Blast Mitigation

Arun Shukla • Yapa D.S. Rajapakse
Mary Ellen Hynes
Editors

Blast Mitigation

Experimental and Numerical Studies

Springer

Editors
Arun Shukla
Department of Mechanical, Industrial
and Systems Engineering
University of Rhode Island
Kingston, RI, USA

Yapa D.S. Rajapakse
Office of Naval Research
Solid Mechanics
Arlington, VA, USA

Mary Ellen Hynes
Infrastructure Protection & Disaster
Management
Science and Technology Directorate
U.S. Department of Homeland Security
Washington, DC, USA

ISBN 978-1-4614-7266-7 ISBN 978-1-4614-7267-4 (eBook)
DOI 10.1007/978-1-4614-7267-4
Springer New York Heidelberg Dordrecht London

Library of Congress Control Number: 2013941355

Printed on acid-free paper

Springer is part of Springer Science+Business Media (www.springer.com)

Preface

The aim of this book is to present a collected source of recent knowledge to graduate students and researchers working in the field of blast loadings and their effect on material and structural failure. This book covers experimental, numerical, and analytical aspects of material and structural response to dynamic blast loads and their mitigation.

Since dynamic blast loading problems are challenging, there is a paucity of mechanics- and physics-based solutions to mitigate the effect of these loadings on structures. The nation has a large, intricate, varied, yet vulnerable infrastructure system that is in turn composed of several smaller systems. The requirements to protect such a complex system and its components from a threat, including an explosive threat, can be examined in terms of systematic *resiliency*. Resiliency here is defined as the ability of a threatened system to prevent, sustain, and recover functionality in time, preferably very rapidly, after successful execution of the threat. Thus, a more resilient infrastructure is better able to recover from both natural and man-made disasters, is less susceptible to disruption, and is, ultimately, less attractive to attack. The material improvements for mitigation should be optimized across as wide a range of threat and vulnerability configurations as possible to improve a structural system's overall resiliency.

The idea for preparing this book originated as a result of two workshops on "Blast Mitigation" organized at the University of Rhode Island in 2009 and 2010 under the sponsorship of the Department of Homeland Security. The thrust of these workshops was in identifying areas of research, yet unexplored, and in recognizing hurdles that hinder progress in the field of blast mitigation.

The presentations at these workshops addressed various means of mitigating explosive threats through the study of composites, as well as a better understanding of construction materials, scaling, blast phenomenology, and of structural and system performance in response to blast loading. As a result of the workshop, several data gaps were identified and some are listed below

- Blast overpressure dampening by utilizing water mist or other methods.
- Development of resilient self-healing materials.

- Understanding failure mechanisms in and developing engineering failure models for novel composite blast-resistant materials.
- Development of novel metallic and composite sandwich structures for energy absorption and structural integrity under blast loading.
- Understanding structural collapse mechanisms, including the effect of elevated temperatures on structural steel, and developing numerical codes to predict structural collapse.
- Understanding key material attributes that improve the response of structures under simultaneous blast and fragment impact.
- Understanding the shattering of glass and developing glass that is both blast resistant and energy efficient.
- Understanding stress wave scattering and attenuation due to particulates for better dampening and channeling of energy.
- Addressing scaling issues to take lab experiments to full scale.
- Development of expendable or sacrificial shields.
- Development of active shielding systems, including triggering through radio frequency broadcast.

The book consists of 11 chapters from leading researchers in the world and broadly covers the field of blast mitigation. It includes chapters on dampening of blast overpressure, energy absorption, and reflection from structures during blast loading, fluid–structure interaction, structural collapse during post-blast and design of novel and better structural configurations. The topic of implosion during blast loading is also discussed. In addition, the structural and material response to air blast as well as underwater blast is addressed.

This book would not have been possible without the timely cooperation of the authors of the chapters contained in this book, and we want to thank them for their contributions. The editors thank the staff at Springer for their exemplary help, patience, and understanding in bringing this book to fruition. The help of Mr. Nicholas Heeder, Mr. Sachin Gupta, and Mr. Frank LiVolsi in proofreading the manuscript is gratefully acknowledged. Finally, we would like to note that the proceeds from this book will go to the educational fund of the Society for Experimental Mechanics (SEM) to help in encouraging students to pursue research in experimental mechanics, which has much to contribute to the society.

Kingston, RI, USA — Arun Shukla
Arlington, VA, USA — Yapa D.S. Rajapakse
Washington, DC, USA — Mary Ellen Hynes

Contents

Chapter 1
Experimental Investigation of Blast Mitigation for Target Protection

S.F. Son, A.J. Zakrajsek, E.J. Miklaszewski, D.E. Kittell, J.L. Wagner, and D.R. Guildenbecher

Abstract An explosion yielding a blast wave can cause catastrophic damage to a building and its personnel. This threat defines an immediate importance for understanding blast mitigation techniques via readily available materials. An unconfined mass of water in the form of a free flowing sheet is one possible readily available mitigant. This chapter focuses narrowly on the protection of high-valued structures and other large targets where removal from the threat zone is often impossible. In these situations, methods are needed to dissipate and reflect incoming blast waves and mitigate damage potential. Any proposed mitigation method must be evaluated for effectiveness, and while steady advances in computational physics have been made in this area, experimentation remains crucial. Therefore, this chapter emphasizes experimental methods for evaluation of blast mitigation, both from a practical and fundamental standpoint. In addition, some of the capabilities of current computational methods are highlighted. The chapter begins with a review of the underlying physics. This is followed by a brief overview of experimental methods. Finally, the remainder of the chapter is dedicated to recent experimental and computational results for a potential configuration involving protective water sheets.

S.F. Son (✉) • A.J. Zakrajsek • E.J. Miklaszewski • D.E. Kittell
Department of Mechanical Engineering, Purdue University,
500 Allison Road, West Lafayette, IN 47906, USA
e-mail: sson@purdue.edu

J.L. Wagner • D.R. Guildenbecher
Sandia National Laboratories, Albuquerque, NM 87185, USA

A. Shukla et al. (eds.), *Blast Mitigation: Experimental and Numerical Studies*,
DOI 10.1007/978-1-4614-7267-4_1,

1.1 Introduction

Blast waves, whether from conventional munitions or improvised explosive devices, pose a major risk to people and property throughout the world (National Research Council 1995). Engineering methods to protect individuals and structures vary widely due to the diverse range of threats and potential situations. This chapter focuses narrowly on the protection of high-valued structures and other large targets where removal from the threat zone is often impossible. In these situations, methods are needed to dissipate and reflect incoming blast waves and mitigate damage potential.

Any proposed mitigation method must be evaluated for effectiveness, and while steady advances in computational physics have been made in this area, experimentation remains crucial. Therefore, this chapter emphasizes experimental methods for evaluation of blast mitigation, both from a practical and fundamental standpoint. In addition, some of the capabilities of current computational methods are highlighted. The chapter begins with a review of the underlying physics. This is followed by a brief overview of experimental methods. Finally, the remainder of the chapter is dedicated to recent experimental and computational results for a potential configuration involving protective water sheets.

1.2 Physics of Blast Waves

Detonation releases a large amount of energy in a small volume, and in an unconfined gaseous environment, such as air, energy interactions quickly give rise to a blast wave expanding in all directions. At a fixed point away from the detonation, a near-instantaneous increase in the static pressure is followed by a period of rapid pressure decay. Depending on the conditions of the blast and the distance from the source, the static pressure may eventually fall below atmospheric. Finally, with sufficient time, pressure equilibrates back to atmospheric.

Figure 1.1 illustrates a typical blast profile. The initial pressure rise is due to a shock wave whose properties are well approximated using the Rankine–Hugoniot relations (Needham 2010). The shock *overpressure*, Δp, defined as the instantaneous peak static pressure minus the ambient pressure, typically peaks during the passage of the leading shock. The time period of positive static pressure is defined as the *positive pulse duration*. The *impulse* is defined as the area under the pressure curve during the positive pulse duration. Finally, the negative pulse duration and negative impulse are typically much longer and smaller than the positive pulse. Effects of the negative pulse are often assumed negligible for blast mitigation design.

In general, a lower overpressure and impulse leads to greater blast-wave survivability. Theoretical analyses of blast waves have been completed by a number of authors, with classic analytic expressions provided by Taylor (1950a, b), Sedov (1959), and others. In agreement with common experience, all of these analyses show a strong decrease in overpressure and impulse with increasing distance from the source. Therefore, it is clear that an effective defense against the damaging

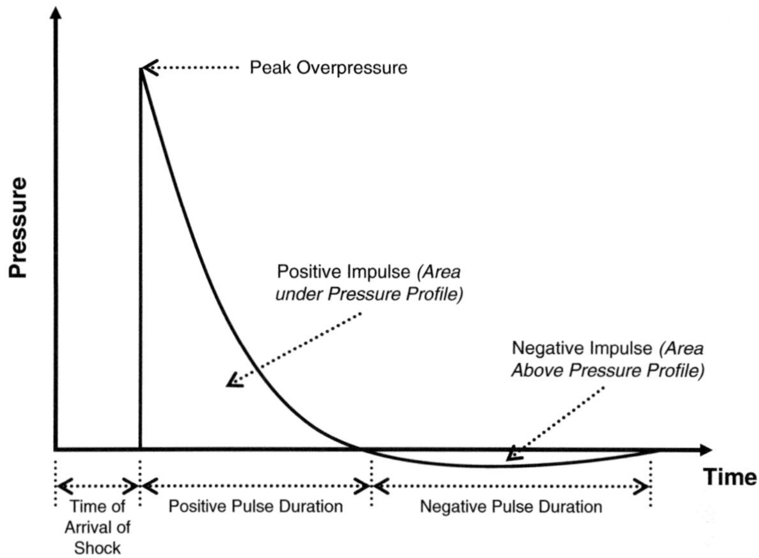

Fig. 1.1 Typical static pressure profile of a blast wave measured at a fixed point in space

effects of a blast wave is to separate the source of the blast and the target. For this reason, high-value targets, such as the White House, embassies, large ships, and so on, often have established perimeters with restricted access for individuals and vehicles. However, such perimeters cannot be made infinitely large and are particularly challenging to implement for existing structures in urban settings. In some cases, adequate standoff distances are simply not possible. For this reason, novel and cost-effective methods are needed to protect such structures and their inhabitants. This is the focus of the remainder of the chapter.

1.3 Experimental Methods

1.3.1 Explosive Field-Tests

The literature includes a large number of proposed blast mitigation methods. Examples include barriers of soil, concrete, foams, sprays, polycarbonate, steel, etc. Evaluation of the effectiveness of these methods requires quantification of the change in overpressure and impulse to the target compared to an un-mitigated blast. This is most often achieved via field-tests involving actual explosives, which is the focus of the first half of this section. However, explosive blast testing is expensive and lacks repeatability necessary to resolve fundamental phenomena. Diagnostics that can be applied in the field are also limited. For this reason, fundamental investigations of mitigation mechanisms have recently been performed in more controlled environments, such as pressure-driven shock tubes, which is the focus of the second half of this discussion.

At a test range, an explosive charge of known quantity is placed some distance away from a target, and blast loading experiments are conducted with and without the proposed mitigant in place. Pressure transducers are used to measure the overpressure, positive pulse duration, and impulse. The time scales of interest are often on the order of milliseconds to microseconds, during which time the thermal and mechanical loading can be significant. Fast pressure transducers are required, which can stand up to this harsh environment. Among the various commercially available configurations, pressure transducers based on the piezoelectric principle are most often utilized. These transducers contain piezoelectric crystals that convert mechanical stress into an electrical signal and are sensitive to rapid changes in overpressure.

To obtain quality measurements, the experimentalist must take care to properly select and configure pressure transducers. In any flow measurement, proper placement and orientation of pressure transducers are crucial. Where the flow is parallel to the sensing surface, the transducer measures the static pressure. In blast measurements, this is most often achieved with a ruggedized probe shaped like a pencil with the pointed end facing the incoming blast wave. The design allows the wave to propagate with minimal distortions along the side of the probe, which contains the sensing element. Alternatively, a pressure transducer may be orientated with its sensing element facing the incoming blast wave. In this orientation, the sensor quantifies the reflected overpressure, which can be related to the incident overpressure assuming idealized shock relations (Needham 2010). Regardless of the orientation, additional care must be taken to ensure that measurements are unaffected by mechanical vibrations, thermal stress, and line losses. Industrial transducer suppliers can assist with these issues, which tend to be transducer specific.

In addition to pressure measurements, optical techniques, such as high-speed imaging, Schlieren, and shadowgraphy, have been applied to image the shock waves in an experiment (Settles 2006a). These techniques are equally applicable to shock tube experiments or blast experiments.

1.3.2 Shock Tube Experiments

The shock tube is a basic instrument often used for shock attenuation research relevant to blast mitigation (Monti 1970; Gel'fand et al. 1983; Skews et al. 1993; Levy et al. 1993; Radulescu and Lee 2002; Teodorczyk and Lee 1995; Sommerfeld 1985; Aizik et al. 1995; Wagner et al. 2012; Britan et al. 2001; Meekunnasombat et al. 2006). In addition, shock tubes are commonly utilized in aerodynamics (Selberg and Nicholls 1968; Takayama and Itoh 1986; Igra and Takayama 1993; Jourdan et al. 2007; Wagner et al. 2011), fluid dynamics (Haas and Sturtevant 1987; Brouillette 2002; Jacobs and Krivets 2005; Leinov et al. 2009; Motl et al. 2009; Orlicz et al. 2009), shock physics (Skews 1967; Sun and Takayama 2003; Tanno et al. 2003), and combustion experiments (Michael and Sutherland 1986; Masten et al. 1990). As is

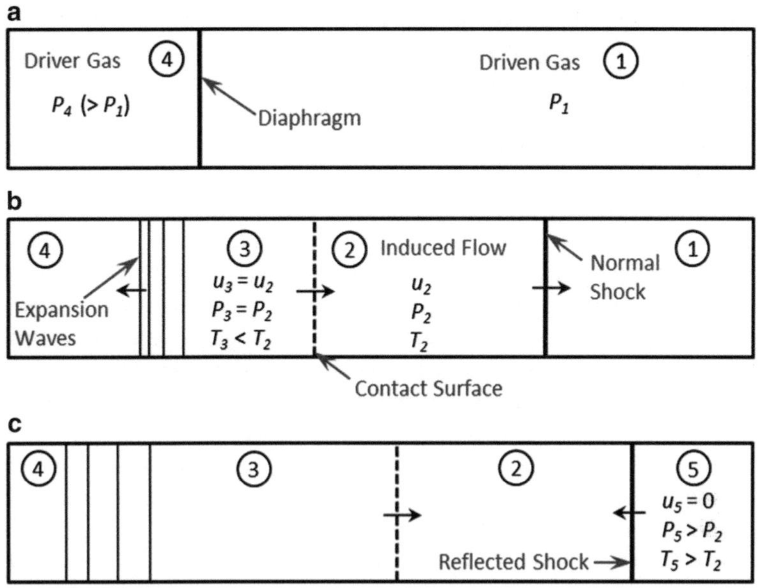

Fig. 1.2 Illustration of basic shock tube principles. Both the figure and accompanying text are after Anderson (Anderson 1990): (**a**) initial conditions, (**b**) following the rupture of the diaphragm, and (**c**) following reflection of the normal shock from the end-wall

shown in Fig. 1.2a, the basic components of a shock tube include a high pressure driver section (region 4) and a low pressure driven section (region 1), which are initially divided by a diaphragm (Anderson 1990). Upon rupture of the diaphragm (Fig. 1.2b), a normal shock wave is formed in response to the instantaneous pressure difference between the driver and driven sections. With continuing time, the shock wave propagates downstream resulting in a high-speed, shock-induced flow with velocity u_2, which is often utilized in fluid dynamics experiments. Upstream of region 2, a contact surface separates the driver and driven gases. The contact surface moves at a velocity u_3 equal to u_2. Across the contact surface, the pressure is equal, but there is a discontinuity in temperature and entropy. Expansion waves form causing pressure in the driver gas to continuously vary between regions 3 and 4. Depending on the initial conditions in the shock tube, the expansion waves may propagate to the left or to the right. Eventually, the normal shock will reach the end of the tube resulting in a reflected shock wave as is shown in Fig. 1.2c. To match the end-wall boundary condition, the reflected shock must completely slow the flow in region 5 to a velocity of zero (Anderson 1990). To obtain high temperatures, chemical kinetics experiments are commonly performed in this region of twice-shocked flow.

Many studies have used shock tubes to evaluate the use of porous materials for blast mitigation (Monti 1970; Gel'fand et al. 1983; Skews et al. 1993; Levy et al. 1993; Radulescu and Lee 2002; Teodorczyk and Lee 1995). Several other experiments have measured the effects of compressible foam placed upstream of a wall in the path of a normal shock wave (Monti 1970; Gel'fand et al. 1983; Skews et al.

1993). The foam results in a decrease in reflected shock strength, although, importantly, the pressure at the back wall is amplified significantly in comparison to experiments without foam. Tests with rigid porous materials placed upstream of shock tube end-walls also indicate an increase in end-wall pressure compared to the baseline case (Levy et al. 1993). However, this increase appears to be less than that in deformable foam experiments (Skews et al. 1993). In addition, experiments have demonstrated the attenuation of detonation waves traveling along porous walls (Radulescu and Lee 2002) and walls lined with foams (Teodorczyk and Lee 1995).

The use of solid particles for blast mitigation has also been tested in shock tube experiments (Sommerfeld 1985; Aizik et al. 1995; Wagner et al. 2012; Britan et al. 2001), where the general configuration involves placing gas-solid mixtures upstream of incident normal shock waves. Based on the data of Sommerfeld (1985), Aizik et al. (1995) gave a correlation for shock attenuation in dilute gas-solid mixtures, which they claimed to be valid for particle volume fractions less than 1 %. The use of denser particle mixtures has also been evaluated. For example, streamwise-thin particle curtains with dense particle volume fractions of 20 % have resulted in modest shock attenuation (Wagner et al. 2012) and streamwise-thick, granular filters consisting of packed particle beds have led to significant shock attenuation (Britan et al. 2001).

Water sheets in the path of normal shock waves have also been tried for mitigation purposes (Meekunnasombat et al. 2006). In comparison to undisturbed shocks, the shocks transmitted through the water sheets were diminished. However, as a result of subsequent compression waves and impinging water droplets, the peak end-wall pressures in experiments with water sheets were up to ten times higher than those without water and the impulses were up to two times higher (Meekunnasombat et al. 2006).

The above discussion demonstrates the complexities and tradeoffs associated with shock attenuation techniques. For example, a reduction in shock strength is often accompanied with an increase in peak pressure that occurs more gradually.

Although many studies have focused on the attenuation of planar shock waves in shock tubes, little work has been aimed at understanding the physical mechanisms involved in mitigating the expanding blast waves associated with real detonations. New data are required to enable effective blast mitigation methods. Towards this end, experiments have been conducted for water sheets subjected to blasts from an explosively driven shock tube, which is the subject of the remainder of this chapter.

1.4 Mitigation Using Unconfined Water Sheets

As discussed above, an explosion yielding a blast wave can cause catastrophic damage to people and property. To mitigate damage from such an event, a number of investigators have proposed the use of water in various configurations. Water sprays, attached barriers, and sheets could be explored. In this investigation, an unconfined free-flowing water sheet, with an approximate thickness of 0.3 cm, is experimentally examined using an explosively driven shock tube at three different standoff distances. This differs somewhat from the shock tube experiments described above and is meant to simulate an explosive blast profile. The results presented in this section show that the water sheet mitigates the initial peak overpressure and impulse of the blast. Further insights into the underlying physics are revealed by a numerical simulation using Sandia's CTH hydrocode (Crawford 2012).

Water is an attractive material for use as a blast mitigant because it is often readily available (Kailasanath et al. 2002). Previous investigations have focused on blast mitigation by water sprays or confined masses of water (Kailasanath et al. 2002; Schwer and Kailasanath 2006) with mixed success as quantified by the peak overpressure and impulse mitigation. An alternative geometry is that of a free-flowing water sheet forming a shield around the object to be protected. A few numerical models have considered this geometry (Cheng et al. 2005; Shin et al. 1998), and their results indicate that water sheets may effectively limit the peak overpressure of the blast. However, there are limited experimental data to validate these findings. Meekunnasombat et al. (Meekunnasombat et al. 2006) considered liquid layers in a confined vertical shock tube, and their experiments highlighted the advantageous behavior of multiple layers of water. However, it is unclear if their results can be extended to the case of the unconfined water sheet, which is the likely geometry for practical blast mitigation. In addition, Bremond and Villermaux (2005) considered the breakup of a thin soap film under normally incident shock loading. However, it is unclear if the physical breakup mechanisms revealed in their investigation can be extended to the thicker water sheets considered here.

The aim of this work is to study the blast mitigating potential associated with a sheet of water. To elucidate the fundamental physics, an experimental investigation is coupled with a numerical model. The following two sections focus on the experimental configuration and results, respectively. Numerical results are presented in subsequent section. The results presented here form an initial starting point to determine the feasibility of water sheets as a possible blast mitigant.

1.5 Experimental Configuration for Water Sheets

1.5.1 Explosively Driven Shock Tube

Experiments were conducted in an open field, with an explosively driven shock tube, and a custom fabricated water sheet generator. A schematic of the experimental configuration is shown in Fig. 1.3. The explosively driven shock tube is used to produce a relevant laboratory size blast wave. Previous work has shown that this configuration yields a blast profile similar to open field explosive tests, and the shock tube directs the energy from the blast in one direction allowing for the use of less explosive as compared to a conventional open field explosive test (Freiwald 1972; Alley 2009).

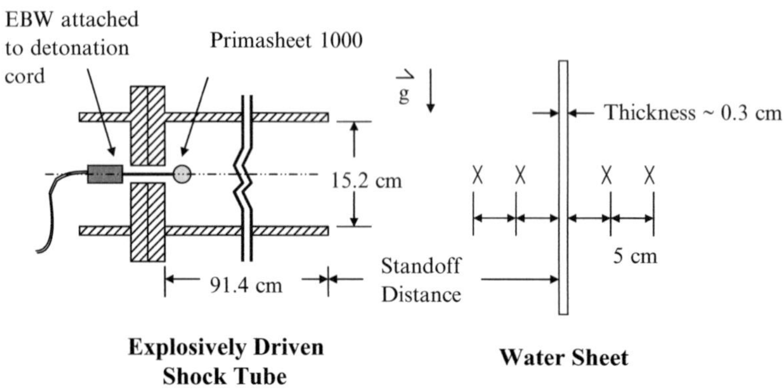

Fig. 1.3 Experimental configuration for investigation of blast loading of an unconfined water sheet. The out of plane water sheet dimension is roughly 70 cm, while the in plane height is 30 cm. The "X's" indicate the location of the static pressure gauges which are positioned along the centerline of the shock tube

In an experiment, the explosively driven shock tube is loaded with 3 g of Primasheet 1000. Primasheet 1000 is a pentaerythritol tetranitrate (PETN) based plastic explosive consisting of approximately 63 % PETN powder. Initiation of the PETN plastic sheet explosive is achieved by a combination of detonation cord (PETN powder) and an explosive bridge wire (EBW) detonator (RP-502 EBW) charged by a firing set, see Fig. 1.3.

The shock tube consists of a detonation chamber and a high explosive chamber. The two chambers are bolted together to protect the cables, test apparatus, and other equipment from the fragments produced by the aluminum detonator cap. A small hole links the two chambers and allows for passage of detonation cord. By changing the distance of the shock tube from the water sheet (defined as the *standoff distance*), the characteristics of the incident shock wave are altered. Decreasing the standoff distance increases the overpressure at the water sheet.

1.5.2 Water Sheet Generator and Pressure Gauges

A water sheet is generated with a custom fabricated water sheet generator, operating at a constant flow rate of approximately 56 L/min, producing a sheet approximately 0.3 cm thick. In the absence of blast loading, the sheet is continuous and displays surface perturbations, which likely arise from turbulent or capillary instabilities. In an experiment, the incoming blast wave is approximately normal to the water sheet.

Due to space constraints near the water sheet, typical pencil gauges cannot be placed at the locations marked in Fig. 1.3 without disrupting the water flow. Instead, PCB 113A22/113B22 piezoelectric dynamic pressure sensors were threaded into custom fabricated plates orientated to measure the static pressure of the blast. The experimental configuration includes two pressure gauges placed in front and two behind the water sheet, as shown in Fig. 1.3. As the water sheet standoff distance is varied from one experiment to the next, the pressure gauges are repositioned to maintain the distances with respect to the water sheet shown in Fig. 1.3.

1.5.3 Shadowgraphy Visualization

In select experiments, the shock wave is visualized using the high-speed shadowgraphy technique described in (Settles 2006b). Videos are recorded at 11,494 fps and an exposure of 26 μs using a Vision Research Phantom v7.3 digital high-speed camera and an Oriel 1000 W xenon arc lamp. The field of view is approximately 53 × 53 cm and is recorded on a 171 × 171 pixel region of the CCD.

1.6 Water Sheet Experimental Results

Experiments were performed at three different standoff distances: 20, 31, and 41 cm (distance from the exit of shock tube to the water sheet). To verify repeatability, all experiments were performed three times. Further repetition of experimental conditions was cost prohibitive, as is often the case in explosive field measurements.

1.6.1 Pressure Measurements

Figure 1.4 shows the free field pressure traces taken without the water sheet. Additionally, tabulated values for all free field distances measured are shown in Table 1.1. Figure 1.5 displays the pressure traces from an experiment with a water sheet at a 31 cm standoff distance. Note the significantly lower peak pressures at distances behind the water sheet (36 and 41 cm). Other standoff distances produce similar results. Table 1.2 summarizes the pressure measurements after the water sheets, for all three standoff distances considered. In this table, the peak overpressure is taken as the highest recorded static pressure during the transient experiment. The standard error is calculated between the three experiments performed at each condition.

1.6.2 Pressure Trace After Water Sheet

The experimental results show that the unconstrained free flowing water sheet significantly reduces the initial overpressure and impulse of the blast. This is best illustrated in Fig. 1.5. At the 31 cm water-sheet standoff distance, the peak overpressure is reduced by 82 % and the impulse is reduced by 75 % as measured by the pressure

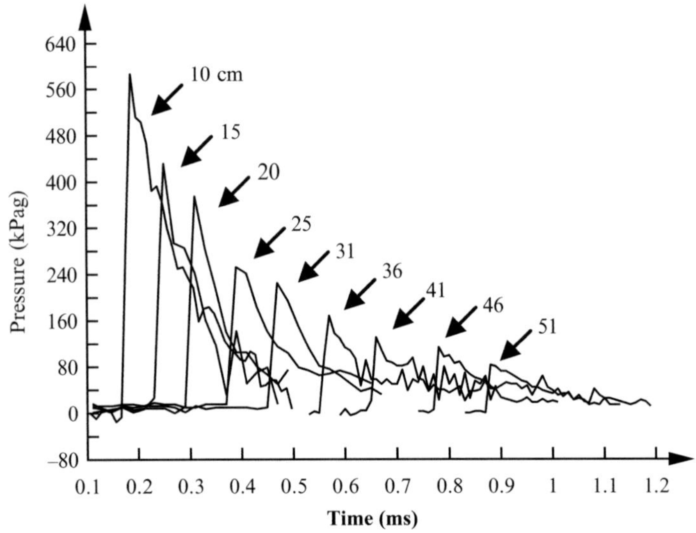

Fig. 1.4 Free field experimental pressure traces. *Numbers* indicate the distances in centimeters between the exit of the shock tube and the pressure gauge

Table 1.1 Free field (no water sheet) experimental shock wave parameters

Pressure gauge standoff distance (cm)	Peak overpressure (kPa)	Impulse (kPa-ms)	Positive pulse (ms)
10	586±29	60±9	0.2±0.1
15	436±18	35±8	0.2±0.1
20	389±15	32±5	0.3±0.1
25	253±21	32±3	0.3±0.1
31	203±19	19±5	0.2±0.1
36	169±3	19±4	0.3±0.1
41	130±1	18±2	0.3±0.1
46	117±4	18±1	0.4±0.1
51	88±6	12±1	0.3±0.1

Uncertainties indicate the standard error

gauge placed 5 cm behind the water sheet. These results show qualitative agreement with previous work with water shields (Cheng et al. 2005; Shin et al. 1998).

The pressures measured downstream of the water sheet also show a somewhat unexpected increase in the pressure at a finite time following the passage of an initial shock wave. Some pressure traces showed this rise in the pressure more distinctly. For example, Fig. 1.6 highlights the pressures measured downstream of the water sheet at 31 cm. Based on the distances between the measurement points and the delay time between the initial pressure rise, it is found that the initial transmitted wave propagates at approximately sonic conditions. The second pressure increase,

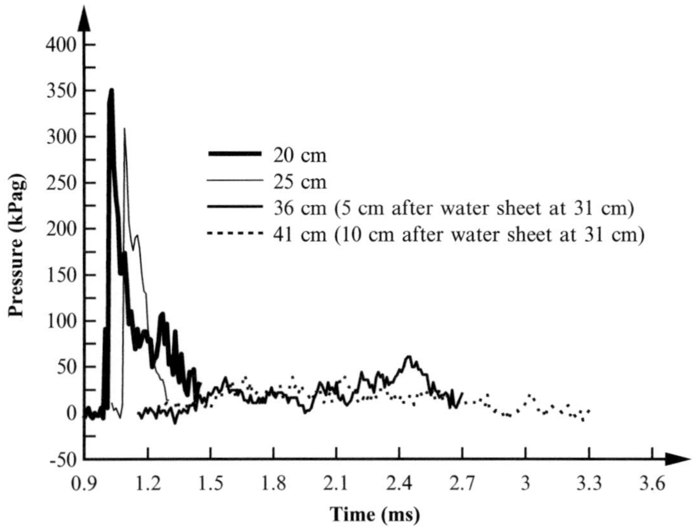

Fig. 1.5 Pressure trace of a blast test with a water sheet at a 31 cm standoff distance. *Numbers* indicate the distances in centimeters between the exit of the shock tube and the pressure gauge

Table 1.2 Initial experimental shock wave parameters 5 and 10 cm after the water sheets

Water sheet standoff distance (cm)	5 cm After water sheet		
	Peak overpressure (kPa)	Impulse (kPa-ms)	Positive pulse (ms)
20	60±8	18±3	0.3±0.1
31	31±4	5±2	0.3±0.1
41	26±5	6±2	0.4±0.1
Water sheet standoff distance (cm)	10 cm After water sheet		
	Peak overpressure (kPa)	Impulse (kPa-ms)	Positive pulse (ms)
20	53±6	12±3	0.3±0.1
31	27±8	4±0.3	0.3±0.1
41	23±3	6±0.4	0.4±0.1

Uncertainties indicate the standard error

which occurs sometime after the passage of the sonic wave, travels at subsonic velocity. The source of this behavior is currently not understood. Possible causes may be impact of liquid water on the pressure gauges or passage of high-pressure products released after the sheet fragments.

This phenomenon is somewhat similar to the rise in pressure sometimes observed during shock wave reflections from porous foams (Skews et al. 1993). However, unlike foam, the water sheet does not contain regular voids. For this reason, wave propagation mechanisms are likely to be distinctly different between the foam and water sheet configurations, and more work is needed to understand this phenomena.

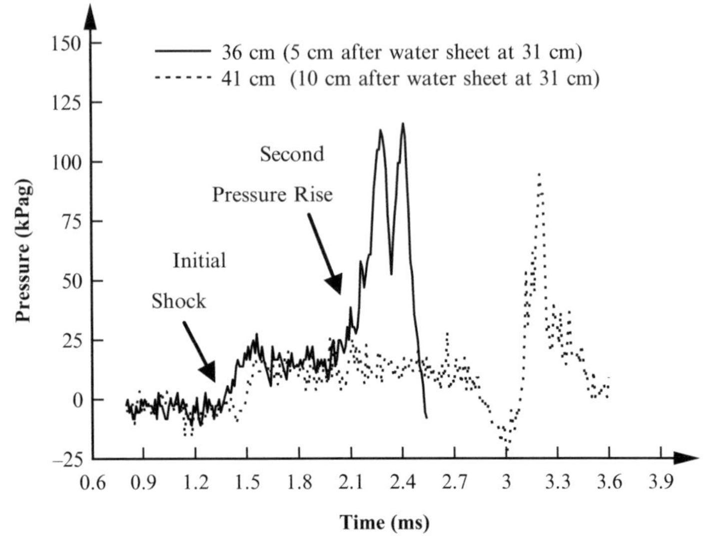

Fig. 1.6 Pressure trace of gauges 5 and 10 cm behind the water sheet at a 31 cm standoff distance

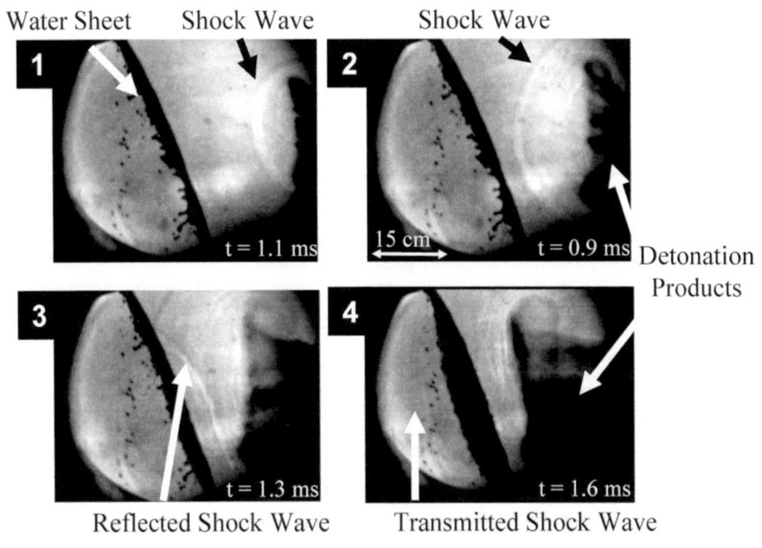

Fig. 1.7 High speed shadowgraphy of a blast loaded water sheet at a 31 cm standoff distance

1.6.3 Shadowgraphy Visualization

The shock wave interaction with the water sheet was visualized using shadowgraphy. Select images from a high-speed shadowgraphy video are shown in Fig. 1.7, where the flow is right to left. The video is taken without pressure gauges which tend to obstruct the observation of the incident, reflected, and transmitted shock waves. The reflected wave seen in the third image is due to the impedance differences between the air and water (Henderson et al. 1990). In the fourth image a weak transmitted shock wave is observed. The disruption of the water sheet is also seen.

1.7 Numerical Simulation

Exploratory numerical simulations were completed utilizing Sandia's hydrocode, CTH. CTH is a multi-material, large deformation, strong shock wave, solid mechanics code developed at Sandia National Laboratories (Crawford 2012). In what follows, the CTH model is first validated using the free field conditions and is then used to qualitatively study the interaction of the shock wave and water sheet. This is meant to demonstrate what is possible with these types of simulations and does not represent a final study. The utility of these types of simulations to interpret the experimental results is demonstrated.

For simplicity, the explosively driven shock tube was not modeled in this study. Rather, an open field charge model was employed; nevertheless, as shown below the model qualitatively captures the physical phenomena. The geometry and boundary conditions are shown in Fig. 1.8. The model is axisymetric with a spherical open field charge of PETN. The bottom boundary condition is set to allow the pressure to be zero in the ghost cells and to later remove all material from that ghost cell. This boundary condition ensures that mass does not enter the mesh but is allowed to leave. The remaining three boundary conditions use a sound speed based absorbing/transmitting condition to approximate an infinite or semi-infinite medium. Here mass can flow into and out of the mesh.

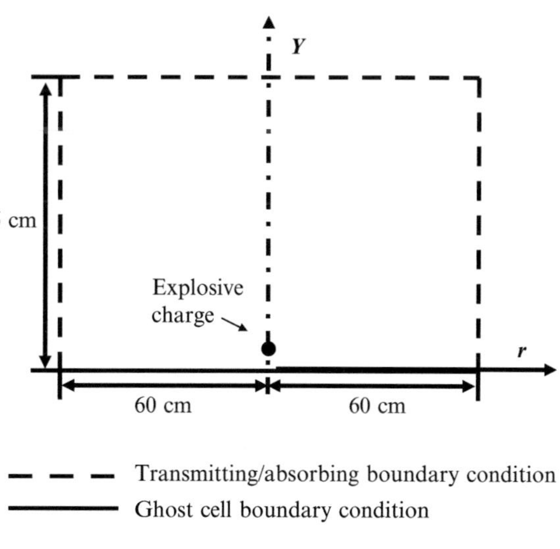

Fig. 1.8 CTH simulation geometry and boundary conditions

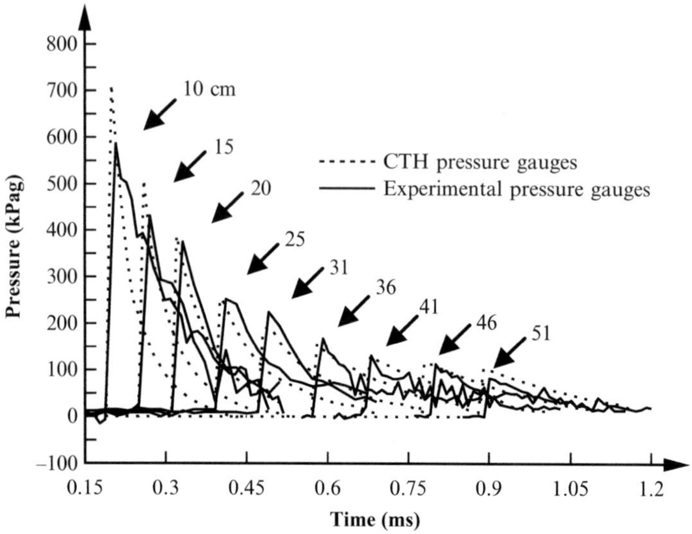

Fig. 1.9 Comparison of CTH pressure traces versus experimental pressure traces. *Numbers* indicate the distances in centimeters between the exit of the shock tube and the pressure gauge

The explosive charge is assumed to follow the Jones–Wilkins–Lee (JWL) equation-of-state (EOS) for PETN. Since the experimental explosive is comprised of 63 % PETN, the model mass is scaled to allow the use of the JWL EOS. The explosive charge is detonated using a history variable reactive burn (HVRB) model. The HVRB is a pressure-based model used to treat shock induced initiation that grows to a detonation for heterogeneous explosive material (Teodorczyk and Lee 1995). When the HVRB is used, the equations of state for the un-reacted and reacted phases are usually the Mie-Grüneisen and JWL equations of state (Todd et al. 2011). Atmospheric air was modeled at an initial absolute pressure of 100 kPa using a tabulated SESAME EOS. The SESAME EOS Library is a standardized, computer-based library of thermodynamic properties developed by Los Alamos National Laboratory (Lyon et al. 2011). Fixed nodes are included in the CTH model to match the experimental free field pressure gauge locations. The mass of PETN used in the model (48.4 g) was determined by comparing the predicted pressure to the experimental pressures and adjusting the mass until reasonable agreement was obtained. It should be noted that the mass of the PETN used in the model is an order of magnitude larger than the experiment value. This difference can be attributed to the fact that a shock tubes is used in the experiment to focus the blast in one direction. Figure 1.9 compares the free field pressure traces between the final CTH model and the experimental results. The average percent difference between the peak overpressure is 7.8 % and the average percent difference between the impulses is around 20 %. Perhaps some of the difference from experiment and calculations observed in the width of the positive pulse close to the water sheet could be due to not modeling the focusing effect of the shock tube.

To model the water sheet, adaptive mesh refinement was used with a radially symmetric mesh. Spatial blocks were divided into 256 zones and had up to four

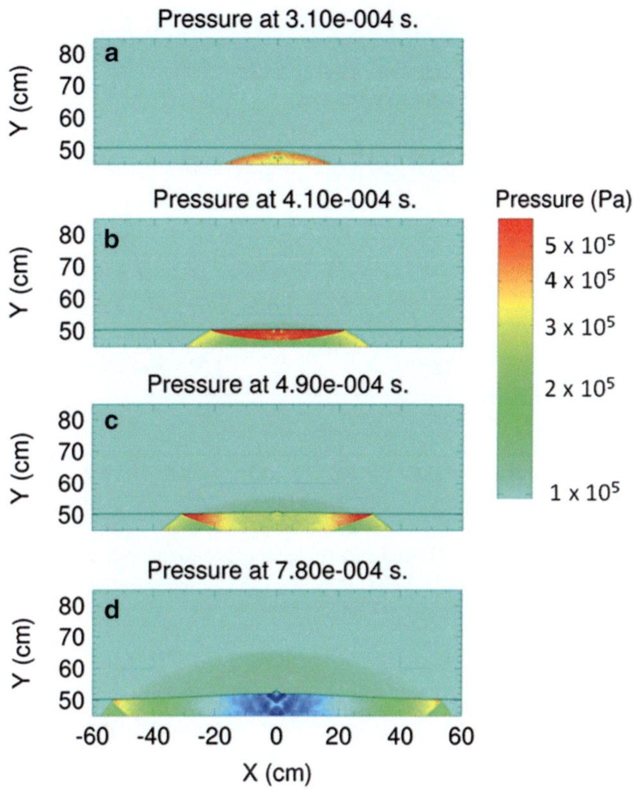

Fig. 1.10 Simulation results showing absolute pressure for shock wave interaction with the water sheet at the 20 cm standoff distance. (**a**) Initial shock wave reaches the water sheet, (**b**) Reflection of a shock wave off the surface of the water sheet, (**c**) Development of weak transmitted shock wave behind the intact water sheet, and (**d**) Water-sheet breakup

levels of refinement. This resulted in a maximum resolution of 195 μm near any moving interface, reaction zone, or region of steep pressure gradient. Note that the results from the adaptive mesh were similar to initial calculations using a fixed Eulerian mesh, except that the water sheet became more jagged during the later stages of breakup with adaptive meshing.

Liquid water is modeled with the Mie-Grüneisen EOS. The water sheets are assumed to be initially smooth with a thickness of 0.3 cm. Furthermore, the sheets are assumed to be infinite (stretching across CTH domain) to eliminate effects caused by diffraction of the shock wave around the water sheet, and to better isolate the interaction of the shock wave and water sheet. In addition, it should be noted that the CTH models do not include surface tension effects which may play a significant role in the physical breakup process. Because of these simplifications, the simulation results are considered for qualitative insight into the physical phenomena, rather than quantitative predictions.

Figure 1.10 shows predicted contours of pressure at select time intervals. Due to the impendence mismatch at the air-water interface, a significant portion of the

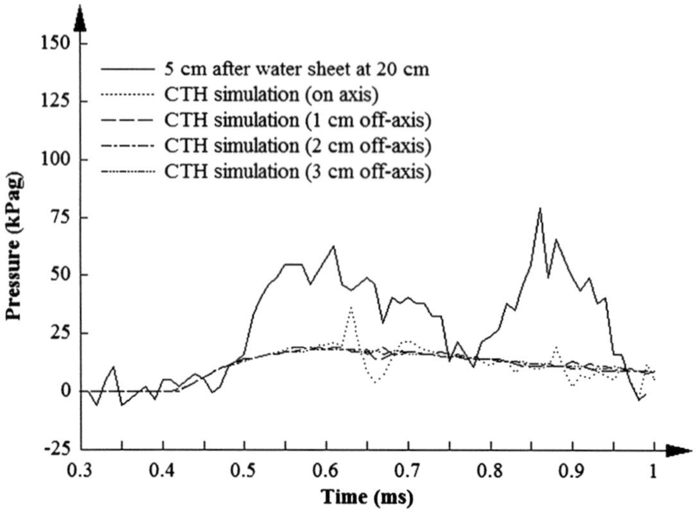

Fig. 1.11 Comparison between CTH and experimental pressure traces 5 cm after the water sheet at a 20 cm standoff distance

initial shock reflects off the water sheet and only a weak shock wave is initially transmitted. Comparison with Fig. 1.8 reveals many qualitative similarities between simulation and experiment including the reflected wave, transmission of a weak shock wave, and fragmentation of the sheet.

Figure 1.11 shows pressure results from the CTH model at 5 and 10 cm after the water sheet at a 20 cm standoff distance. This can be qualitatively compared to experimental results presented in Fig. 1.6. Both the experimental and numerical results show the initially transmitted pressure wave. However, the simulation does not capture a distinct second rise in pressure. Note, the pressure fluctuations seen in the simulation results, particularly near the centerline, are thought to be a result of water passing through the domain resulting in an unphysical, numerical artifact. These differences between simulations and experiments highlight the importance of coupled experimental and numerical efforts. Further work is needed to fully understand the experimentally observed pressure profile.

1.8 Discussion of Water Sheet Mitigation

As discussed by Henderson et al. (1990), when a blast wave first contacts a water sheet, a large portion of the incident energy is reflected back towards the source due to the impedance mismatch at the air-water interface. The remaining energy, which is transmitted through the water sheet, forms the observed weak shock wave and the resulting initial pressure rise. For the experimental conditions considered here, the initial blast overpressure is mitigated by as much as 80 %. This indicates that a water sheet may be an effective emergency blast mitigant.

However, a second rise in pressure is observed, resulting in a greater overall peak overpressure behind the water sheet. As discussed above, experimental and numerical results are insufficient to elucidate the source of this behavior and further work is needed. Such work could include investigations in a pressure-driven shock tube as discussed in the first half of this chapter. Such an investigation would be particularly advantageous due to the well-controlled environment and the resulting ability for quantitative optical diagnostics.

1.9 Conclusions

Blast loaded objects undergo complex fluid–structure interactions. These interactions introduce a large range of length-scales and physical phenomena, which complicate first-principle models and numerical simulation. For this reason, experimental investigations of blast mitigants are vital to the determination of their effectiveness. In addition, fundamental experimental results can be used to guide model improvements.

This chapter has discussed the most relevant experimental methods, including large-scale blast loading experiments and fundamental shock-tube experiments. In addition, an example investigation of blast mitigation via a water sheet is presented. Initial results from the experimental investigation reveal significant blast mitigation at certain conditions. However, more work is needed to elucidate the relevant physical phenomena such that predictive models can be developed. This is highlighted by a comparison with numerical simulation.

The further development of this, and other blast mitigation methods, requires a three pronged approach. Further blast loading experiments are required to identify optimal conditions and quantify effectiveness. Furthermore, fundamental experiments, such as those conducted in pressure-driven shock-tubes, are needed to visualize and quantify the details of the complex interaction of a shock-wave with the proposed mitigant. Finally, these efforts must be closely integrated with numerical simulations to aid in development of predictive models. Simulations may possibly be improved with true fluid modeling of the water sheet and full simulation of the shock tube.

Acknowledgments We would like to thank the Department of Homeland Security and the Center of Excellence for Explosive Detection, Mitigation and Response, Sponsor Award No. 080409/0002251. Additionally, special thanks to Matthew Massaro and Jesus Mares who assisted with experiments.

Sandia National Laboratories is a multi-program laboratory managed and operated by Sandia Corporation, a wholly owned subsidiary of Lockheed Martin Corporation, for the US Department of Energy's National Nuclear Security Administration under contract DE-AC04-94AL85000.

References

Aizik F, Ben-Dor G, Elperin T, Igra O, Mond M, Groenig H (1995) Attenuation law of planar shock waves propagating through dust-gas suspensions. AIAA J 33(5):953–955

Alley MD (2009) Explosive blast loading experiments for TBI scenarios: characterization and mitigation. M.S. Thesis, Purdue University, West Lafayette

Anderson JA (1990) Modern compressible flow: with historical perspective. Mc-Graw Hill, New York

Bremond N, Villermaux E (2005) Bursting thin liquid films. J Fluid Mech 524:121–130

Britan A, Ben-Dor G, Igra O, Shapiro H (2001) Shock waves attenuation by granular filters. Int J Multiphase Flow 27(4):617–634

Brouillette M (2002) The RICHTMYER-MESHKOV instability. Annu Rev Fluid Mech 34(1):445–468

Cheng M, Hung K, Chong O (2005) Numerical study of water mitigation effects on blast wave. Shock Waves 14(3):217–223

Crawford D (2012) CTH shock physics, Sandia National Laboratories. Accessed 14 Dec 2011. http://www.sandia.gov/CTH/

Freiwald DA (1972) Approximate blast wave theory and experimental data for shock trajectories in linear explosive-driven shock tubes. J Appl Phys 43(5):2224–2226

Gel'fand BE, Gubanov AV, Timofeev EI (1983) Interaction of shock waves in air with a porous screen. Fluid Dyn 18(4):561–566

Haas J-F, Sturtevant B (1987) Interaction of weak shock waves with cylindrical and spherical gas inhomogeneities. J Fluid Mech 181:41–76

Henderson LF, Jia-Huan M, Akira S, Kazuyoshi T (1990) Refraction of a shock wave at an air–water interface. Fluid Dyn Res 5(5–6):337–350

Igra O, Takayama K (1993) Shock tube study of the drag coefficient of a sphere in a non-stationary flow. Proc Math Phys Eng Sci 442(1915):231–247

Jacobs JW, Krivets VV (2005) Experiments on the late-time development of single-mode Richtmyer–Meshkov instability. Phys Fluids 17(3):034105

Jourdan G, Houas L, Igra O, Estivalezes J-L, Devals C, Meshkov EE (2007) Drag coefficient of a sphere in a non-stationary flow: new results. Proc Math Phys Eng Sci 463(2088):3323–3345

Kailasanath K, Tatem PA, Mawhinney J (2002) Blast mitigation using water – A status report

Leinov E, Malamud G, Elbaz Y, Levin LA, Ben-dor G, Shvarts D, Sadot O (2009) Experimental and numerical investigation of the Richtmyer–Meshkov instability under re-shock conditions. J Fluid Mech 626:449–475

Levy A, Ben-Dor G, Skews BW, Sorek S (1993) Head-on collision of normal shock waves with rigid porous materials. Exp Fluids 15(3):183–190

Lyon SP, Johnson JD, (1992) Group T-1, SESAME: The Los Alamos National Laboratory equation of state database, Report number LA-UR-92-3407 http://t1web.lanl.gov/doc/SESAME_3Ddatabase_1992.html.

Masten DA, Hanson RK, Bowman CT (1990) Shock tube study of the reaction hydrogen atom + oxygen.fwdarw. hydroxyl + oxygen atom using hydroxyl laser absorption. J Phys Chem 94(18):7119–7128

Meekunnasombat P, Oakley J, Anderson M, Bonazza R (2006) Experimental study of shock-accelerated liquid layers. Shock Waves 15(6):383–397

Michael JV, Sutherland JW (1986) The thermodynamic state of the hot gas behind reflected shock waves: implication to chemical kinetics. Int J Chem Kinet 18(4):409–436

Monti R (1970) Normal shock wave reflection on deformable solid walls. Meccanica 5(4):285–296

Motl B, Oakley J, Ranjan D, Weber C, Anderson M, Bonazza R (2009) Experimental validation of a Richtmyer–Meshkov scaling law over large density ratio and shock strength ranges. Phys Fluids 21(12):126102

National Research Council (1995) Protecting buildings from bomb damage: transfer of blast-effects mitigation technologies from military to civilian applications. The National Academies Press, Washington, DC

Needham CE (2010) Blast waves. Springer, Heidelberg

Orlicz GC, Balakumar BJ, Tomkins CD, Prestridge KP (2009) A Mach number study of the Richtmyer–Meshkov instability in a varicose, heavy-gas curtain. Phys Fluids 21(6):064102

Radulescu MI, Lee JHS (2002) The failure mechanism of gaseous detonations: experiments in porous wall tubes. Combust Flame 131(1–2):29–46

Schwer D, Kailasanath K (2006) Blast mitigation by water mist (3) Mitigation of confined and unconfined blasts. NRL memorandom Report 6410–06–8976

Sedov LI (1959) Similarity and dimensional methods in mechanics. Academic, New York

Selberg B, Nicholls J (1968) Drag coefficient of small spherical particles. AIAA J 6(3):401–408

Settles GS (2006a) Schlieren and shadowgraph techniques: visualizing phenomena in transparent media. Springer, Berlin

Settles GS (2006b) High-speed imaging of shock waves, explosives, and gunshots. Am Sci 94(1):22–31

Shin YS, Lee M, Lam KY, Yeo KS (1998) Modeling mitigation effects of watershield on shock waves. Shock Vib 5(4):225–234

Skews BW (1967) The shape of a diffracting shock wave. J Fluid Mech 29(02):297–304

Skews BW, Atkins MD, Seitz MW (1993) The impact of a shock wave on porous compressible foams. J Fluid Mech 253:245–265

Sommerfeld M (1985) The unsteadiness of shock waves propagating through gas-particle mixtures. Exp Fluids 3(4):197–206

Sun M, Takayama K (2003) Vorticity production in shock diffraction. J Fluid Mech 478:237–256

Takayama K, Itoh K (1986) Shock waves and shock tubes. Proceedings of the fifteenth international symposium. Stanford University Press, Stanford, CA

Tanno H, Itoh K, Saito T, Abe A, Takayama K (2003) Interaction of a shock with a sphere suspended in a vertical shock tube. Shock Waves 13(3):191–200

Taylor G (1950a) The formation of a blast wave by a very intense explosion. I. Theoritical discussion. Proc Math Phys Eng Sci 201(1065):159–174

Taylor G (1950b) The formation of a blast wave by a very intense explosion. II. The atomic explosion of 1945. Proc Math Phys Eng Sci 201(1065):175–186

Teodorczyk A, Lee JHS (1995) Detonation attenuation by foams and wire meshes lining the walls. Shock Waves 4(4):225–236

Todd SN, Anderson MU, Caipen TL (2011) Model ms for Non-shock Initiation. In Dynamic Behavior of materials, Vol.1, proceedings of the 2010 Annual conference on Experimental and Applied Mechanics. Springer: New York

Wagner J, Beresh S, Kearney S, Pruett B, Wright E (2011) Shock tube investigation of unsteady drag in shock-particle interactions. AIAA 2011–3910, 41 st AIAA Flold Dynamic conference and Exhibit

Wagner J, Beresh S, Kearney S, Trott W, Castaneda J, Pruett B, Baer M (2012) A multiphase shock tube for shock wave interactions with dense particle fields. Exp Fluids 52(6):1507–1517

Chapter 2
Application of High Performance Computing to Rapid Assessment of Tunnel Vulnerability to Explosive Blast and Mitigation Strategies

L. Glascoe and T. Antoun

Abstract To credibly assess tunnel vulnerabilities to explosives and the mitigation of these vulnerabilities, one must consider the energetic source of the shock, the geophysical response of the surrounding soil or rock under strong dynamic loading, and the structural and material response of the tunnel to the explosion. Tools used for such assessments often require significant computational resources and expert analyses. This chapter provides a concise overview of some recent high fidelity multiphysics approaches to modeling tunnel structures and the need for careful experimental validation of these models. Presented also is an example of how the building of reduced order models from these higher fidelity capabilities can help to quickly estimate the effects of explosions within carefully defined tunnel scenarios.

2.1 Introduction

High performance computational multiphysics analysis tools, e.g., the ALE3D, GEODYN, LDEC, DYNA3D, and CTH codes (Nichols and The ALE3D team 2009; Lomov et al. 2005; Lin 2005; McGlaun et al. 1990; Morris and Johnson 2009), are useful for describing and predicting the response of geology and structures to shock loads, and, if properly validated and caveated, can be used for vulnerability assessments and corrective measures to reduce overall risk (e.g., Noble et al. 2008; Glascoe et al. 2009, 2011; McMichael et al. 2009). To credibly assess tunnel vulnerability and mitigation, one must consider the energetic source of the shock, the physics of geologic response to strong dynamic loads, and, ultimately, the physics of structural and material response and failure resulting from dynamic shock loading events.

L. Glascoe, Ph.D., P.E. (✉) • T. Antoun
Lawrence Livermore National Laboratory, Box 808, L-126 7000
East Avenue, Livermore, CA 94550, USA
e-mail: Glascoe1@llnl.gov; Antoun1@llnl.gov

A. Shukla et al. (eds.), *Blast Mitigation: Experimental and Numerical Studies*,
DOI 10.1007/978-1-4614-7267-4_2,

As these tools require significant computational resources, timely assessment involving a large range of threats and threat locations is difficult. Further, high fidelity tools are typically deterministic in nature and, therefore, must assume a specified set of material properties for the structure itself. Consequently, there is a need for (1) careful experimental validation of relevant phenomena used in the high fidelity models, and (2) reduced order capabilities that can rapidly evaluate effects associated with various threat sizes, threat locations, and system states while quickly highlighting uncertainties associated with structural and system response. This chapter presents an overview of high fidelity multiphysics modeling for tunnel structures and the need for careful experimental validation, as well as the building of reduced order models (ROMs) from these higher fidelity capabilities to quickly estimate the effects of explosions within a contiguous system of confined spaces such as tunnels.

2.2 Multiphysics Models and High Performance Computing

Most civil structures, and, in particular, tunnels are not designed to withstand the intense dynamic loads produced by an explosive detonation in the vicinity of the structure. These impulsive loads are significantly different in magnitude and duration from those expected under service conditions, and consequently they induce deformation modes and damage mechanisms not normally experienced by the structure. Conventional engineering mechanics and structural analysis techniques are not adequate for assessing tunnel vulnerability under these severe loading conditions. Novel methods that can account for the prevalent dynamic loading phenomena and their nonlinear interactions are required to assess tunnel vulnerability.

Scenarios of interest are initiated by an explosive detonation either inside or outside the structure but usually in close proximity. This is followed by the propagation of a blast wave in the case of an internal detonation, or a ground shock in the case where the explosive charge is placed outside the structure. The energy coupling associated with the interaction of those waves with the tunnel can cause damage to various elements of the structure, sometimes leading to catastrophic collapse. Assessing tunnel vulnerability therefore requires the ability to analyze the explosive detonation, the propagation of the stress waves, and the response of the tunnel. In short, one must be able to solve a multiphysics problem involving several nonlinear, interacting phenomena. Multiphysics hydrocodes are ideally suited for addressing this class of problems, and the remainder of this section is dedicated to the description of the modeling approach and the challenges associated with the various phases of the analysis.

2.2.1 Modeling the Energetic Material Source

Capturing the response of energetic materials in simulations of tunnel vulnerability is essential to ensure accurate representation of the time-dependent release of chemical energy, the ensuing blast effects, and the eventual coupling of the released energy into the structure. A wide array of explosives with widely varying energy densities and thermodynamic properties are usually considered in these simulations, ranging from military explosives and ANFO (ammonium nitrate–fuel oil), to improvised and homemade explosives.

For events triggered by explosives, the most important aspects of the behavior of the explosive are the energy release characteristics of the explosive and the pressure–volume response of the detonation product. For ideal explosives, these can be adequately described with a phenomenological equation of state calibrated using empirical data. The equation of state (EOS) most often used for this purpose is the Jones-Wilkins-Lee (JWL) EOS (Lee et al. 1968). The JWL EOS is an empirical pressure–volume equation that describes the adiabatic expansion of the detonation products. This equation contains several fitting parameters that are calibrated using experimental data. Due to its relative simplicity, and also because it has been demonstrated to accurately represent the behavior of many explosives, the JWL EOS is widely available in practically all the hydrodynamic and structural dynamics codes used in simulations involving explosive detonations. Standardized procedures have also been developed to determine the JWL EOS parameters using Chapman–Jouguet state parameters and data from cylinder expansion tests (Lawrence Livermore National 2012).

JWL is an appropriate EOS for ideal explosives, i.e., explosives with a relatively high reaction rate, and once calibrated, this EOS can be used to describe detonation phenomena under a wide range of conditions and for practically all explosive configurations of interest in tunnel vulnerability assessments. However, the behavior of nonideal, i.e., relatively slower reacting, explosives is typically more complicated than that of an ideal explosive, with the response often varying with the size and geometric configuration of the explosive charge. Characterizing the behavior of nonideal explosives therefore requires the use of more complex models and more sophisticated testing procedures (e.g., Fig. 2.1), with the data used to improve the model and the model used to better characterize the response of the explosive during detonation.

Most improvised and homemade explosives are nonideal and some of their response features during and after detonation cannot be adequately represented using the JWL EOS formulation. For those types of explosives where kinetic and other nonideal effects are important, more sophisticated models like the JWL++ (Souers et al. 2000), which accounts for time dependent reaction in prompt detonations, or the ignition and growth model (Lee and Tarver 1980), which accounts for heterogeneous mesoscopic mechanisms (friction, pore collapse, strain localization, etc.) using a phenomenological continuum approach, can be used for more accurate representation in hydrodynamic simulations. Advanced computer codes have also been developed to calculate the properties and thermodynamic state in nonideal heterogeneous energetic mixtures. CHEETAH (Fried 1995) and its predecessor TIGER (Cowperthwaite and Zwisler 1974) are two codes that fall into this category, with CHEETAH being the current industry standard. CHEETAH is a thermochemical code which uses physical and chemical properties to predict the performance of ideal and nonideal high explosives and explosive formulations. It employs Chapman–Jouguet detonation theory and performs molecular calculations to determine bulk explosive properties like pressure and detonation velocity. The output from CHEETAH can be used to characterize explosive EOS properties in hydrodynamic calculations.

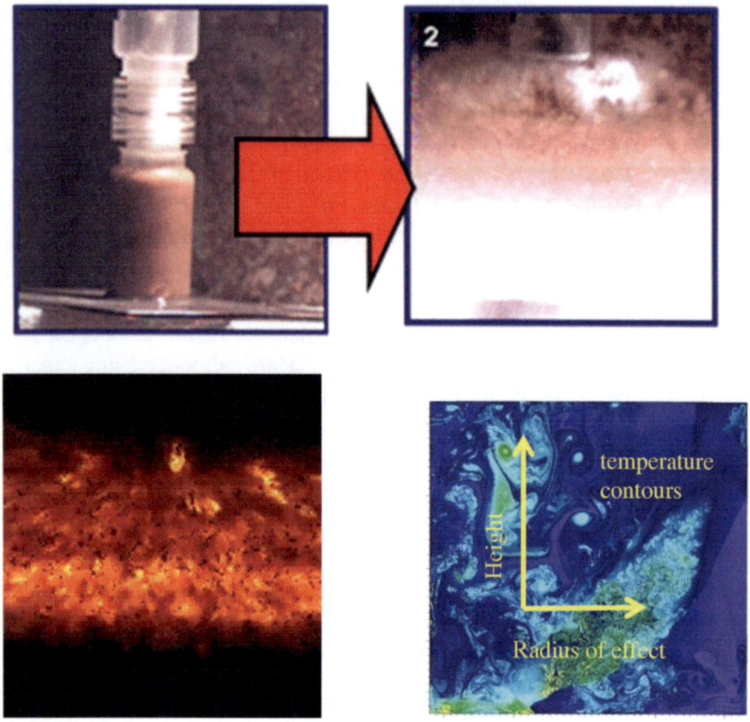

Fig. 2.1 Experimental (*top and lower left*) and numerical characterization (*lower right*) of nonideal explosives

Circumstances arise where the response of the explosive is highly nonideal, the reaction rates are variable and dependent on geometry and thermochemical state, and the EOS evolves with the rapidly changing local conditions. Under these circumstances, the thermochemical and hydrodynamic behaviors are fully coupled, and the notion of an EOS must be replaced with an approach that allows accurate representation of the coupled chemical and physical processes. Dynamic coupling of a thermochemical code like CHEETAH and a hydrodynamic code like ALE3D enables this important capability and allows accurate representation of the complex, interacting physical and chemical processes associated with the explosive detonation.

2.2.2 Modeling Geologic Materials

Earthen materials, including rock, sand and soils are the geologic settings in which tunnels are constructed. For this reason, the ability to accurately simulate the propagation of stress waves in geologic media, and the interaction of those waves with buried structures is of significant importance in assessing tunnel vulnerability and associated mitigation strategies. This requires the use of constitutive and failure models that capture the behavior of geologic materials under dynamic loading conditions. Continuum mechanics based constitutive models are currently the only practical models for use in large scale (field scale) simulations of the dynamic response of geologic materials (e.g., Rubin et al. 2000). These models rely on fitting parameters to phenomenologically describe the dependence of processes like yielding, bulking, damage and porous compaction on the deformation history and the rate of load application.

The material response in these models is generally correlated to laboratory measured properties like density, elastic sound speed, stress–strain data under a range of loading conditions, and measurements of the strength and failure envelopes. Models calibrated in this manner can be shown to reproduce the behavior of the geologic medium under dynamic loading conditions, as shown in Fig. 2.2. The figure compares simulated and measured particle velocity and displacement histories at two different ranges away from an underground nuclear explosion in granite (Antoun et al. 1999, 2001). The data shown in the figure is from the PILE DRIVER underground nuclear test (Perret 1968), a 61 kt nuclear explosion detonated at a depth of 462.8 m in the Climax Stock granitic outcrop of Area 15 at the Nevada National Security Site (NNSS), formerly known as the Nevada Test Site (NTS). The data in Fig. 2.2a correspond to ground motion 204 m away from the center of detonation (ground zero) along the shot horizon. The data in Fig. 2.2b was collected along the same radial, at a distance of 470 m away from ground zero. The material behavior at both these ranges is inelastic, and as shown the model closely reproduces the response of the medium at the locations of these measurements. Once calibrated, the model can be used to represent the geologic medium in large scale simulations aimed at assessing the dynamic response of buried structures.

Geologic materials are highly heterogeneous, and when those heterogeneities are not properly represented in the small laboratory samples used to extract material property data for model calibration, the resulting constitutive model will not adequately represent field scale behavior where such heterogeneities may influence, and even dominate, the dynamic response. A common manifestation of this effect is that models developed using laboratory data have been shown to overpredict the free field environment in large scale field experiments in fractured granite. This is largely due to the effects of joints and other heterogeneities that are present at the field scale, but not in pristine laboratory samples. Indeed, the good agreement between the simulated and measured motions shown in Fig. 2.2 required modifications to the model parameters originally calibrated using the static laboratory data. Specifically, a scaling law was introduced to degrade the yield and strength surfaces as shown in Fig. 2.3.

Phenomenologically, this is in line with experimental data that show the strength of granite and other geologic materials to be size-dependent, decreasing with increasing specimen dimensions (e.g., Hoek 1994; Hoek and Brown 1980). However, the manner in which this size scaling was introduced into the model was empirical. A function was added to the model to represent this effect, and its parameters were calibrated to achieve good agreement with the free field wave propagation data.

This modeling approach, whereby mechanistic data are used to calibrate the model, and the calibrated model is used to shed insight into the behavior of the

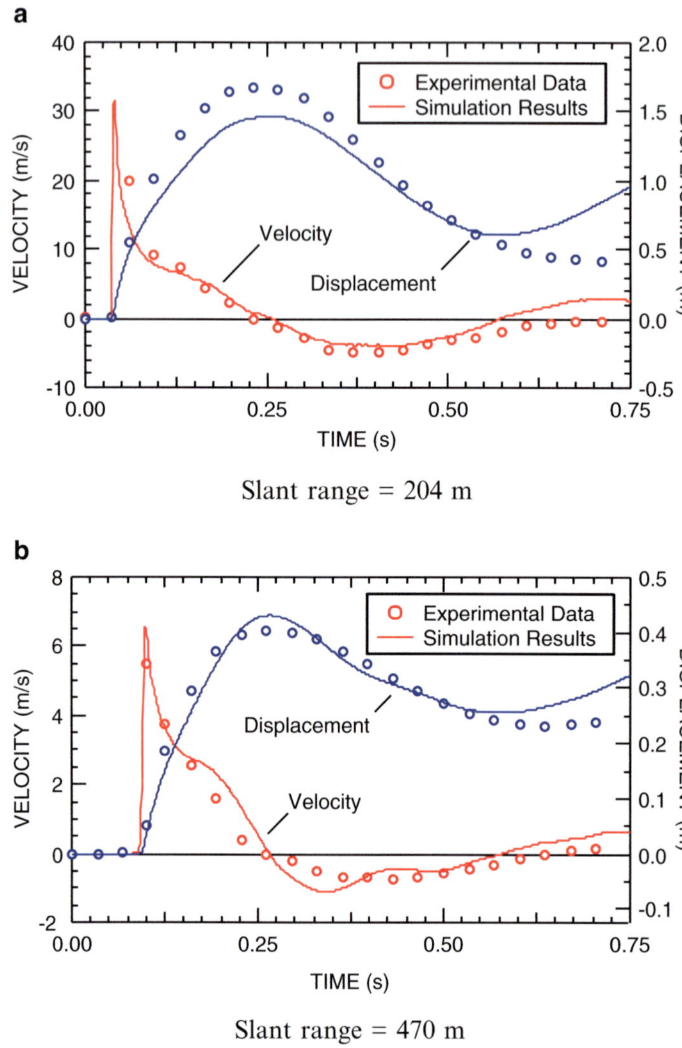

Fig. 2.2 Comparison of simulated and measured radial velocity and displacement histories at two different slant ranges away from the PILE DRIVER underground nuclear detonation

geologic medium during wave propagation, is reasonable when sufficient data are available to calibrate the model. When calibrated in this manner, the applicability of the model is limited to the geologic medium for which it was calibrated. Extrapolation to other locations where the joint spacing and/or orientation may be different from the original location will require recalibration of the constitutive model. This is illustrated in Fig. 2.4 which shows snapshots of the velocity field from two different 3D simulations of wave propagation in a jointed medium (Heuze and Morris 2007). The two snapshots on the left (Fig. 2.4a) correspond to a randomly jointed medium,

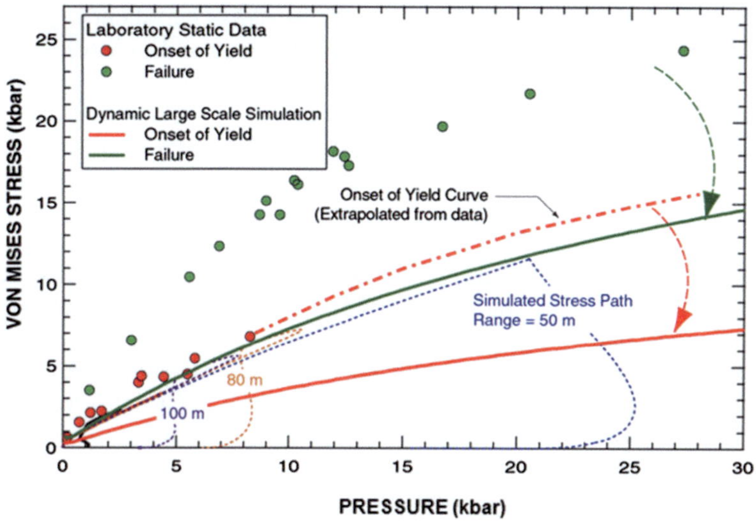

Fig. 2.3 Yield and failure surfaces for granodiorite used in the PILE DRIVER calculations (*solid curves*) together with the data from static experiments (Schock et al. 1973). *Dashed lines* are the stress path trajectories experienced by the material at different ranges away from the source of explosion

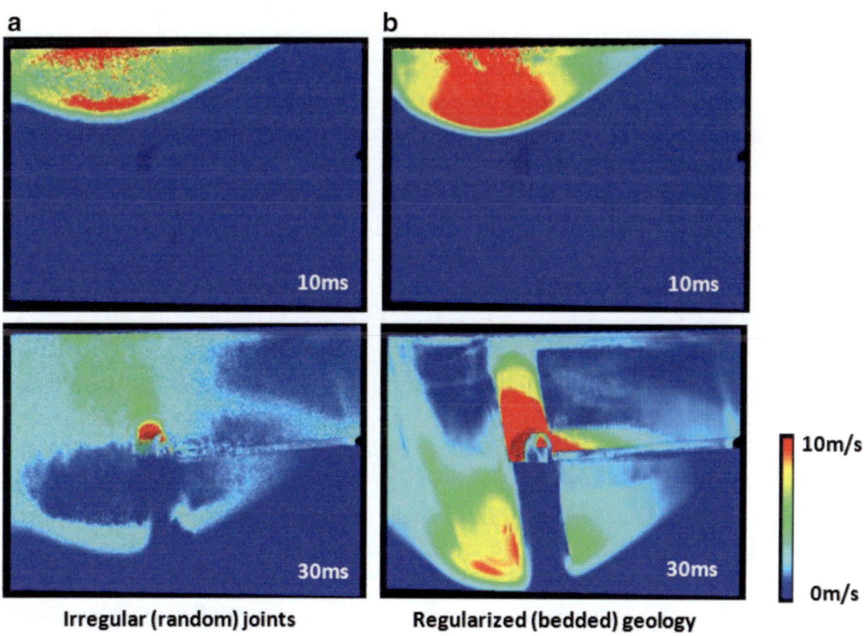

Fig. 2.4 Snapshots of the velocity field from two different LDEC simulations of wave propagation in a jointed geologic medium

while the snapshots on the right (Fig. 2.4b) correspond to a medium with bedding planes and regularized joints. Although the joint spacing and the contact properties used in the two calculations are the same, it is clear from Fig. 2.4 that joint orientation has a pronounced effect on the velocity distribution in the free field, and particularly in the vicinity of the tunnel. The constitutive model described above, which represents the state of the art in the groundshock modeling community, does not account for joint orientation and therefore is incapable of reproducing the kinds of effects shown in Fig. 2.4.

The currently used modeling approach is best described as "descriptive," and is distinguished from the sought-after "predictive" modeling approach wherein the effects of joints and other heterogeneities are taken into account using a rational physics based approach that relates the macroscopic continuum response of the medium to elemental properties like joint spacing, joint orientation, and interface properties. Recent advances in modeling capabilities coupled with modern high performance computing platforms enable physics-based simulations of jointed geologic media with unprecedented details, offering a prospect for significant advances in the state of the art. Initial steps demonstrating the feasibility of this approach have already been taken (e.g., Vorobiev et al. 2005; Vorobiev and Antoun 2011). As these novel approaches mature, they will be more tightly integrated in vulnerability assessment simulations thus reducing uncertainty in the simulation results and leading to more efficient and economical mitigation strategies.

Porous geologic materials present a unique set of challenges that must be addressed as part of a comprehensive constitutive model development strategy. Aside from the need for pressure–volume relations that take into account porous compaction and dilation, the question of saturation must also be addressed because under dynamic loading conditions, the response of dry porous geologic materials is markedly different from that of the saturated material. Furthermore, saturated soils, especially those with low cohesion, could undergo liquefaction, a phenomenon widely associated with structural damage during earthquakes. Liquefaction can also occur under shock loading conditions, leading to increased vulnerability and heightened risk of catastrophic structural failure.

Liquefaction occurs in loosely consolidated saturated soils when the shear resistance of the soil decreases as a result of increased pore pressure (or fluid pressure) during dynamic loading. Since the fluid only occupies the interstitial pore space, the pressure in the fluid increases in response to the decreasing volume brought about by porous compaction during loading. Under dynamic loading conditions like earthquakes and explosive applications, the rapid loading rate does not allow sufficient time for fluid migration to accommodate the transient increase in fluid pressure leading to an instantaneous loss of shear strength. This is accompanied by a transition from a solid-like response characterized by pressure-dependent strength to a liquid-like response governed by viscosity and hydrodynamic motion. Even if liquefaction does not occur, the decrease in strength associated with increased pore pressure in saturated geologic media during dynamic loading can have significant implications for the response of buried tunnels. Therefore, this effect must be incorporated in the constitutive model used to describe the behavior of the geologic medium in simulations of tunnel vulnerability.

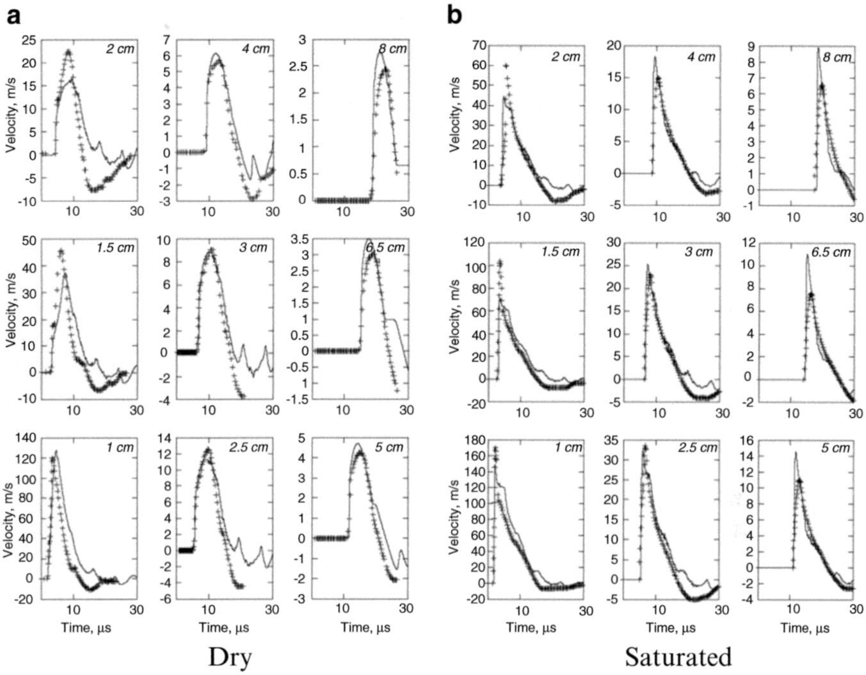

Fig. 2.5 Comparison of velocity waveforms for dry and saturated limestone at varying distances from the explosive charge. Experimental data (+) are from ref. (Gefken and Florence 1993) and simulation results (—) are from ref. (Liu et al. 2005)

The effective stress model, first introduced by Terzaghi for geotechnical applications, is used to account for the effect of saturation on the response of geologic media. This model has been used extensively to model the behavior of a variety of porous geologic materials including soils, concrete, and rock. In this formulation the "effective pressure" is introduced and defined as the difference between the total pressure and the fluid pressure (e.g., Jaeger and Cook 1976). The effective pressure is then used, instead of the total pressure, to determine the deformational and strength characteristics of the saturated material response.

The effective stress concept is particularly attractive because the saturated models can be trivially generated from the corresponding dry material models by providing a fluid equation of state. The model treats both fully and partially saturated materials and allows the use of arbitrary equations of state (from analytical equations, tabular data, or run-time databases) to be used for both the solid and the pore fluid. This is particularly useful at the high pressures associated with shock loading.

This effective stress model was used to simulate the behavior of porous limestone under shock loading (Liu et al. 2005). Experimental spherical wave data (Gefken and Florence 1993) for dry and saturated limestone were compared with results using the effective stress model. The model was first calibrated to fit the dry data as shown in Fig. 2.5a. The calibrated model was then used to predict the fully saturated response as shown in Fig. 2.5b. The significant differences in behavior between the

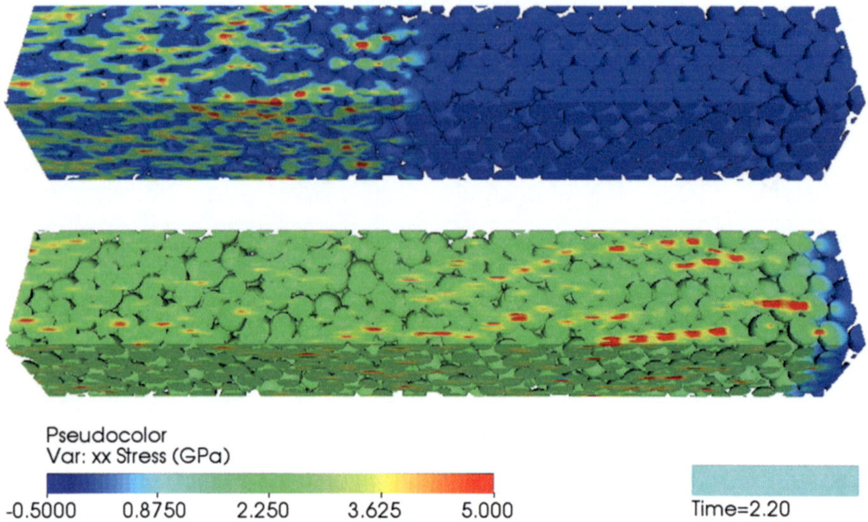

Fig. 2.6 Stress distribution in dry (*top*) and saturated (*bottom*) granular silica during shock loading. Significant differences in stress distribution and effective sound speeds are observed. The simulations were performed using GEODYN (2, Private Communication)

dry and saturated limestone are noteworthy. Also noteworthy is the fact that the effective stress model reproduces both the dry and saturated responses under shock loading with reasonable fidelity without changing any of the model parameters, thus demonstrating that the model can be reasonably expected to represent the behavior of variably saturated geologic materials in large scale simulations.

Terzaghi's effective stress model has been shown to work relatively well for most geotechnical applications, though significant deviations have been observed under pressures in excess of 100 MPa (Lade and de Boer 1997). These observations have also been corroborated with grain scale simulations (Fig. 2.6) of the response of representative volume elements of dry and saturated granular silica (SiO_2) (i.e., sand). At low pressures, the grain scale simulation results agree with the effective stress model predictions, but deviate significantly at high pressures. The development of improved modeling capabilities for saturated geologic materials is fertile ground for future research. An approach which combines grain scale simulations with experimental observations to elucidate the deformation mechanisms, leading to the development of a constitutive model that captures the main features of the material response, seems particularly promising and is being actively pursued.

2.2.3 Simulating Dynamic Structural Response

Dynamic simulations of the response of tunnels to explosive loading are complex multiphysics simulations that involve several nonlinear interacting phenomena with widely varying length and time scales. End to end vulnerability assessments often utilize one or more multiphysics codes capable of representing the events and interactions of interest. Simulations using high-performance structural and fluid mechanics finite element codes are run on teraflop-class supercomputers to characterize structural response to explosive shock. Modeling frameworks of varying complexity are used to capture computationally demanding fully coupled soil-fluid–structure interactions. Explosive threats under different conditions of confinement or placement have been modeled using the ALE3D (Nichols and The ALE3D team 2009) and Paradyn (DeGroo et al. 2008) finite element codes, as well as discrete element and smoothed particle hydrodynamics (SPH) approaches (Morris and Johnson 2009).

Explicit structural dynamics codes like ParaDyn (DeGroo et al. 2008) are often used for system level response to structural failures, and multiphysics hydrodynamics codes like ALE3D (Nichols and The ALE3D team 2009) are used for fully coupled blast analysis. ParaDyn, the parallel version of DYNA3D (Lin 2005), is an explicit, transient dynamic analysis code for solid and structural mechanics which uses a Lagrangian formulation. Component failures can be accommodated through element erosion and the conversion of failed elements into SPH particles which preserve the system's mass and momentum balances. The SPH particles can interact with other particles and intact elements which provides a means to transfer the kinetic energy of failed components into other parts of the structure.

ALE3D is an arbitrary-Lagrangian–Eulerian hydrodynamic analysis code that simulates the fluid and elastic–plastic material response. The code incorporates continuum mechanics, thermal diffusion, chemistry, incompressible flow, multiphase flow, and magneto-hydrodynamics. The detonation energy released by the explosive can be represented using a JWL equation of state or other more complex models as described previously. The resulting shock wave is propagated through the surrounding medium (e.g., air, water, or a geologic medium) using advection and mesh relaxation techniques to allow material flow through the mesh without element tangling, which could compromise the quality of the numerical solution. The code tracks the wave propagation and interactions (reflections), so the blast pressures applied to the structure vary temporally (arrival time) and spatially (stand-off distance) as the blast wave travels over and around the structure. The materials representing the structure can be held Lagrangian to best preserve material history parameters or allowed to advect if severely deformed or extensively damaged.

Energy coupling into the structure often involves propagation of a blast wave either in air, in the case of an internal blast, or in water or a geologic medium, in the case of an external blast. Among the blast wave parameters of interest in assessing structural damage and tunnel vulnerability are peak material stress, positive phase impulse, and peak particle velocity. Resolving these quantities adequately in a hydrodynamic simulation places stringent requirements on the mesh size. Additional resolution requirements arise from the need to represent construction details (e.g., rebar in a reinforced concrete tunnel liner) that are likely to affect the dynamic response of the structure. Meeting these requirements necessitate a cell size on the order of 10^{-2} m. Additionally, problems of interest usually span spatial domains on the order of tens of meters in size. The three-dimensional multiphysics nature of the problem, coupled with the requirements for high resolution over large spatial domains lead to simulations that are of such size as to require the use of high

Fig. 2.7 High performance simulation of dynamic structural-failure of a complex soil/structure system; illustrated are the plastic strain of steel liner and rebar (*left*) and the material strain in soil and structure (*right*)

performance computing platforms to perform physically reasonable simulations in realistic wall clock time.

Several material models are available to represent the component materials' deviatoric strength and pressure–volume relationship (i.e., equation of state). Structural components can be modeled using a combination of beam, shell, and solid elements. Frame members (be they steel, iron, etc.) are often modeled using beam elements with user-defined integration rules to represent the joist, girder, and column cross-sections and account for partial yielding through a cross-section. The constitutive response is represented by large deformation inelastic models that include plasticity models with options for representing hardening, damage and effect of strain rate on material response. Concrete members are typically modeled with solid elements and often use the Karagozian & Case (K&C) concrete model (Malvar et al. 1997, 2000). The K&C model determines concrete strength with respect to confining pressure, damage, and strain rate. The K&C model also allows a homogenized representation of reinforcement using volume fractions. Dry and saturated geologic materials are modeled using the GEODYN material model as described in a previous section. Figure 2.7 shows a simulation of an underground reinforced concrete tunnel subjected to explosive loading. In this simulation, the tunnel structure was modeled with the K&C concrete model, the steel rebar within the concrete were modeled using an elastic–plastic model with a strain to failure criterion for rupture, and the saturated soil medium was represented using the GEODYN model with effective stress. As shown in the figure, the concrete tunnel experiences severe damage, including tensile rupture of several rebar, but the overall structural integrity of the tunnel is not compromised.

Structural damage is time-dependent and often requires several levels of analyses over different time-scales to determine the ultimate failure or survival of a structure. During an explosive event, structural damage in the form of a breach may be apparent within milliseconds after the blast and may be best modeled using specific fully coupled codes such as described above and shown in Fig. 2.7. Longer term failure modes may be indirectly associated with a shock event and may occur at timescales on the order of seconds, e.g., longer term structural collapse, soil liquefaction, and the progressive collapse of a structure due to the failure of a critical support. Such longer term failure modes are likely to be realized by employing less computationally demanding capabilities that allow for longer temporal considerations.

Tunnel construction methods vary widely depending on the medium in which the tunnel is constructed and the intended function of the tunnel. Construction techniques vary from the reinforced concrete tunnel shown above, to cast iron and brick tunnels, to even unlined tunnels in competent rock where the strength of the rock is sufficient to maintain the stability of the rock mass. Assessing vulnerability of unlined tunnels presents a unique set of challenges which require highly specialized code capabilities to perform the analysis. As was previously discussed, rock masses are highly fractured, and those fractures have a controlling effect on wave propagation in the rock medium and on the response of structures buried therein. Assessing the stability of unlined tunnels therefore requires the utilization of a numerical method capable of explicitly accounting for the effect on joints (or fractures) on the response of the tunnel. Morris and Johnson (2009) developed the LDEC code specifically for performing this type of calculations. LDEC is a fully 3-dimensional, massively parallel code for simulating the stability of openings in fractured rock masses. LDEC represents the rock mass using a large number of polyhedral blocks that interact at their points of contact according to experimentally validated contact force laws. By directly simulating the discrete, blocky nature of rock masses, LDEC takes a fundamental approach to simulating the behavior of these systems while limiting the number of empirically derived model features.

LDEC was used to simulate the response of a tunnel complex to a near surface explosion in the vicinity of the tunnel (Morris 2004). The simulation was the largest of its kind and included a detailed facility spanning approximately 50 m with multiple tunnels and junctions. The fracture spacing in the rock was 30 cm, resulting in a computational domain of ~8 million polyhedral blocks and ~100 million computational elements. The facility was subjected to loading from a near-surface explosion, resulting in collapsed portions of the tunnels (Fig. 2.8). Simulations of the type shown in the figure are made even more daunting when geologic uncertainties are taken into account. In this case multiple runs are needed to assess the effect of variability in properties and rock texture (joint spacing and orientation) on the stability of the tunnel.

The closer a code is founded upon first principles of physics and chemistry (i.e., the less reliant on empirical calibration) the better the model is suited for problems of different scales and different applications. However, theoretical computational models can only approximate reality. Building confidence in the computational models requires extensive comparison to experimental data under loading and response regimes relevant to the application. Appropriate and carefully considered validation experiments can build confidence in the use of these computational tools. Ideally, validation experiments are true representations of the problem under consideration. This ideal is usually compromised: true experimental representations are costly or otherwise prohibitive, particularly when considering the importance of

Fig. 2.8 Sequence of snapshots showing the progressive collapse of a section of a complex where an access drift intersect one of the main functional areas of the facility

experimental repeatability. Such is the case for structural systems of interest (tunnels, bridges, etc.) particularly when considering different failure modes spanning large timescales (e.g., blast failure taking milliseconds versus structural progressive collapse taking seconds to minutes to hours). Confidence in a model's ability to represent these large, and difficult-to-test, structural systems can be enhanced by carefully considered component tests, or by carefully and appropriately scaled phenomenological tests.

In building confidence in modeling the blast vulnerability of a large structural system, e.g., an aircraft fuselage or a tunnel structure, one must establish the level of detail necessary to answer the question at hand. For instance, if an order of magnitude threat-size is sufficient for making a decision, then less model validation will be required to establish model confidence, whereas other programs may require higher resolution of threat particularly if a threshold is important. The following items must then be considered in any study: (1) model construction and assumptions as well as parametric sensitivity, (2) verification of solution (model-to-model comparison or comparison to analytic solution is often sufficient here), and (3) a testing plan to validate models that can help provide metrics of confidence. A testing plan could include the following:

- Full-scale structure tests. These nearly "full system" tests are expensive, difficult to control and repeat, and may be prohibitive for structures of most interest, i.e., for a fleet of aircraft, or specific tunnels. Nonetheless, these types of tests are closest to the actual scenario of concern and highlight, at the very least, the primary mechanisms of system failure that must be captured, for instance, connection failure due to blast followed by the longer time progressive failure due to external loading. Any historical tests of this nature are valuable but limited in number. Full-scale tests are most effective if coordinated with numerical model analysts to optimize both the test design and to provide a set of tests that lend themselves best to model comparison.

- Component tests. These most basic component tests are relatively inexpensive and controllable providing confidence in the most fundamental aspect of modeling a structure under blast loads, for example, plate deflection and plate failure. Modifications of plates to represent structural complications, e.g., connections or rebar, can be included in a test to study specific failure modes. Careful consideration of plate testing must include the geometry of the plate relative to blast size as well as the influence of plate boundaries.
- Scaled bridging system tests. A scaled representation of the important structural components of a large structure could be a relatively inexpensive bridge between the small-scale component testing and the full-scale testing. Testing under loading conditions, for example at different pressurization levels, would build confidence in the ability of the models to capture consistently the relevant failure modes and timescales.

Following development of a validation process (test and modeling plan), impartial judgment by, e.g., a review board of experts or another independent organization with the relevant expertise, of the proposed process as well as the results is helpful. A difficulty, of course, is finding the right experts (expertise in blast-structure response is not commonly found) who are available to review the validation effort.

Verification and validation of the code and the models at the appropriate scale and under representative loading conditions are essential to ensure accuracy of the numerical results. Considerable efforts have been made to validate the blast responses predicted by these simulation tools. Figure 2.9 shows representative examples of the code benchmarking and validation efforts performed to increase confidence in the ability of the codes to reproduce important aspects of the experimental observations. Efforts such as the US Army Engineering Research and Development Center's Precision Test Wall Study, the Defense Threat Reduction Agency's Divine Buffalo and Discrete Gemini test series, and a host of smaller lab and field studies are used to partially validate simulations of air and underwater blasts. Such validation efforts also provide opportunity to evaluate the utility of, for example, faster running simplified approaches to structural response incorporating, for instance, homogenized material assumptions. Sample validation studies, illustrated in Fig. 2.9, include full-scale concrete testing (e.g., Crawford et al. 2004; Noble et al. 2005), and water-tamping/bubble-collapse testing (e.g., Thrun et al. 1993; Couch and Faux 1996).

Scale model testing can be an effective complement to large scale simulations in the assessment of structure vulnerability. Such an approach was taken by Glascoe et al. (2010) to evaluate a mitigation strategy for the protection of submerged structures from a standoff underwater explosion. Underwater explosions are efficient at the propagation of energy. Coupling the relatively high density with the relative incompressibility of water makes for enhanced shock transmission when compared to an air-blast of the same threat size (Cole 1948). For enclosed air-filled structures or ship hulls below the waterline, damage associated with shocks from a nearby underwater blast can be much more severe than damage associated with an air-blast from a similar sized threat.

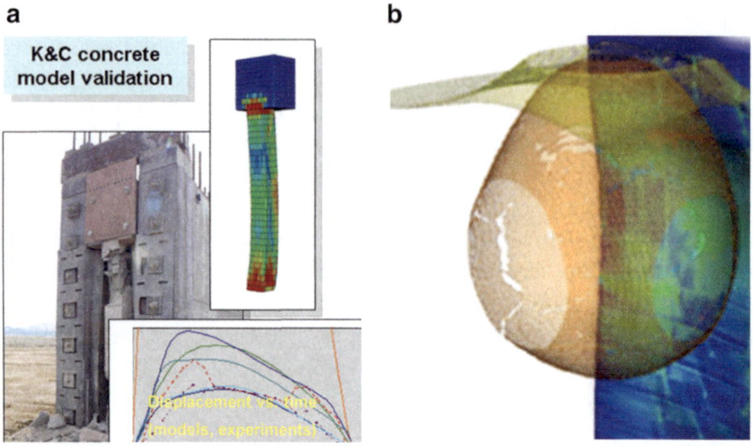

Concrete model validation and benchmarking

Underwater explosion validation and benchmarking

Fig. 2.9 Simulations of relevant experimental configurations are used to validate the codes and models used in numerical analysis of tunnel vulnerability

Effective mitigation of vulnerable structures against underwater shock is often best achieved by forcing increased standoff distance from the structure and by redistributing and/or breaking up the incident shock using strategically placed air pockets to maximize the impedance mismatch between materials, a function of density and sound speed. It is towards this end that the relative performance of a proposed mitigation option for a large concrete and steel structure was investigated: a belt of air-filled tubes encasing the waterside of an enclosed submerged structure.

A series of small (3-gallon) aquarium tests were used to evaluate the response of an instrumented aluminum plate to underwater explosions with and without mitigation near the charge. Several scaled 3-gallon aquarium experiments were designed including water-tamped tests without mitigation, water-tamped tests with mitigation, and an air-blast (control) test. A non-rupturing scenario was chosen for ease of model validation. The aquarium experiments were coupled closely with numerical simulation for pretest prediction and post-test comparison. The primary goal of the study is to couple the experimental results to numerical simulation for code validation and to build confidence in the proposed mitigation scheme. The ALE3D hydrodynamics code was used for experiment design, preexperiment deformation predictions, and post-experiment sensitivity studies.

The experiments were instrumented with a suite of diagnostic measurements that included pressure gages to measure the pressure history in the water, as well as strain gages and high speed photography to measure transient plate deformation during loading. Half of the aluminum plate was speckled for digital image correlation measurements. Figures 2.10 and 2.11 compare the temporal and full spatial field response of the aluminum plate's relative displacement from the DIC measurements. Figure 2.10 shows the response of the plate in the unmitigated test and Fig. 2.11 shows the response in the mitigated test. As shown in the figures, these measurements compare well with numerical predictions.

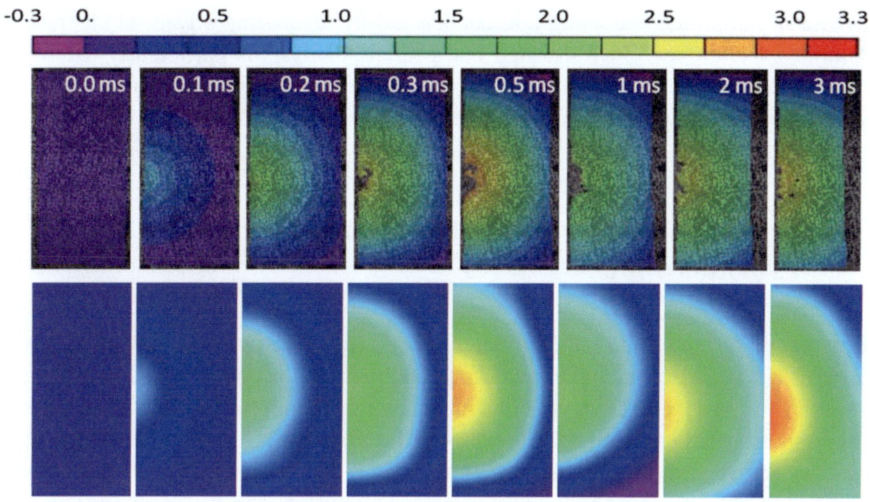

Fig. 2.10 Half-symmetry digital image correlation (DIC) experimental measurements in centimeters (*see color bar*) of relative plate deflection (*top images*) compare well with ALE3D simulation of plate deflection (*lower images*) for the unmitigated test over the first 3 ms

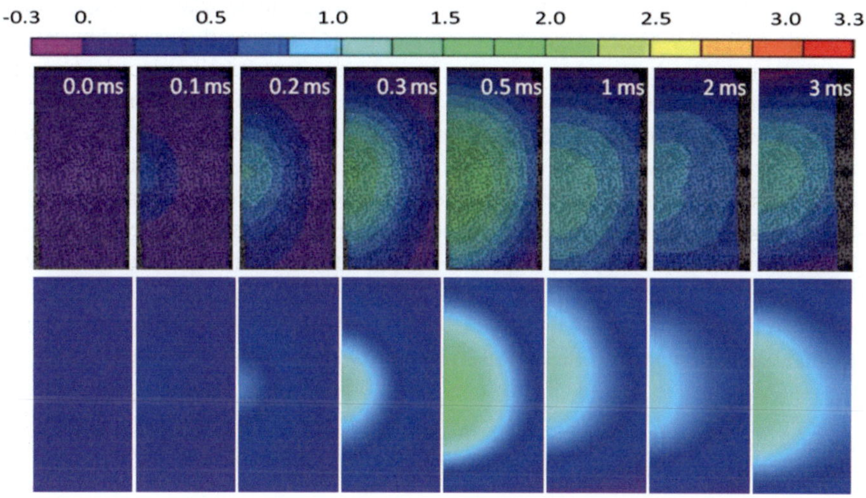

Fig. 2.11 Half-symmetry DIC plate deflection measurements (*top images*) in centimeters (*see color bar*) compare well with ALE3D simulation of plate deflection (*lower images*) for the mitigated test over the first 3 ms

Informed use of multiphysics modeling built upon experimental validation can form the basis of an end-to-end capability for analyzing and correcting structural vulnerabilities associated with explosive blast. These techniques are inherently computationally expensive due to the multi-dimensional nature of the problem space and the often necessary requirement for full coupling between fluid, solid and soil media phases. Computational expense can be minimized and results optimized by use of simplified modeling approaches and by leveraging advanced stochastic sampling techniques. Such detailed but a timely analysis is particularly useful to accommodate the needs of government agencies and infrastructure owners to guide security efforts for critical infrastructure nationwide.

2.3 A Reduced Order Model for Tunnels: Pressure/Impulse Curves

Emergency response and national defense needs demand fast-running yet accurate tools as a first line of technical reachback and planning for both infrastructure protection and target defeat. Expert-constructed reduced order models (ROM) can form the basis for such tools. In this chapter, we discuss two types of tunnel ROMs that quickly and accurately simulate the blast environment and associated tunnel damage by portably leveraging high performance computing. The validated approach is robust, stable and lends itself to greater usability and interpretation by an engineer or scientist competent in explosives and structural response.

It is difficult to conduct direct simulation of blast propagation in tunnels or pipelines due to a combination of factors. Tunnels have high length to diameter ratios, resulting in a lengthy computational domain relative to tunnel diameter. Such long domains coupled with the high resolution required to capture the boundary effects that dominate the flow make direct three-dimensional simulation of flow in long tunnels prohibitively expensive. Even two-dimensional simulations of tunnels becomes computationally expensive when considering multi-parametric study and cannot be utilized if the tunnel system in question contains inherently three-dimensional features such as bends or intersections. One ROM that focuses on pressure and impulse (P/I) employs a simpler algorithm originally developed for gas guns (Glenn 1990, 1997) and captures the essential physics of blasts in tunnels (Glenn 2001; Neuscamman et al. 2011a, b), but runs in minutes on standard personal computing hardware. The approach taken for building this particular ROM for internal blasts in tunnels is described in detail below.

2.3.1 Building a Physics-Based P/I ROM

The P/I ROM for internal tunnel blast solves the one-dimensional fluid flow equations of mass, momentum and energy. The effects of wall drag are accounted for in the momentum equation using a friction factor, applicable to both laminar and turbulent flows. For simulation of an in-tunnel blast, the P/I ROM couples several one-dimensional representations of the tunnel and blast into a higher dimensional representation. Specifically, the code includes a spherical flow region for the detonation that is coupled to one-dimensional axial flow through the tunnel segments. By varying the cross section of the tunnel along its length, it is possible to account for the effect of platforms (larger cross section) and trains (restricted cross section) upon the blast wave propagation and attenuation. The code can predict the effect of an arbitrary number of bends in a tunnel and supports coupling to additional one-dimensional segments to simulate the effect of tunnel intersections upon the shock wave. This ROM is ideally suited for providing pressure and impulse histories along a complex tunnel/station configuration to examine, for example, down-tunnel blast effects on personnel or impulse loading of structural components.

2.3.2 *Validation of a P/I ROM*

As discussed above, validation of any code is necessary before confident use. The P/I ROM has been validated against several sets of data involving a range of threat sizes and tunnel/pipe configurations (Glenn 2001). We compare the output of the P/I ROM to higher-fidelity two-dimensional axisymmetric hydrocode calculations using the ALE3D code as well as results from two small-scale high-explosive experiments done in Vicksburg at the now ERDC Laboratory.

Blast pressure data were obtained from a series of small-scale high explosive experiments of confined airblast (Lunderman and Ohrt 1997). In the experiment, a straight pipe section was simulated with a 243 mm inner diameter steel pipe (10 in. schedule 80 steel pipe). Pipe sections were connected with heavy bolted flanges to form an assembled tunnel about 7.75 m in length. The inlet was constructed with steel plates to emulate the entrance to a real tunnel in a mountainside. The far end of the pipe was left open. Pressure profiles were obtained using diaphragm-type air blast gages flush-mounted to the wall of the pipe. Spherical explosive charges were formed from hand-packed hemispheres of composition C-4. The hemispheres were packed around a detonator to approximate a center-detonated charge. The charge was suspended by the firing line of the detonator to the centerline of the pipe. A series of detonations were conducted, with varying charge size and location. We focused on the 15.7 g C-4 charge located inside the pipe, four pipe diameters (D) from the inlet. This configuration was simulated with the two codes assuming rigid walls and the ALE3D simulation calculating the 2D axisymmetric solution.

Peak pressure and impulse results from each simulation and the experiment are shown in Fig. 2.12. Both simulations show good agreement with the experimental data. The STUN calculation performs better than ALE3D for this metric, and agreement to experiment increases for peak pressure at x/D locations greater than one. Very close to the explosive source, both models under-predict the peak pressure measured in the experiment. However, further downpipe, when the flow has developed into normal shock wave propagation, we see excellent agreement. ALE3D peak pressure results also improve with distance down the pipe, but impulse is consistently under-predicted. Full pressure and impulse profiles at 15.7 diameters down the tunnel are shown in Fig. 2.12; the STUN profiles at this measurement position are virtually congruent with the experimental ones.

The second set of simulations compares to data from an experiment involving an explosion in a steel pipe where the blast wave encounters a sudden decrease in cross-section (Lunderman et al. 1993). In this case, a 340 g charge of C-4 explosive

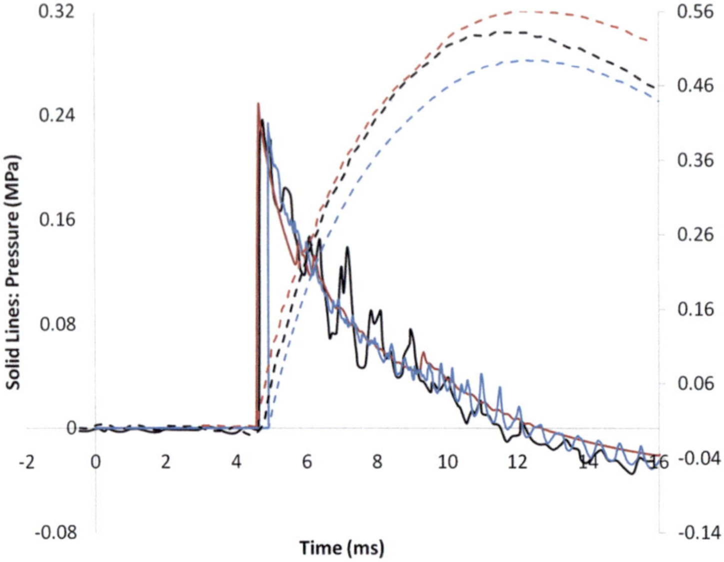

Fig. 2.12 Peak pressure (*left axes*) and impulse (*right axes*) comparison of 1D ROM prediction (*red lines*), a 2D hydrocode prediction (*blue lines*) and the experiment of Lunderman and Ohrt (*black lines*)

(5.04 cm × 5.04 cm × 8.23 cm) was centered in a 1.8 m long section of heavy duty steel pipe with an internal diameter of 14.6 cm (the detonation chamber). Connected to one side of the detonation chamber were a set of 4 m long segments of smooth Schedule 80 steel pipe with an internal diameter of 7.36 cm. On the other side, a similar arrangement was used except that the center two segments were replaced by grooved pipe. Pressure profiles were obtained using diaphragm-type air blast gages flush-mounted to the wall of the pipe; no pressure gages were used in the detonation chamber. For the purposes of this study, the focus is on the smooth side and the fourfold decrease in cross section that occurs at the end of the detonation chamber. The experimental setup was again simulated assuming rigid walls with the ROM and ALE3D codes, and the ALE3D simulation ran in 2D. The numerical results from both codes are in good agreement in terms of both pressure and impulse (Fig. 2.13).

The reduced order physics code for estimating blast effects in pipelines or tunnels performs exceptionally well when comparing to experimental data for a smooth, straight pipe. Down tunnel from a contraction, STUN overestimated peak pressure and impulse compared to both the experiment and 2D axisymmetric ALE3D simulation. However, since the goal is to create a fast-running tool, the relative calculation times should be considered. The 1D calculation in each study was carried out on a single processor PC in less than a minute of clock time whereas the 2D calculations required 2,300 h (96 h on 24 processors) for the single pipe simulation and 18 h (6 h of clock time with 3 processors) for the two pipe with a contraction simulation.

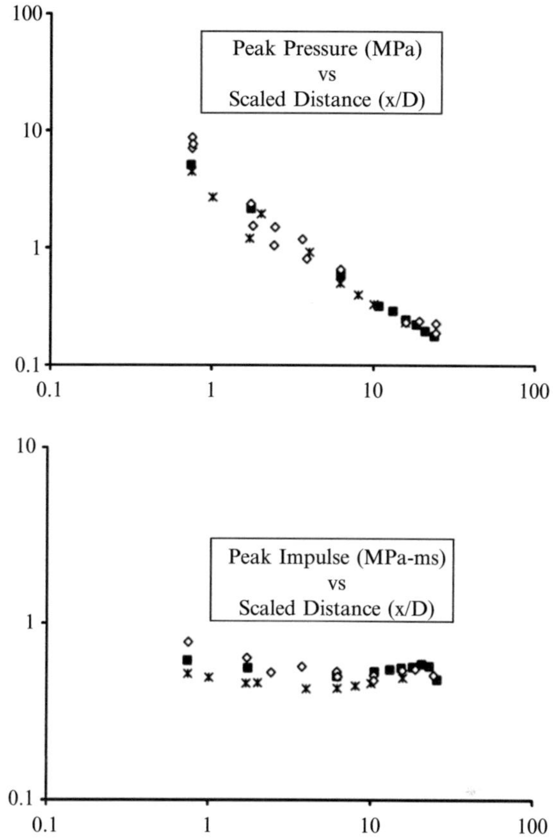

Fig. 2.13 Modeled peak pressure and impulse (*top and bottom plots*) by the 1D ROM (*filled square*), the 2D hydrocode (*open circle*), and experimental results (*multiplication sign*)

2.3.3 P/I ROM Example: An Explosive in a Complex Tunnel System

An example of a complex hypothetical single tunnel system is useful for illustrating the utility of the P/I ROM for tunnels. Figure 2.14 illustrates a tunnel with a 1,000 m straight section running to the left of the explosive source and a complex system of partial blockages (contractions), expansions, and multiple drifts to the right of the explosive source. The black dots indicate points where time varying details of pressure and impulse per unit area can be acquired after simulation. This particular example is computed quickly (tens of seconds on a 2012 single processor laptop) producing both pressure and impulse per unit area histories at various locations. Figure 2.15 illustrates peak overpressure as a function of distance from the charge both down the straight section (left) and the complex section (right). Peak pressure observed across the tunnel is useful information when considering life-safety in a tunnel away from a blast, or when considering the design of secondary structures in

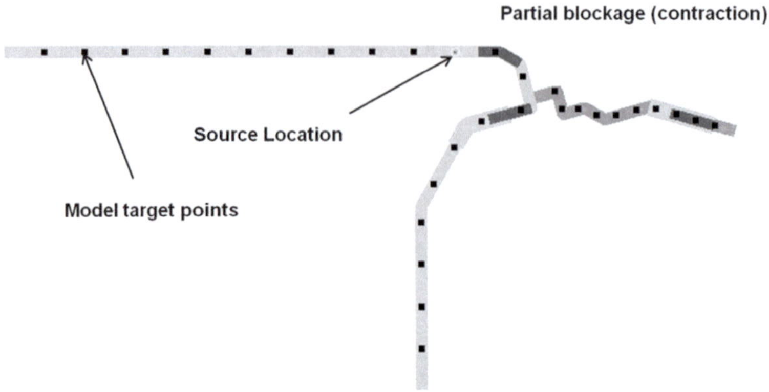

Fig. 2.14 A hypothetical complex tunnel system is constructed to illustrate an example use of the P/I ROM. Shown here is a tunnel with a 1,000 m "straight tunnel" running to the left of the explosive source and a "complex tunnel" system of partial blockages (contractions), expansions, and multiple drifts to the right of the explosive source

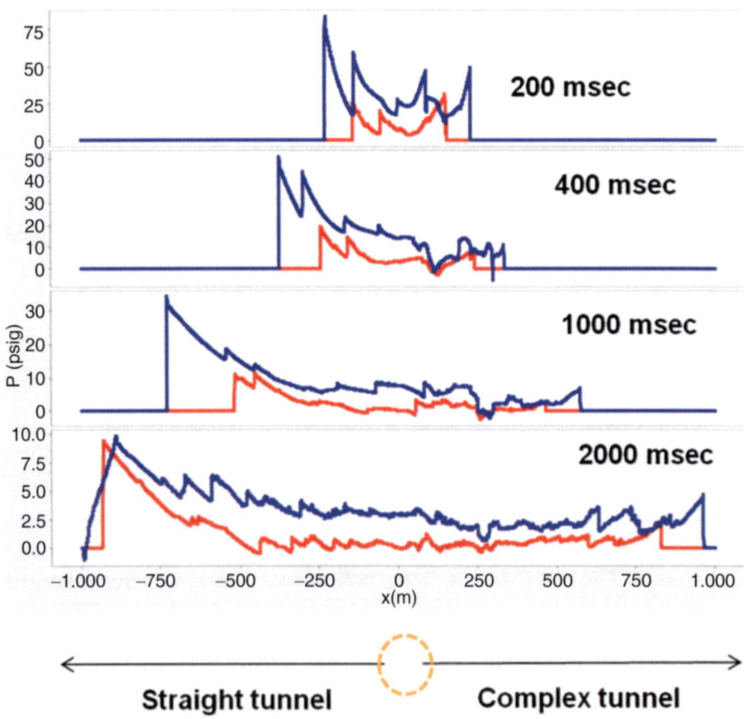

Fig. 2.15 Pressure profiles in time propagating down the left-hand "straight tunnel" and right-hand "complex tunnel" of Fig. 2.14

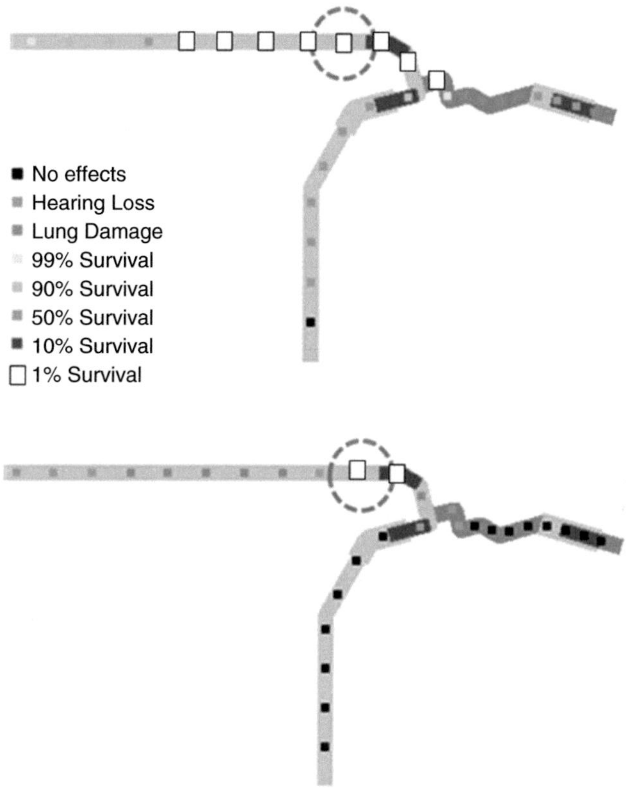

Fig. 2.16 Down-tunnel blast effects on people using P/I fragility curves on scenarios of the system in Fig. 2.14; shown are the effects of a large explosive (*top*) and a small explosive (*bottom*). Results are illustrated on a point by point basis for output purposes only

a tunnel system (e.g., emergency lighting, ventilation structures). It is notable that the pressures could be significantly higher if the explosion were tamped by the presence of a train or other large object located near the charge.

The severity and extent of personnel injuries along the tunnel, as well as in the stations and drifts, can be estimated by using fragility curves of Bowen et al. (1968) relating human survivability to the peak overpressure and duration of exposure to the blast overpressure as plotted in Fig. 2.16. Survivability estimates are evaluated at the same target points specified for the pressure time histories (refer to Fig. 2.14). The target points are gray-scale coded according to the survival probability. In this example, the blast overpressures have dropped below the threshold for 1 % survival in the complex section of the tunnel (right) sooner than in the straight section of the tunnel (left) for both a large and a small explosive. Such information can be used in conjunction with system usage data to estimate the number of casualties that might occur for a particular scenario to assist stakeholders in response planning.

2.4 A Reduced Order Model for Structures: Complex Failure

The simplified physics code discussed above allows tool users to determine the characteristics of the blast propagation along a tunnel; however, a fully 3D calculation is required to estimate near field damage, e.g., structural failure, for a given threat and tunnel configuration and these calculations cannot be accommodated in fast-running mode. Accordingly, we have used a statistical learning approach to deal with this problem. Using ALE3D, a series of parametric calculations was performed to identify bounds on the values of charge size, standoff distance, and wall thickness that may result in structural breach. First, a high fidelity multiphysics model of a generic but representative structure is created that can be run and rerun under different threat and tunnel conditions. Careful consideration of the trends gleaned from these runs can be used to generalize multi-parameter "response curves" for rapid assessment (Fig. 2.17).

2.4.1 Building a Complex Failure ROM: HPC Response Analysis

Numerical analyses are conducted for a hypothetical steel- (rebar) reinforced concrete tunnel or pipe system of limited length. Constrained boundary conditions are imposed at the bottom of the soil surrounding the tunnel, the side boundary, and the

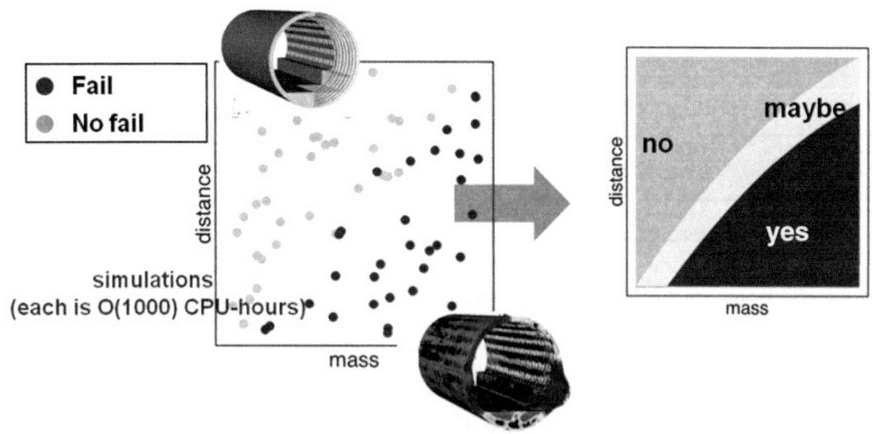

Fig. 2.17 Careful deliberate selection of high fidelity calculations of a full three-dimensional soil/fluid/structure blast calculation can be used to create an interpolated lookup of a specific mode of structural response (in this case "breach") under a range of variable conditions, e.g., variable charge mass and charge standoff distance

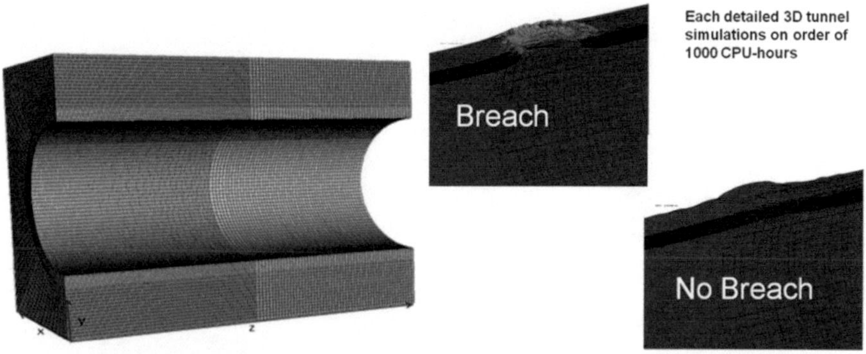

Fig. 2.18 An example of a breached concrete structure; the breached zone is defined by a dual criteria capturing tensile and compressive failure. Note that this figure shows "breach" by removing elements considered to be failed by our dual criteria

vertical section of soil and concrete portion where nodal velocities are assumed to be near zero. Continuous pressure boundary and zero fluxes of energy and mass are assumed for the vertical face of the air portion within the structure's inner radius. The center of the spherical charge is located, and its radius is calculated, based on the standoff, charge weight and fixed density of charge. Standoff is defined as the distance from the tunnel wall to the closest point on the charge's surface. Structural damage is quantified using a combined index of shear dilation and density change in the concrete. Both metrics indicate the separation of aggregate and mortar necessary for the formation of rubble.

Damage evaluation must take place in two steps because the continuum representation of the structure is incapable of developing a breach hole. Elements that begin the simulation connected must remain so throughout the duration of the simulation. To determine breach, a description must be made of the portions of the concrete material, if any, that have sustained heavy damage (Fig. 2.18). Information must be extracted from the phenomenological plasticity model to determine damaged sections. First, portions of the structure that have been determined to contain rubble are highlighted. Compressive damage is tracked well by material dilation. Dilation is a natural and observed phenomenon in actual concrete and other geological materials that occur when the concrete is experiencing shear stress and aggregate begins to "roll over" the neighboring particles. Large dilation strains emulate a real-world loading scenario where aggregate has been separated violently from the mortar. Based on Split Hopkinson pressure bar tests and expert analysis, a 0.85 % volumetric strain *from dilation* is chosen to represent damage (see Fig. 2.19). Another indication of concrete damage severe enough to cause breach is a decrease in density which can be similarly compared to dynamic Brazilian split cylinder tests (Fig. 2.20). Density is tracked by the constitutive material model and its decrease can be attributed to cracking and spall failure.

Next, the calculated breached volumes based on the dilation and density results are inspected by an analyst to see if they are likely, in his or her opinion, to

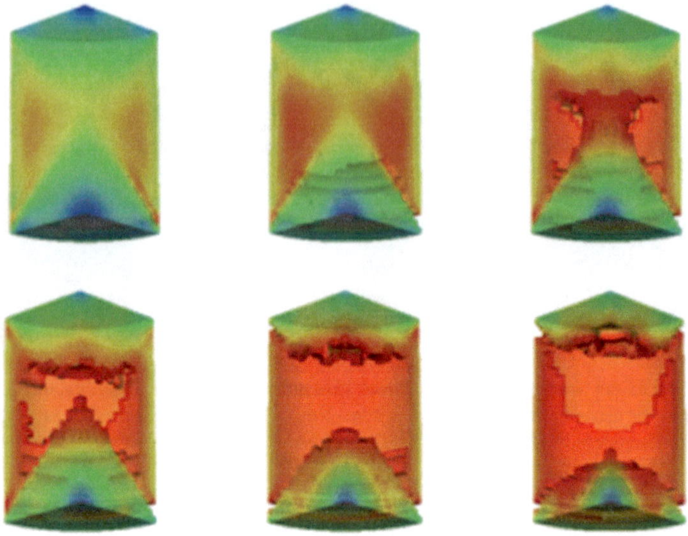

Fig. 2.19 For breach determination, split-Hopkinson pressure bar (SHPB) tests were modeled for dilation and rubble formation; shown are modeled SHPB concrete specimens in quarter symmetry and the evolution of modeled failure cones from top left to bottom right under ever increasing strain rate. Note with increasing strain, failure cones begin to form illustrating a dilation failure model represented in the breach

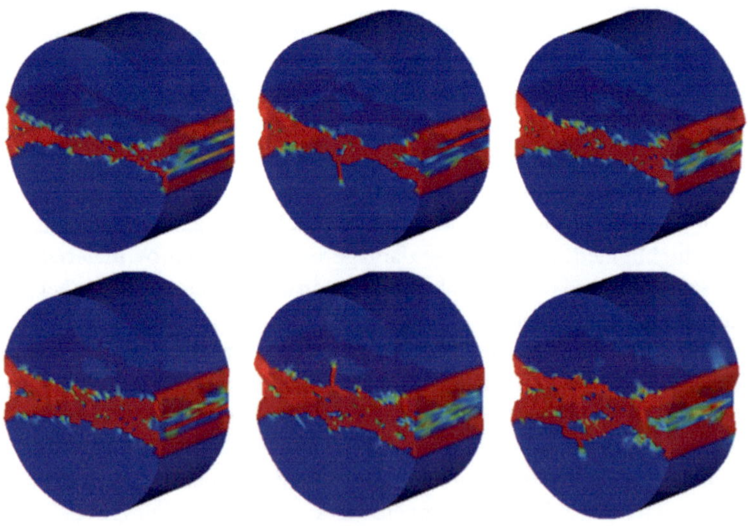

Fig. 2.20 Breach criteria for damage in tension can be represented as a density drop and compared to dynamic "Brazilian" split cylinder tests. Shown here starting from top left moving to bottom right is a modeled decrease in density (*lighter portions*) for a concrete sample under a direct side impact of the specimen representing the split cylinder test

constitute a breach. Finally, the analyst's decisions about breach, which include an option for indeterminate or "possible breach," are fed into a statistical algorithm that produces breach curves, as described in the following section.

2.4.2 Building a Complex Failure ROM: Statistical Emulation

Due to the high computational cost, breach determination via the full 3D physics model is not feasible for every structure/charge configuration of interest. Statistical emulation can be used to supplement a limited number of model runs and provide breach predictions for the full parameter space. Such emulators can also be used as a part of an adaptive sampling strategy, to ensure that costly 3D simulations are run only at points with high predictive value. Once we have adequately characterized the parameter space through such adaptive sampling, we use our emulator to generate "breach curves" as shown in Fig. 2.22. These curves, generated from a relatively small number of precomputed simulation runs, are included in our fast-running tool, and can provide users with breach risk assessments for a wide variety of threats almost instantly.

A support vector machine (SVM) is a classifier that generalizes a maximal margin hyper plane: the linear surface that divides two classes with the largest "margin" from existing data. Only a few points (support vectors) define the classification boundary. The SVM extends this framework to nonlinear boundaries by using kernel functions to map data into a higher dimensional space, and by introducing slack variables to allow for classification errors. SVMs can be used for adaptive sampling. Choosing new points which lie along the current boundary will maximize the information gained. This is critical for computer experiments, where sampling may be costly.

Our methodology was developed to address several challenges inherent, and in some cases unique, to our breach prediction problem: a priori constraints on our decision boundary, uncertainty in breach classification, and the need for adaptive sampling. Our first constraint relates to the need to incorporate knowledge of breach behavior into our classifier. In particular, our classifier needed to exhibit proper monotone behavior (increasing charge weight increases risk, increasing wall thickness decreases risk, etc.) and the fact that there is no risk of breach below a known threshold value. We also had to address the uncertainty in breach characterization. Recall that, in addition to simulation runs that clearly failed or clearly did not, there were a number of "possible breach" cases. Finally, we needed a method that had the ability to improve its own results through adaptive sampling. While we began this stage of the project with several hundred existing simulation runs, they were not necessarily in the neighborhood where the boundary for our final breach criterion would be. Therefore, we needed a way to identify "high value" points for new simulation runs, which would allow us to further refine our classifier.

All of these factors lead to the decision to use parametric support vector machines (PSVMs; Lennox and Glascoe 2011). The PSVM is a form of the standard support vector machine (Cortes and Vapnik 1995) that allows the form of the classification boundary to be constrained by expert knowledge (Fig. 2.21).

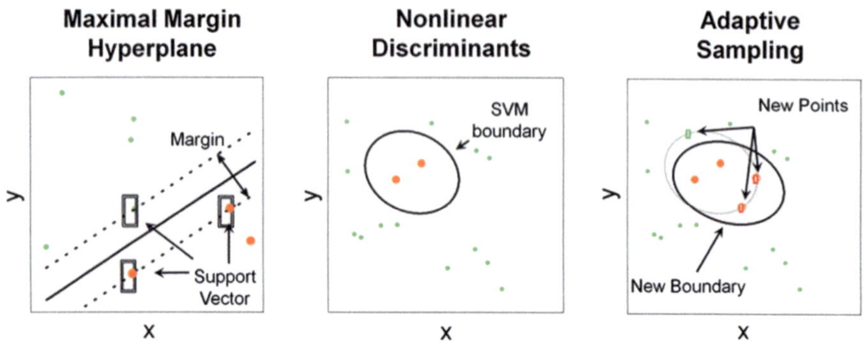

Fig. 2.21 Support vector machine approach with adaptive sampling

Our problem involves three classes: breach (also referred to as "red" for the color used in the breach curve in our tool), no breach (green), and possible breach (yellow). We define the yellow region by fitting two separate classifiers. The first classifier combined all possible breach points with no breach points, and constituted the boundary between red and yellow areas on our plots. The second made the more conservative assumption that all possible breaches should be counted as breaches, and this classifier was used to generate the green/yellow boundary. As is typical for active learning with SVMs (Basudhar and Missoum 2008), our strategy focuses on selecting points on these classification boundaries.

To select a new set of test points for boundary refinement we generated a candidate set of points located on either the conservative (green/yellow) or anti-conservative (yellow/red) SVM boundary. For every point in the candidate set, we determined the closest neighbor in the current data, and the distance to this closest neighbor. The point with the maximum distance from its nearest neighbor ("maximin" criterion) was chosen to be added to the dataset, see Fig. 2.22 (right plot). All variables were scaled to fall between zero and one to ensure that those with the largest ranges did not dominate this point selection process. The process was then repeated for the remaining candidates, until we found a reasonable test set size (on the order of 25–50 points), and then these were processed using our ALE3D model. The new points were then added into the dataset, new SVM boundaries were calculated, and the adaptive sampling process repeated (Fig. 2.23). The finalized breach curves are accessed through the same user interface as the 1D model and take a form similar to Fig. 2.22.

2.4.3 A Failure ROM Example: With and Without Liner Mitigation

An example of a hypothetically failing structure is useful for illustrating the utility of a structural failure ROM. Here we consider two sets of hypothetical reinforced

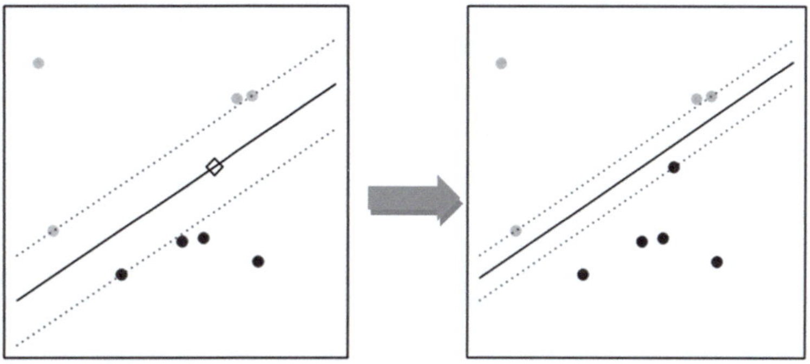

Fig. 2.22 Notional standoff versus threat size breach curve (*left plot*). The *right plot* shows a linear SVM separating two classes. The *diamond* represents a highly informative point (located on the decision boundary and distant from existing points) that would be selected by our adaptive sampling method for evaluation

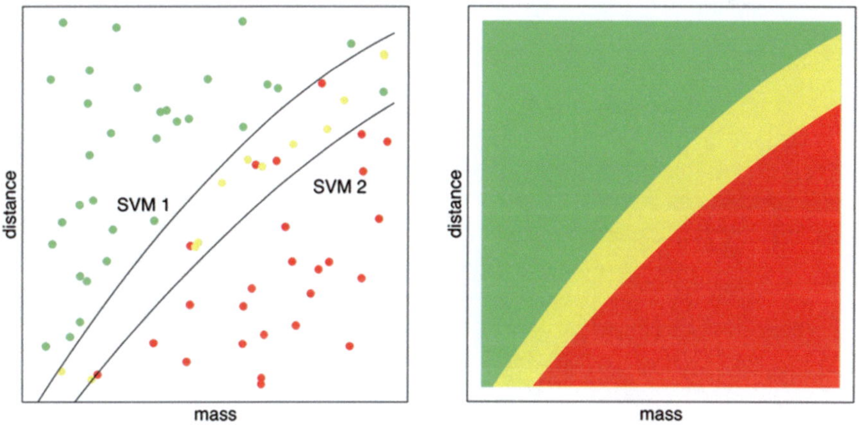

Fig. 2.23 Structural response output in terms of breach (*red*), no breach (*green*), and indeterminate (*yellow*). *Left* is a plot of the discrete three-dimensional output and the use of sequential vector machine (SVM) curves separating no breach (SVM 1) and separating breach (SVM 2). The resulting continuum shown *right* is used as output to the fast running tool

concrete structures with and without a ¼″ thick steel plate on the outside part of the structure. Breach is considered if the structure "daylights" after the modeled explosive event, i.e., a clear hole exists through both concrete and plate. The response curves discussed in the subsection above are employed and generated using the SVM methodology; two SVMs segregate the breach into likely breach, possible breach, and unlikely breach. The two sets of response curves in terms of standoff and threat size are illustrated for a baseline concrete structure (Fig. 2.24) and for a structure of twice the concrete thickness (Fig. 2.25). All results are expectedly

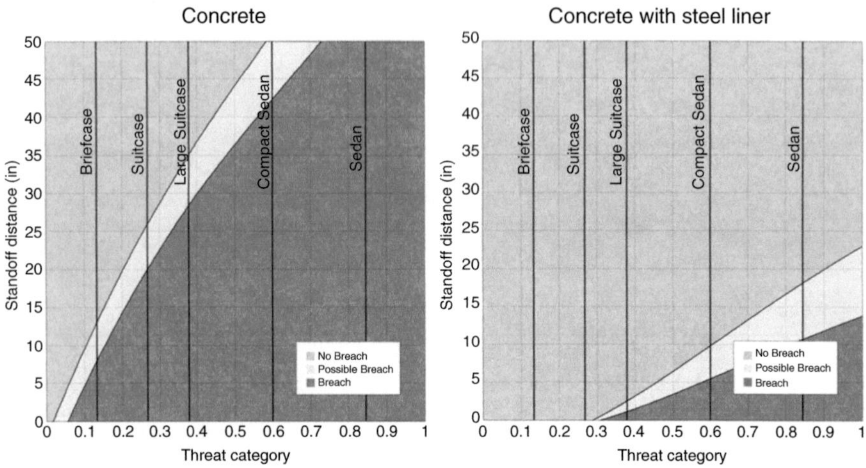

Fig. 2.24 A hypothetical reinforced concrete structure as represented by the complex failure ROM. On the *left* is the structure without an external steel liner; on the *right* is the structure with a ¼″ thick steel liner

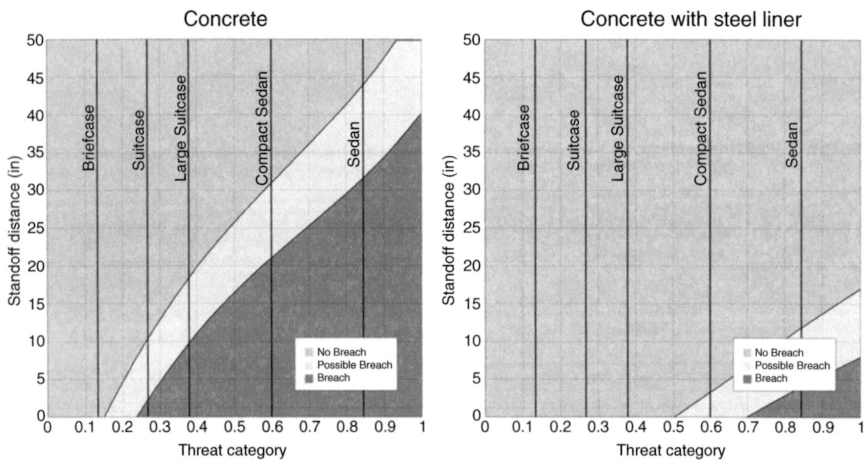

Fig. 2.25 A hypothetical reinforced concrete structure twice as thick as the structure in Fig. 2.24. On the *left* is the structure without an external steel liner; on the *right* is the structure with a ¼″ thick steel liner

monotonic with respect to breach failure and standoff and threat size. The thick concrete structure requires a much larger charge at, for example, 25 in. of standoff (a vehicle bomb) than the thinner concrete structure (a suitcase bomb). The efficacy of the steel mitigation is obvious: neither structure is likely to be breached at 25 in. of standoff if a ¼″ steel liner is present.

2.5 Summary

High performance computational hydrodynamic and structural analysis tools are useful for predicting the response of structures to impulsive dynamic loads, assisting in the implementation of vulnerability corrective measures to reduce overall risk. As these tools require significant computational resources, timely assessment involving a large range of threats and threat locations is difficult. Consequently, there is a need for capabilities that can rapidly evaluate effects associated with various threat sizes, threat locations, and system states while quickly highlighting uncertainties associated with structural and system response. In this chapter we highlighted the importance for high fidelity models of tunnels with considerations of the energetic source of the shock, the physics of geologic response to strong dynamic loads, and, ultimately, the physics of structural and material response and failure resulting from dynamic shock loading events. Careful validation of these high fidelity tools is necessary to provide confidence in model prediction. Faster running, phenomena-focused, reduced order models can be constructed from these validated higher fidelity capabilities to quickly estimate the effects of explosions within a contiguous system of confined spaces such as tunnels. Emergency response and national defense needs demand fast-running yet accurate tools as a first line of technical reachback and planning for both infrastructure protection and target defeat. Expertly constructed reduced order models can form the basis of such tools whether the simplification is by using a 1D limited physics code instead of a 3D code when assumptions permit, or by providing a response curve fitted to previously analyzed high fidelity multiphysics analysis.

Acknowledgments This work is performed under the auspices of the US Department of Energy by Lawrence Livermore National Laboratory under Contract DE-AC52-07NA27344. We would like to thank the Department of Homeland Security (DHS) Science and Technology (S&T) Directorate and the Transportation Security Administration for their sponsorship and assistance in these efforts. We particularly acknowledge the assistance and leadership of Dr. John Fortune (S&T), Dr. Mary Ellen Hynes (S&T), and Mr. Christopher McKay (TSA). We would also like to acknowledge the technical contributions of Lew Glenn, Larry McMichael, Chad Noble, Ilya Lomov, Stephanie Neuscamman, Oleg Vorobiev, Joe Morris, Greg Schebler, Allen Kuhl, Ben Liu, Ed Kokko, Ana Kupresanin, and Will Elmer. This document is LLNL-BOOK-589112.

References

Antoun TH, Vorobiev OY, Lomov IN, Glenn LA (1999) Numerical simulations of an underground explosion in granite. 11th topical conference on shock compression of condensed matter, Snowbird, UT, June 27–July 2, 1999

Antoun TH, Lomov IN, Glenn LA (2001) Development and application of a strength and damage model for rock under dynamic loading. In: Elsworth D, Tinucci J, Heasley K (eds) Proceedings of the 38th U.S. rock mechanics symposium, Rock mechanics in the national interest. A.A. Balkema, Lisse, The Netherlands, pp 369–374

Basudhar A, Missoum S (2008) Adaptive explicit decision functions for probabilistic design and optimization using support vector machines. Comput Struct 86(19–20):1904–1917

Bowen LG, Fletcher ER, Richmond DR (1968) Estimate of man's tolerance to the direct effects of air blast, Defense Atomic Support Agency Report DASA-2113

Cole H (1948) Underwater explosions. Princeton University Press, Princeton, NJ

Cortes C, Vapnik V (1995) Support-vector networks. Mach Learn 20:273–297

Couch R, Faux D (1996) Simulation of underwater explosion benchmark experiments with ALE3D. Lawrence Livermore National Laboratory, UCRL-CR-123819, 1996

Cowperthwaite M, Zwisler W (1974) TIGER computer program documentation. Stanford Research Institute Technical Report, AD/A-02 791, March 1974

Crawford, Malvar JL, Shi Y (2004) Verification and validation for DYNA3D models of RC components subjected to cased and uncased munitions. Karagozian & Case, TR-99-10.3, October 26, 2004

DeGroo A, Sherwood R, Hoover C (2008) ParaDyn, a parallel nonlinear explicit, three-dimensional finite element code for solid and structural mechanics. Version 8.1 user manual, LLNL-SM-407000, Lawrence Livermore National Laboratory

Fried L (1995) CHEETAH 1.22 user's manual. Lawrence Livermore National Laboratory, Livermore, CA, UCRL-MA-117541, Rev. 2

Gefken PR, Florence AL (1993) Spherical waves in jointed limestone. SRI International, Defense Nuclear Agency Report No. DNA-TR-92-122

Glascoe L, Noble C, Reynolds JG, Kuhl A, Morris J (2009) Securing transportation systems from high explosive threats. Proceedings of the 2009 IEEE international conference on technologies for homeland security, Waltham, MA, LLNL-PROC-411500, May 11–12, 2009

Glascoe L, McMichael L, Vandersall K, Margraf J (2010) Underwater blast experiments and modeling for shock mitigation. 14th Det Symposium

Glascoe L, Elmer W, Glenn L, Kupresanin A, McMichael L, Minkoff S (2011) Proceedings for DHS University summit, Washington, DC, 7 February

Glenn LA (1990) Design limitations on ultra-high velocity projectile launchers. Int J Impact Eng 10:185

Glenn LA (1997) On how to make the fastest gun in the West. Proceedings of the international workshop on new models and numerical codes for shock wave processes in condensed media, St. Catherine's College, Oxford, UK, September 15–19, 1997

Glenn LA (2001) Simulating MARVEL with the STUN Code. LLNL UCRL-ID-143993, June 6, 2001

Heuze F, Morris J (2007) Insights into ground shock in jointed rocks and the response of structures there-in. Int J Rock Mech Min Sci 44:647–676

Hoek E (1994) Strength of rock and rock masses. ISRM News J 2(2):4–16

Hoek E, Brown ET (1980) Underground excavations in rock. The Institution of Mining and Metallurgy, London

Jaeger JC, Cook NGW (1976) Fundamentals of rock mechanics. Halsted, New York

Lade PV, de Boer R (1997) The concept of effective stress for soil, concrete and rock. Géotechnique 47:61

Lawrence Livermore National Laboratory (2012) LLNL explosives reference guide, Lawrence Livermore National Laboratory, online limited access website, October 2012

Lee E, Tarver C (1980) Phenomenological model of shock initiation in heterogeneous explosives. Phys Fluids 23:2362–2372

Lee E, Hornig H, Kury J (1968) Adiabatic expansion of high explosive detonation products, UCRL 50422, Lawrence Livermore Laboratory, May 1968

Lennox KP, Glascoe L (2011) Constrained classification for infrastructure threat assessment. In: Proceedings for the IEEE international conference on technologies for homeland security, 15–17 November, Waltham, MA

Lin J (2005) DYNA3D: a nonlinear, explicit, three-dimensional finite element code for solid and structural mechanics user's manual. UCRL-MA-107254

Liu BT, Lomov I, Vorobiev O (2005) Simulation of shock loading in saturated geologic materials. In: Khan AS, Khoei AR (eds) Dislocations, plasticity, damage and metal forming: material

response and multiscale modeling. Proceedings of plasticity '05: the 11th annual symposium on plasticity and its current applications. NEAT, Fulton, Maryland, pp 346–348
Lomov I, Pember R, Greenough, Liu B (2005) Patch-based adaptive mesh refinement for multi-material hydrodynamics. In: Five-Laboratory Conference on Computational Mathematics, Vienna, 2005
Lunderman C, Ohrt AP (1997) Small-scale experiments of in-tunnel airblast from internal and external detonations. 8th International symposium on interactions of the effects of munitions with structures, April 21–25, McLean, Virginia
Lunderman CV, Welch CR, Joachim CE, Furey JS (1993) Effects of tunnel wall roughness on airblast from internal detonations – results from small-scale experiments. 64th Shock & vibration symposium, October 25–28, Fort Walton Beach, Florida
Malvar L, Crawford J, Wesevich J, Simons D (1997) A plasticity concrete material model for DYNA3D. Int J Impact Eng 19(9–10):847–873
Malvar LJ, Crawford JE, Morrill KB (2000) K&C concrete material model release III – automated generation of material model input, TR-99-24.3. Karagozian & Case, Structural Engineers, Glendale, CA
McGlaun JM, Thompson MG, Elrick MG (1990) CTH: a three-dimensional shockwave physics code. Int J Impact Eng 10:351–360
McMichael LM, Noble C, Margraf J, Glascoe L (2009) Fully-coupled blast effects modeling of large critical infrastructure. ASCE fifth congress on forensic engineering, LLNL-CONF-411649, November 10–15
Morris J (2004) Thunder science runs: simulation of hard and deeply buried targets using LDEC, UCRL-PRES-206648
Morris J, Johnson S (2009) Dynamic simulations of geological materials using combined FEM/DEM/SPH analysis. Geomech Geoeng 4(1):91–101
Neuscamman S, Glenn L, Schebler G, McMichael L, Glascoe L (2011a) Simulating small-scale experiments of in-tunnel airblast using STUN and ALE3D. LLNL-TR-498891
Neuscamman S, Glenn L, Glascoe L (2011b) A fast-running, physics-based tool for explosives in tunnels: model validation. Paper presented at 82nd Shock & Vibration Symposium, October 31–November 3, Baltimore, Maryland. LLNL-PRES-509019
Nichols AL, The ALE3D team (2009) Users manual for ALE3D, An arbitrary Lagrange/Eulerian 2D and 3D code system, Volume 2, Material and chemical models, Lawrence Livermore National Laboratory, UCRL-SM-404490
Noble C, Kokko E, Darnell I, Dunn T, Haggler L, Leininger L (2005) Concrete model descriptions and summary of benchmark studies for blast effects simulations. Lawrence Livermore National Laboratory, UCRL-TR-215024
Noble CR, Wemhoff AP, McMichael LD (2008) Thermal-structural analysis of the MacArthur Maze Freeway collapse. ASME 2008 summer heat transfer conference, Jacksonville, FL, August 10–14, 2008
Perret WR (1968) Free field ground Motion in granite. Report No.POR-4001. Sandia Laboratory, Albuquerque, New Mexico
Rubin MB, Vorobiev OY, Glenn LA (2000) Mechanical and numerical modeling of a porous elastic-viscoplastic material with tensile failure. Int J Solids Struct 37:1841–1871
Schock RN, Heard HC, Stephens DR (1973) Stress–strain behavior of a granodiorite and two graywackes on compression to 20 Kilobars. J Geophys Res 78(26):5922–5941
Souers P, Anderson S, Mercer J, McGuire E, Vitello P (2000) JWL++: a simple reactive flow code package for detonation. Propell Explos Pyrotech 25:54–58
Thrun R, Goertner JF, Harris GS (1993) Underwater bubble collapse against a flat plate. 1992 Seneca Lake test series data report, 1993, NSWCDD/TR-92/482
Vorobiev O, Antoun T (2011) Equivalent continuum modeling for nonlinear wave propagation in jointed media. Int J Numer Meth Eng 86:1101–1124. doi:10.1002/nme.3094
Vorobiev O, Lomov I, Antoun T (2005) Numerical simulation of wave propagation in jointed rock media. In: Oñate E, Owen DRJ (eds) Eighth international conference on computational plasticity, Barcelona, 2005

Chapter 3
Numerical Analysis of the Response of Biomimetic Cellular Materials Under Static and Dynamic Loadings

Ashkan Vaziri and Ranajay Ghosh

Abstract Biological structures distinguish themselves in preserving material heterogeneity for better mechanical behavior in contrast to man-made structures where they are deliberately removed through design. In this book chapter, we explore the distinct advantages in mechanical behavior brought about by regular, irregular, and functionally graded heterogeneities inherent in these cellular materials under quasi-static and dynamic crushing as well as impact loading. In particular we investigate energy dissipation and deformation modes during such loadings. We also investigate the effect of increasing self-similar structural hierarchy and cell wall material plasticity on the mechanical properties and failure behavior of these structures. Specifically, we discuss how mechanical response can be tuned using structural variations alone in particular, the tuning of strength to stiffness envelopes. Therefore, we show that material hierarchy can potentially reverse the conventional practice of design of structures using a variety of materials rather than vice versa. We also show how variation of structure can fundamentally change the deformation and damage patterns of these structures under dynamic crushing and impact loading. Finally we discuss some recent advances in fabrication and characterization of novel ultralight structures which exploit these essential self-similar properties of biological structures to create materials with remarkable stress–strain behavior strikingly different from the very materials originally used to make them.

A. Vaziri (✉) • R. Ghosh
Department of Mechanical and Industrial Engineering, Northeastern University, 209 Snell Engineering Building, 360 Huntington Ave, Boston, MA 02115, USA
e-mail: Vaziri@coe.neu.edu

A. Shukla et al. (eds.), *Blast Mitigation: Experimental and Numerical Studies*,
DOI 10.1007/978-1-4614-7267-4_3, © Springer Science+Business Media New York 2014

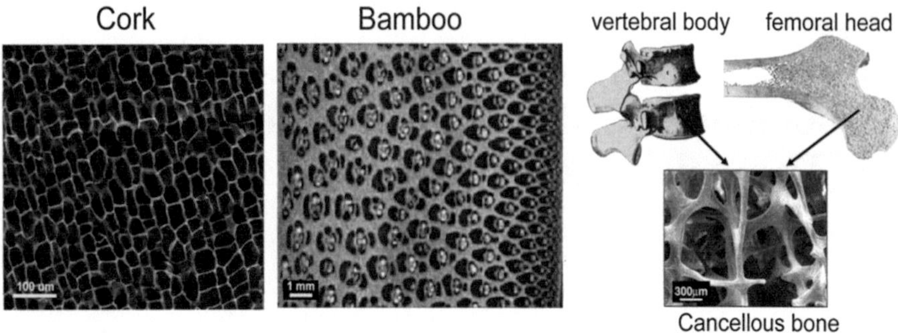

Fig. 3.1 Examples of cellular structures in nature. Cork has random heterogeneity, bamboo has a functionally graded cellular organization, and bone has a highly heterogeneous structure with hierarchical organization

3.1 Introduction

The constituents of biological materials generally have compositions, shapes, sizes, and spatial distribution leading to a multi-scale hierarchical organization as found in wood, vertebrate bones and teeth, mollusk shells, arthropod exoskeletons (Vincent and Currey 1980; Vincent 1990), and so on. In contrast to man-made structural systems—where material heterogeneity is avoided as a nuisance—the hierarchical structure and inherent heterogeneity of the biological materials are known to enhance their fundamental material properties such as elasticity, damage tolerance, and toughness (Vincent and Currey 1980; Vincent 1990; Tai et al. 2007), which are crucial for the overall function and survival of the biological systems. In fact, the overall properties of a biological material only rarely reflect the bulk properties of its constituents, depending rather on the geometry and topological arrangement at different scales (Fabritius et al. 2009). Figure 3.1 displays three biological systems with intricate cellular structures, displayed in order of increasing structural and functional complexity (from left to right). In each system, the toughness and energy absorption characteristics of the cellular structures are crucial for the biological system function and survival. For example, cork is the outer bark of a tree and has intricate heterogeneous cellular organization resulting in remarkable energy absorption. Other examples of cellular biomaterials are elk antlers, banana peel, and bone. Specifically the heterogeneous and hierarchical structural organization have been recognized as key factors in the elk antler's superior energy absorption and toughness crucial to its survival in the wild (Chen et al. 1999).

From the structural point of view, cellular structures have properties that are much superior compared to the properties of the base material, including high strength-to-weight ratio and energy absorption (Gibson and Ashby 1999). Topological irregularities and heterogeneity are well known to affect the behavior of cellular structures. In this book chapter, we provide a selected set of results on the performance and behavior of cellular structures and lattice materials with architectures inspired by the

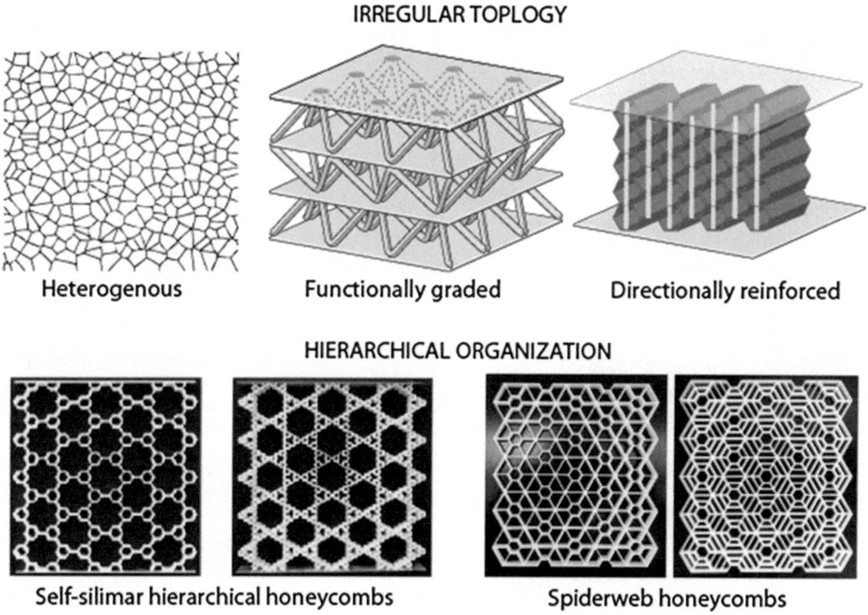

Fig. 3.2 Examples of biomimetic cellular structures

nature. Figure 3.2 displays examples of biomimetic cellular structures: The graded cellular structures are a novel class of material, where variations in cell size, shape, and wall thickness result in a functional variation in the relative density and organization of the cellular structure. These structures are promising candidates for creating novel multifunctional properties with superior properties compared to cellular structures with uniform distribution. Cellular structures with irregular and graded structural organization could promote the materials energy dissipation at the macro-level. The design of hybrid cellular structures that are based on directional reinforcement of the metal cellular structures with elastomer components is inspired by studying the structural organization of cellular materials in *Kingdom Plantae*. In this context, fiber reinforcement is used ubiquitously to provide additional structural support while also serving as a path for nutrients to different parts of the plant. Two examples of self-similar hierarchical cellular structures are also shown.

In this book chapter, first, we discuss the crushing behavior of regular, irregular, and functionally graded cellular structures (Sect. 3.2). This section is divided into three subsections, each dealing with a particular aspect of topological organization starting first from regular honeycombs, moving on to random irregularities, and finally to functionally graded variation in the structure. Another significant feature of biological systems is the inherent hierarchy present in their constitution. We devote Sect. 3.3 to discuss this effect in some detail via two subsections. In the first subsection we discuss the calculation of effective mechanical properties in detail. The dynamic crushing behavior is then left for the next subsection. In Sect. 3.4 we

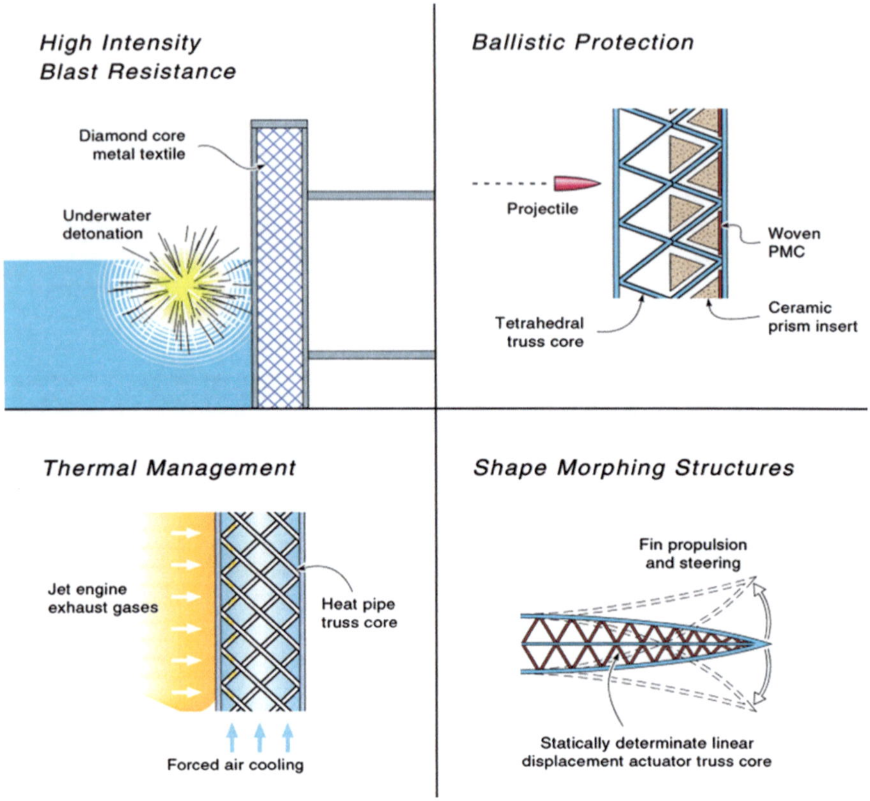

Fig. 3.3 Ultralight cellular material concepts to defeat fragments and the blast effects created by explosions while also providing options for thermal management and shape morphing

highlight some recent advances in other hierarchical structures and close the chapter in Sect. 3.5 with concluding remarks.

3.2 Dynamic Crushing of Regular and Heterogeneous Cellular Structures

The application of cellular structure ranges from the architectural masterpieces of Antonio Gaudi (Nonell and Levick 2001) to thermal insulators (Lu and Chen 1999) and three dimensional scaffolds for tissue engineering (Hollister 2005; Hutmacher 2000). One of the key applications of the cellular structures is the design of structural components with superior energy absorption and impact resistance. The basic applications pertaining to these characteristics are packaging of fragile components (e.g., electronic devices) and various protective products such as helmets and shielding. Another emerging application in this area is usage of cellular structures as the core material for metal sandwich panels, which are shown to have superior performance over the counterpart solid plates of equal mass under shock loading (Dharmasena et al. 2009; Liang et al. 2007; Mori et al. 2007; Mori et al. 2009; Rathbun et al. 2006; Vaziri and Hutchinson 2007; Vaziri et al. 2007; Wadley et al. 2007; Wadley et al. 2008; Wei et al. 2008; Xue and Hutchinson 2003; Xue and Hutchinson 2004). Figure 3.3 shows examples of not only the possibilities for threat protection but also the ability of cellular systems to simultaneously support structural loads, provide thermal management, and be deployable (through shape morphing). Another multifunctional advantage of cellular structures is active cooling. The fluid (e.g., air or water) can flow through the openings of the cellular structures, allowing removing the excessive heat and minimizing thermo-mechanical stresses (Dharmasena et al. 2009; Liang et al. 2007; Mori et al. 2007; Mori et al. 2009; Rathbun et al. 2006; Vaziri and Hutchinson 2007; Vaziri et al. 2007; Wadley et al. 2007; Wadley et al. 2008; Wei et al. 2008; Xue and Hutchinson 2003; Xue and Hutchinson 2004). Thus, cellular structures can be used for simultaneous load bearing, structural protection, and active cooling for spreading the heat in components that undergo extreme heat flux such as compact electronic cooling devices and ultralight actively cooled aerospace structures.

In this section, we discuss the role of heterogeneity and property variation in the framework of a simple 2D cellular structure called honeycomb. We focus our discussion on the dynamic crushing behavior and energy absorption of these structures. As the first step, we briefly discuss the dynamic crushing behavior of a regular hexagonal honeycomb, which is used widely in engineered applications (Sect. 3.2.1). We then introduce topological irregularities in the organization of these honeycombs starting with random variation discussed in Sect. 3.2.2 and then functional variation in the relative density in the crushing direction in Sect. 3.2.3. We discuss the role of each of these topological alternations on the dynamic crushing behavior of honeycombs.

3.2.1 *Dynamic Crushing and Impact Behavior of Regular Honeycombs*

Lu et al. (2005), Rakow and Waas (2004), and Shuang and Boming (2011) carried out numerical simulation of dynamic crushing of regular honeycomb structures. In their work a finite element study of honeycomb structures with regular hexagonal cell shape made from aluminum was used to study their impact and in-plane crushing behavior. In all the calculations presented in this study, the relative density of the honeycombs was varied by changing the thickness of the cell walls since it is well known that for a regular hexagonal honeycomb, the effective yield stress can be calculated from $\sigma_{Yc} = 0.5\rho_c^2\sigma_Y$, where ρ_c is the relative density of the honeycomb, defined as the area fraction of the cell walls with respect to the structure's dimensions (Ajdari et al. 2011). The numerical results showed that

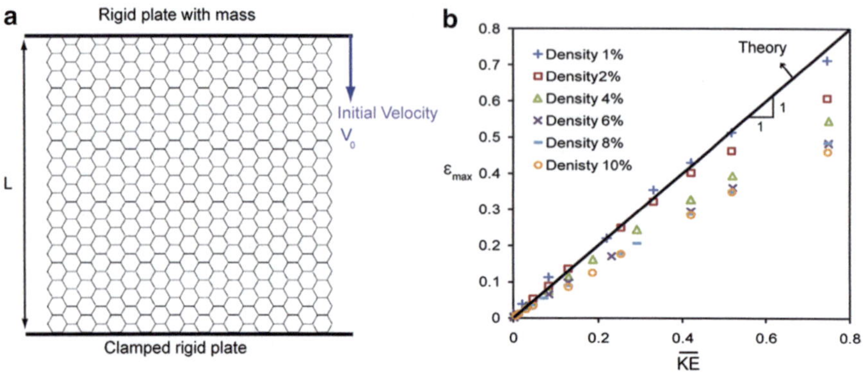

Fig. 3.4 Impact response of a regular hexagonal honeycomb. (**a**) (*Left*) Schematic of the finite element model. (**b**) (*Right*) The maximum crushing strain of honeycomb versus the normalized initial kinetic energy of the rigid plate

varying the cell wall size did not have a significant effect on the crushing response of regular hexagonal honeycombs, as long as the relative density of cellular structure is kept constant.

To simulate the impact response of the hexagonal honeycomb, a rigid flat plate with mass, M, was modeled to impact the honeycomb with an initial velocity, V_0, as shown schematically in Fig. 3.4a (Gibson and Ashby 1997). The honeycomb was taken to be clamped at its bottom surface. Periodic boundary condition was applied on each side of the structure to avoid the influence of the model boundaries on the simulation results (Harders et al. 2005). After the impact, the rigid plate to rest as it crushed the honeycomb structure and finally detached from it with a velocity that was much lower (generally less than 2 % of the impact velocity) than its initial impact velocity. The initial kinetic energy of the rigid plate was $MV_0^2/2$, which was mainly absorbed by the plastic deformation of the cell walls as the honeycomb was crushed by the rigid plate. The maximum crushing strain of the honeycomb, $\varepsilon_{\max}$, and the corresponding crushing time (i.e., at which the rigid plate has zero velocity), t_s, were found to depend on the relative density of the structure and the impact velocity, as quantified in Fig. 3.4b for a wide range of normalized initial kinetic energies of the rigid plate and honeycomb relative densities. In this set of calculations, the initial kinetic energy of the rigid plate was varied by changing the plate initial velocity, while its mass was kept constant. Based on the assumption that the total initial kinetic energy of the rigid plate was dissipated by the plastic deformation of the honeycomb, the maximum crushing of the honeycomb could be estimated from, $MV_0^2/2 = \sigma_{\mathrm{Yc}} AL\varepsilon_{\max}$, where σ_{Yc} is the theoretical effective yield stress of the honeycomb, and A is the cross section area of the structure normal to the longitudinal direction. This yields $\varepsilon_{\max} = \overline{\mathrm{KE}}$, is plotted in Fig. 3.4b as a solid line, along with the results from detailed finite element calculations. For low values of normalized kinetic energy, where the crushing strain was not high, the numerical results were in good agreement with the theoretical estimate. However, for high

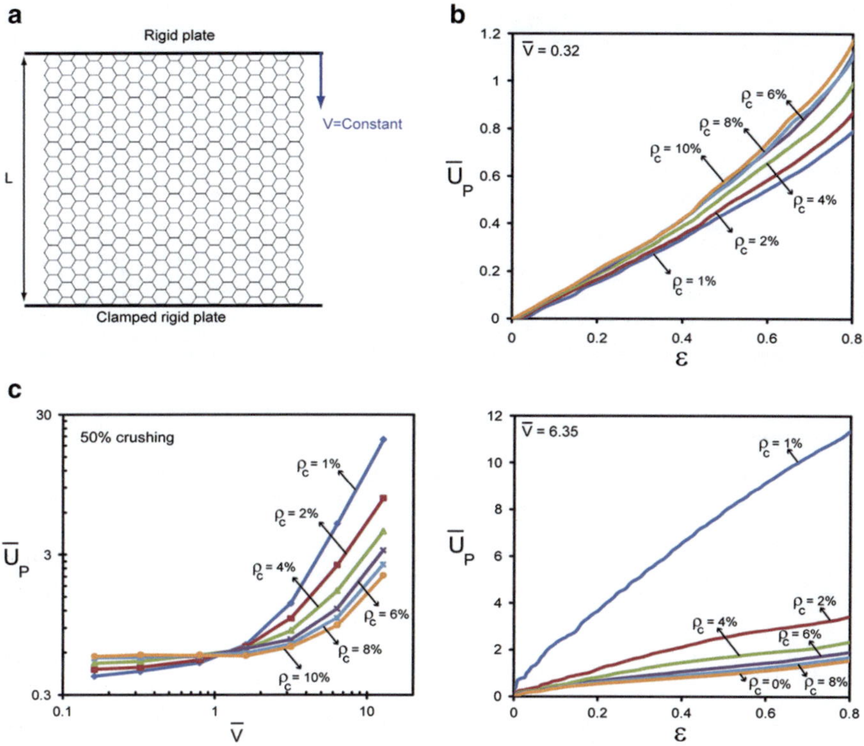

Fig. 3.5 Dynamic crushing of a hexagonal regular honeycomb. (**a**) Schematic of the finite element model. (**b**) Normalized plastic energy dissipation versus the crushing strain for $\bar{V} = 0.32$ and $\bar{V} = 6.35$. (**c**) The normalized plastic energy dissipation of regular honeycombs versus the normalized crushing velocity at 50 % crushing. The results in (**c**) are presented in the log–log scale

values of initial kinetic energy, the numerical simulations predicted $\varepsilon_{max} < \overline{KE}$ due to the dynamic effects and nonlinearity caused by cell wall contact. For a perfectly plastic material, the role of the micro-inertial resistance of core walls is minimal and thus, the effective yield strength of honeycomb was approximately equal to the honeycomb yield strength. However, at high strain rate, the contact between the cell walls increased the effective yield strength of the honeycomb, leading to elevation of the effective yield strength of honeycombs.

To explore the energy absorption of honeycombs under dynamic crushing, the authors simulated the honeycomb response under in-plane dynamic loading at a constant prescribed velocity, V, as shown schematically in Fig. 3.5a. Similar to the previous simulations, the honeycomb was clamped along its bottom edge and periodic boundary conditions were imposed at its sides. An important dimensionless parameter which governs the inertial effects is $\bar{V} = V/(c_0 \varepsilon_Y)$, where $c_0 = (E/\rho)^{0.5}$ is the elastic wave speed in the cell wall material, and $\varepsilon_Y = \sigma_Y / E$ (Xue and Hutchinson 2006). Figure 3.5b shows the normalized plastic energy dissipation, $\bar{U}_P = U_P / \sigma_{Yc} AL$,

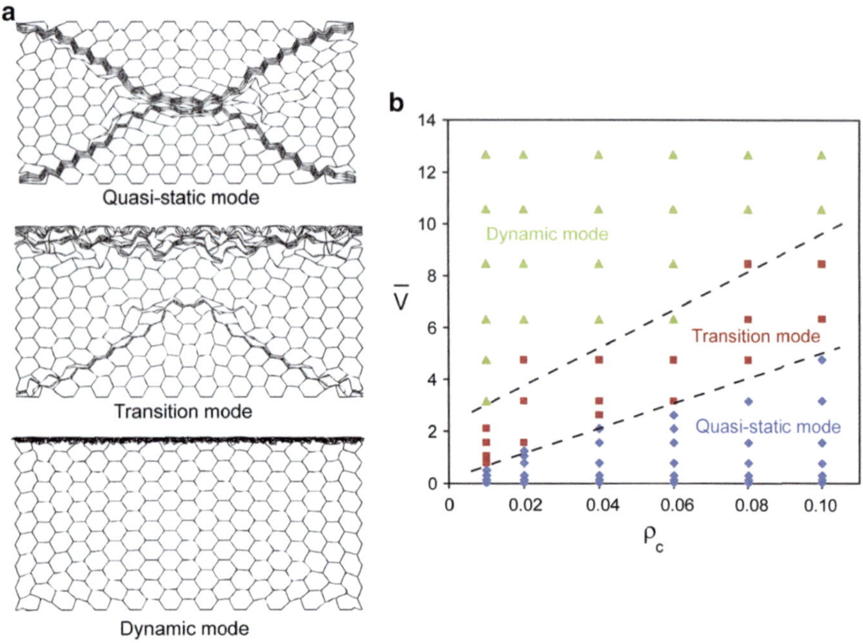

Fig. 3.6 Deformation modes of a hexagonal regular honeycomb. (**a**) Three distinct deformation shapes. (**b**) Deformation map for a regular honeycomb subjected to dynamic crushing. The markers show the finite element results. The *dotted lines* show the approximate limit associated with the transition between different deformation modes

of honeycomb, where U_P is the plastic energy dissipation calculated directly from the numerical simulations for two different crushing velocities, $\bar{V} = 0.32$ and $\bar{V} = 6.35$. At low crushing rates, the normalized dissipated plastic energy increased slightly by increasing the structure's relative density. However, all honeycombs had comparable normalized plastic energy absorption. In contrast, at high crushing velocities, the normalized plastic energy dissipation of a honeycomb strongly depended on its relative density and was higher for a honeycomb with lower relative density, which is due to dynamic effects and the nonlinearity caused by the contact of the cell walls as discussed for honeycombs subjected to rigid plate impact. In Fig. 3.5c, they have quantified the normalized plastic energy dissipation of regular honeycombs with $0.01 \le \rho_c \le 0.10$, under dynamic crushing. The results were presented in the log–log scale and at $\varepsilon = 0.5$ or 50 % crushing. The authors reported that at low crushing rates, $\bar{V} < 1$, the normalized plastic energy dissipation of the honeycomb increased by increasing its relative density. For $\bar{V} > 1$, honeycombs with smaller relative density had remarkably higher normalized plastic energy dissipation at the same level of crushing strain.

The study of deformation modes of the honeycomb under dynamic crushing showed three distinct deformation shapes. These shapes, which depend on the crushing rate and honeycomb relative density, were identified for the regular honeycombs as shown in Fig. 3.6a: quasi-static or X-shape, transition or V-shape, and dynamic shape. Figure 3.6b shows a deformation map for regular honeycomb structures subjected to dynamic crushing. The critical normalized crushing rate associated with the transition between different modes, $\bar{V}_c$, increased approximately linearly with increasing the relative density, as shown in Fig. 3.6b by the dotted lines. This concludes our discussion on the effect of dynamic loading, particularly energy absorption and strain rate, on regular honeycombs.

3.2.2 Dynamic Crushing of Irregular Cellular Structures

Continuing the discussion from previous section we now discuss the role of imperfections in the structural organization of cellular materials. Typical imperfections found are missing walls and variations in the arrangement of cell walls, cell size, and wall thickness. These structural variations are ubiquitous in natural cellular materials (e.g., cork, sponge, and toucan beak) and cellular structures manufactured using random physical processes (e.g., foams). They could also appear in engineered cellular structures that are designed to have uniform cellular arrangement due to fabrication errors. Several studies have been carried out to investigate the effect of different defects and structural variations on the quasi-static behavior of cellular structures (Ajdari et al. 2008; Chen et al. 1999; Fazekas et al. 2002; Fortes and Ashby 1999; Guo and Gibson 1999; Silva and Gibson 1997; Silva et al. 1995; Wang and McDowell 2003; Zhu et al. 2001). Liu and Zhang (2009) studied the dynamic crushing behavior of honeycombs with different cell topologies and arrangements. Their results suggest that the energy absorption of a honeycomb depends not only on the structure relative density but also on the cellular arrangement and irregularity. In the same context, Zheng et al. (2005) showed that cell arrangement irregularity elevates the energy absorption of two-dimensional cellular structures. Somewhat contradictory results were reported by Tan et al. (2005), who suggested that the irregularity in the cell arrangement does not have considerable effect on the plastic energy absorption of a cellular structure.

Ajdari et al. (2011) studied the role of irregularity in the form of variation in arrangement of cell walls on the dynamic behavior and energy absorption of cellular structures under in-plane crushing. Specifically, they considered two types of irregularities: the role of a missing wall cluster on the energy absorption of a regular hexagonal honeycomb and the dynamic crushing of irregular cellular structures. In this section, we review on the results presented (Ajdari et al. 2011) related to the latter topic.

In their work, models of cellular structures were constructed from a set of points located at the centroids of a regular hexagonal honeycomb, where each point is located at distance $\sqrt{3}l$ from its adjacent points—see Fig. 3.7a. To introduce irregularity in the cell arrangement, each point was moved randomly in both in-plane coordinates by

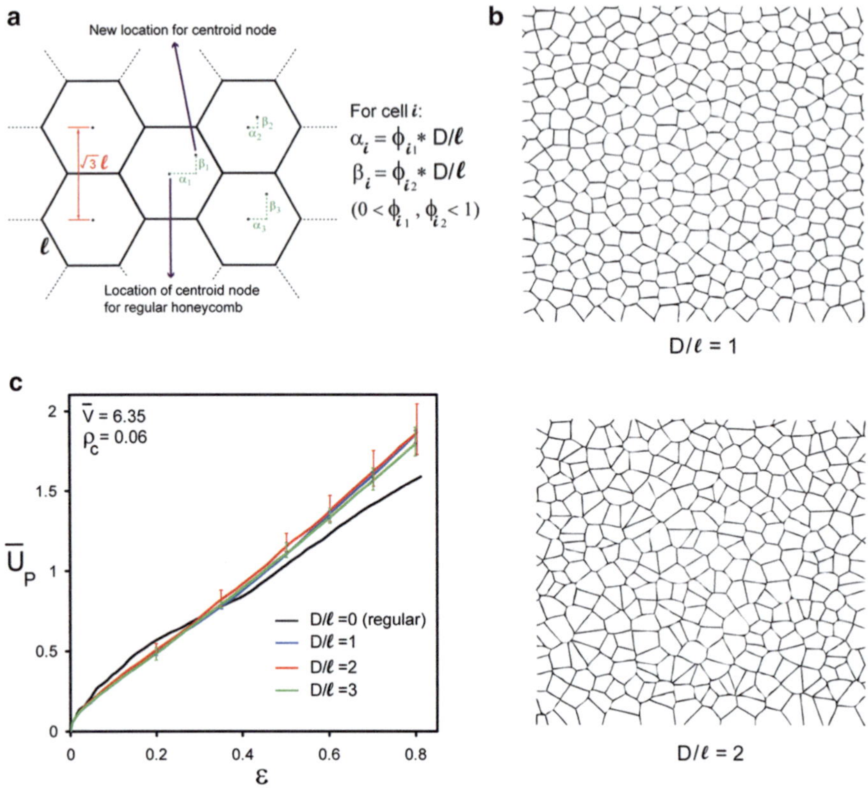

Fig. 3.7 Dynamic crushing of irregular honeycombs. (**a**) Schematic of a regular hexagonal structure and location of centroids for irregular structures. (**b**) Irregular honeycombs with *D*/*l* = *1* and *D*/*l* = *2*. (**c**) Normalized plastic dissipation of irregular structures with different irregularity indexes versus the crushing strain at $\bar{V} = 6.35$

$\alpha_i = \varnothing_{i1} \times D/l$ and $\beta_i = \varnothing_{i2} \times D/l$, where subscript i denoted the point number, D the maximum dislocation of the initial points and was defined as "irregularity index," and $0 \le \varnothing_{i1}, \varnothing_{i2} \le 1$ were random numbers generated for each point separately. Then, the Voronoi diagram which was generated by constructing the perpendicular bisectors of each pair of adjacent points (Silva et al. 1995; Zhu et al. 2001) was used to generate the models of the irregular cellular structures. Application of the Voronoi diagram to the original reference points (i.e., $D/l = 0$) gave the arrangement of a regular hexagonal honeycomb. Examples of structures with $D/l = 1$ and $D/l = 2$ are shown in Fig. 3.7b. It should be noted that the arrangement of a cellular structure with a given cell size and irregularity index changed each time the model was constructed, since $\varnothing_{i1}$ and $\varnothing_{i2}$ are generated in each trial separately. Figure 3.7c shows the normalized plastic energy dissipation versus the crushing strain for three irregular cellular structures with $D/l = 1$, 2 and 3 and relative density $\rho_c = 0.06$. The plot shown for each irregularity index was the average results obtained for six different irregular structure

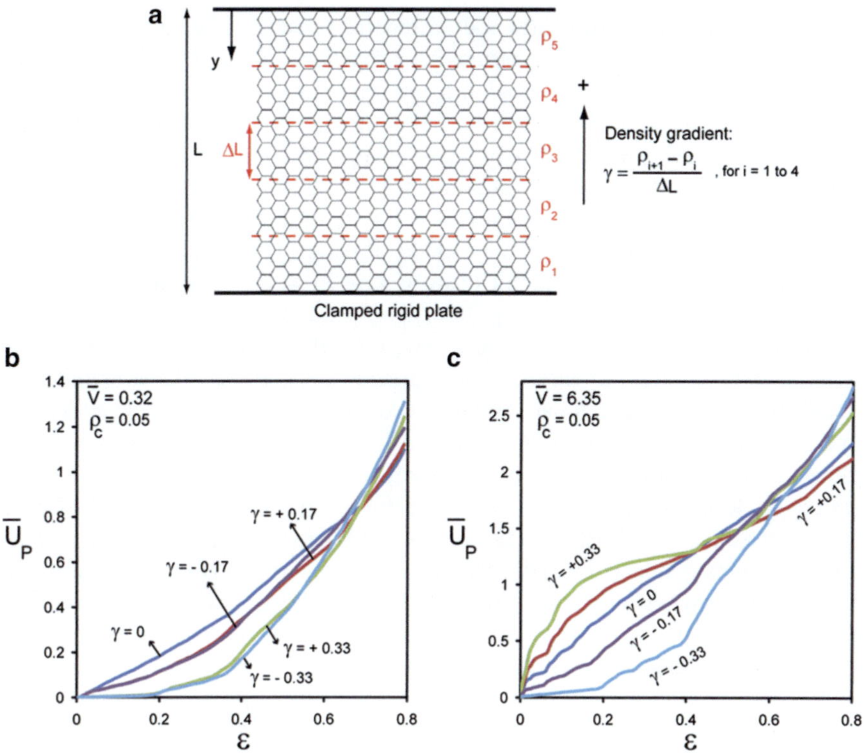

Fig. 3.8 Dynamic crushing of a functionally graded regular hexagonal honeycomb. (**a**) Schematic of the model. (**b**) Normalized plastic energy dissipation versus the crushing strain for honeycombs with different density gradients at low crushing rate, $\bar{V} = 0.32$, and (**c**) at high crushing rate, $\bar{V} = 6.35$. The overall relative density of the honeycombs was kept constant, $\rho_c = 0.05$

models. The standard deviations from the results at several different crushing strains were also shown. The result for a regular honeycomb was also presented for comparison and they show that the irregularity did not have a considerable effect on the normalized energy dissipation of the cellular structures studied.

3.2.3 Dynamic Crushing of Functionally Graded Cellular Structures

In this section, we discuss the consequence of irregularities in the context of functional gradation. Functionally graded cellular structures are a novel class of materials, where variations in cell size, shape, and wall thickness result in a functional variation in the relative density and organization of the cellular structure. Examples of functionally graded cellular structures in nature are bamboo, banana peel, and elk antler (Silva et al. 2006; Chen et al. 2008). Previous studies on impact resistance and energy absorption of functionally graded cellular structures have shown their potential for creating impact-resistant structures and cushioning materials (Cui et al. 2009a; Ali et al. 2008). Cui et al. (2009) suggested that a functionally graded foam can exhibit superior energy absorption compared to a uniform foam with equal mass. In another effort, Wadley et al. (2008) constructed a multilayered pyramidal lattice from stainless steel and investigated the quasi-static and dynamic compressive response of these structures. The developed method allows fabrication of functionally graded cellular structures by varying the relative density of the pyramidal lattice at each layer.

Cui et al. (2009b) constructed finite element models of functionally graded cellular structures by changing the thickness of the cell walls—and thus, the relative density—in the direction of crushing. They divided a cellular structure to five equal-size regions with height $\Delta L = L/5$, and assigned different cell wall thicknesses to each region to introduce a constant density gradient, γ, in the cellular structure. The models were created for cellular structures with both regular hexagonal and irregular cellular arrangements—see Figs. 3.8a and 3.10a, respectively. The density gradient, γ, was defined as $\gamma = (\rho_{i+1} - \rho_i)/\Delta L$, where ρ_i indicates the relative density of i-th region, as shown in Fig. 3.8a. $\gamma = 0$ gives a honeycomb with constant relative density, and a positive density gradient gives a cellular structure with a relative density that gradually decreases in the crushing direction. The total relative density of the structures was kept constant (here, $\rho_c = 0.05$), as the density gradient was introduced. For example, $\gamma = 0.33$ gave a cellular structure with a relative density that functionally decreased from 9 to 1 % in the direction of crushing. However, the average density of this honeycomb structure was $\rho_c = 0.05$. Figure 3.8b, c shows the energy absorption of functionally graded regular hexagonal honeycombs with different density gradients subjected to low-velocity and high-velocity dynamic crushing. As discussed above, the comparison between cellular structures with different density gradients was made at a constant average relative density. At both low and high crushing velocities, the normalized energy absorption of honeycomb with a constant relative density, $\gamma = 0$, changed approximately linearly with the crushing strain. At low crushing velocity (i.e., quasi-static mode), introducing the density gradient decreased the energy absorption of the honeycombs up to crushing strains ~67 %, and the honeycomb with $\gamma = 0$ had the maximum energy absorption for this crushing strain range. There were minimal differences between the response of functionally graded honeycombs with positive and negative density gradients, as quantified in Fig. 3.8b and expected in the quasi-static regime. In contrast at high-velocity crushing, the density gradient had a remarkable influence on the energy absorption of the honeycomb (Fig. 3.8c). Up to ~50 % crushing, the functionally graded cellular structures with positive density gradient, $\gamma > 0$—where the cellular structure relative density was high at the crushing side and changed gradually to its lowest value at the clamped side—had higher energy absorption compared to a honeycomb with $\gamma = 0$. The negative density gradient, $\gamma < 0$, resulted in reduction of the honeycomb energy absorption. The influence of the density gradient on the energy absorption was very significant in early stages of crushing (i.e., crushing strain of up to 25 %). At crushing strain ~56 %, all honeycombs show similar energy absorption capacity ($\bar{U}_P \cong 1.5$ for all honeycombs). At higher crushing strains,

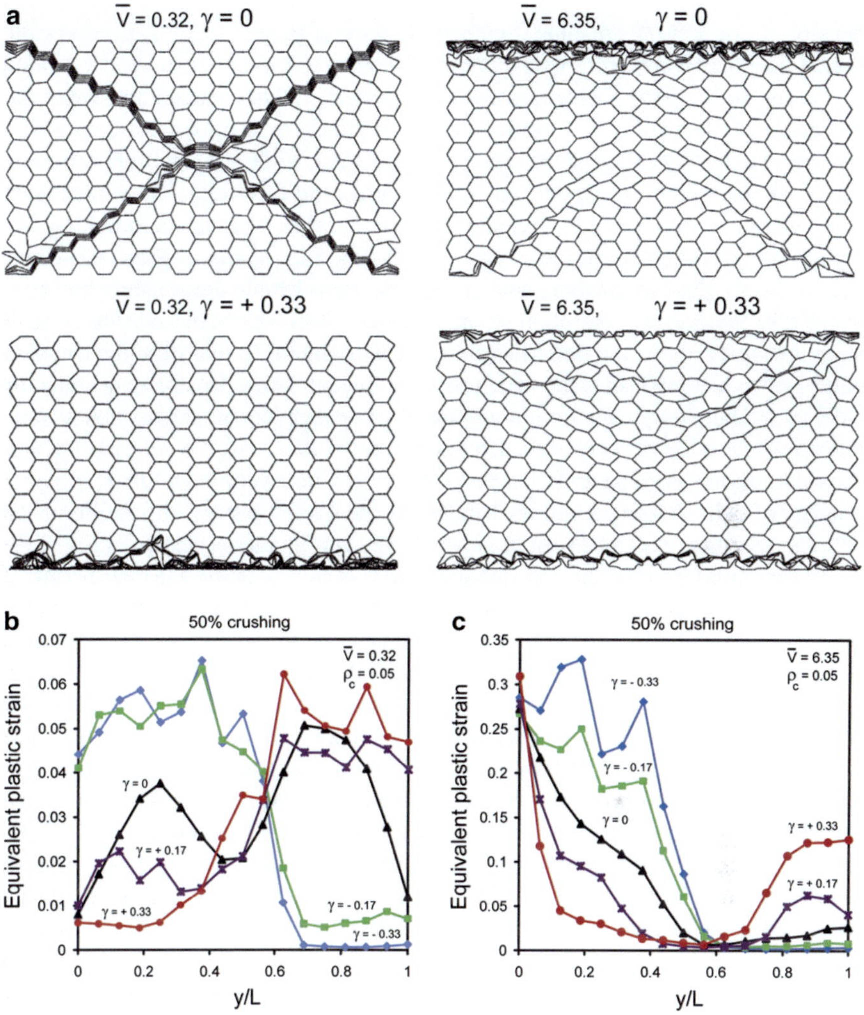

Fig. 3.9 (**a**) Deformation shapes of regular cellular structures with constant and functionally graded relative density at 50 % crushing at low and high crushing rates. (**b**) and (**c**) Equivalent plastic strain through the height of honeycombs with different density gradients subjected to $\bar{V} = 0.32$ and $\bar{V} = 6.35$, respectively, at 50 % crushing. (The overall relative density of the honeycombs was kept constant, $\rho_c = 0.05$)

honeycombs with a negative density gradient showed considerable increase in their energy absorption capacity.

To explain these observations, the authors studied the deformation modes and the distribution of equivalent plastic strain of functionally graded honeycombs with $\rho_c = 0.05$ under crushing. Figure 3.9a compares the deformation modes of a functionally graded honeycomb with $\gamma = 0.33$ and a honeycomb with a constant relative density, $\gamma = 0$, at 50 % crushing. In the quasi-static regime ($\bar{V} = 0.32$), the deformation of the honeycomb with constant relative density was mainly concentrated along two bands, forming the X-shape as explained in Sect. 3.2. In contrast, for a functionally graded honeycomb, the deformation was limited to the part of the structure with low relative density (close to the bottom in Fig. 3.9a), while the rest of the structure stayed almost undeformed. At high-velocity crushing ($\bar{V} = 6.35$), the deformation mode was quite different: For a regular honeycomb with $\gamma = 0$, the deformation was highly localized to the crushing side—as described in details in Sect. 3.2—thus, only the cells close to crushing side underwent considerable deformation and contributed to the energy absorption of the honeycomb as it got crushed. For cellular structures with $\gamma < 0$, the deformation mode was similar to that of regular honeycomb with $\gamma = 0$, and since the structure had a lower relative density at the crushing side, its overall energy absorption was even lower than its counterpart honeycomb with $\gamma = 0$. For a functionally graded honeycomb with $\gamma > 0$, the deformation was focused at both crushing and clamped sides of the honeycomb and thus, a higher number of cells deformed and contributed to the overall energy absorption of the cellular structure, as can be seen in the deformed configurations shown in Fig. 3.9a.

To further understand the deformation modes of honeycombs, the distribution of the equivalent plastic strain along the height of honeycombs with different density gradients was plotted (Fig. 3.9b, c). To calculate the equivalent plastic strain through the height of the structure, the authors averaged the equivalent plastic strains for all elements in each row of the honeycomb from top to bottom. These plots show the relative contribution of cells located at different heights of the structure on the overall energy absorption of the honeycomb. For the honeycomb with $\gamma = 0$ and subjected to $\bar{V} = 0.32$, all cell wall rows contributed to the overall energy absorption of the structure, by forming the X-shape deformation mode shown in Fig. 3.9a. At this crushing rate, for honeycombs with $\gamma > 0$, the cell walls with the lower relative density (wall thickness) were located near the clamped side and were mostly deformed under crushing, while for honeycombs with $\gamma > 0$, cell walls with lower density were located near the crushing side and got more deformed compared to other cell walls. Figure 3.9c shows the results for honeycombs with different density gradients at high crushing rates, $\bar{V} = 6.35$ at 50 % crushing strain. For honeycombs with $\gamma \geq 0$, the cell walls near the crushing side were mainly deformed, while the induced plastic strains in cells located in the bottom half of the structure ($0.5 < y / L \leq 1$) were negligible. For honeycombs with $\gamma > 0$, cells at both sides of the structure (crushing side and clamped side) got strongly deformed under crushing, while the cells in the mid height of the structure contributed minimally to the overall energy dissipation of the cellular structure. It should be noted that the plastic energy dissipation of all honeycombs were comparable at this crushing strain. Figure 3.10 repeated the calculations for an irregular cellular structure. The role of density gradient was more remarkable in this case compared to the regular hexagonal honeycombs. The deformation modes of irregular cellular structures with two different density gradients are displayed in Fig. 3.10c, which show qualitative

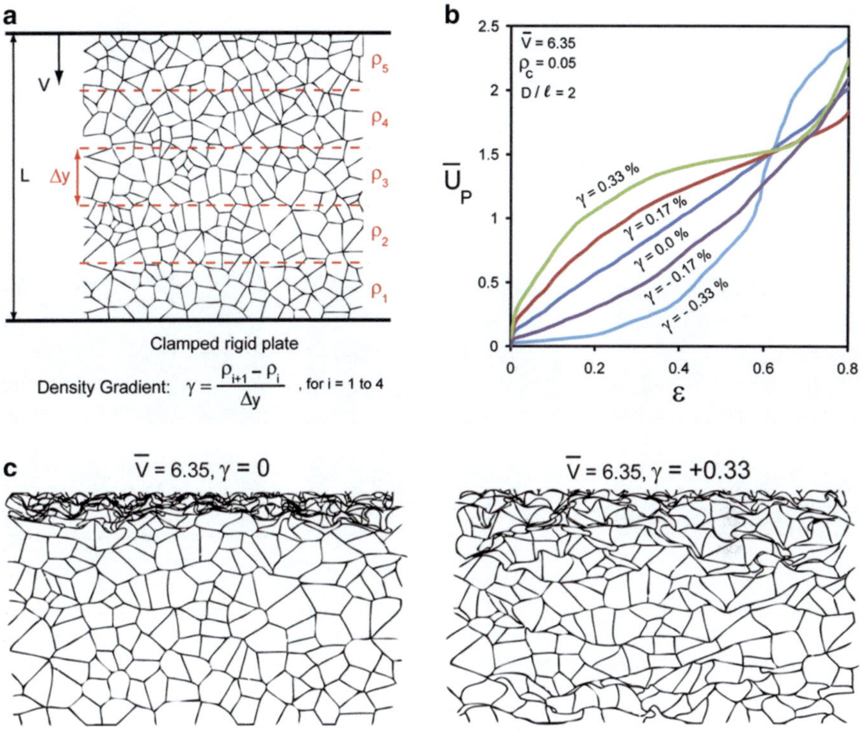

Fig. 3.10 Dynamic crushing of a functionally graded irregular structure. (**a**) Schematic of the model of a functionally graded irregular cellular structure with $D/l = 2$. (**b**) Normalized plastic energy dissipation versus the crushing strain for an irregular cellular structure with different density gradients at the high-velocity crushing, $\bar{V} = 6.35$. (**c**) The deformation shapes of irregular cellular structures

agreement with the deformation mode of their regular hexagonal counterpart. Again, a higher number of cells of the structure with $\gamma = 0.33$ had deformed considerably, compared to the deformation mode of the counterpart irregular structure with $\gamma = 0$, which was highly localized at the crushing side.

3.3 Self-Similar Honeycombs with Structural Hierarchy

In this section we carry forward the discussion from simple honeycombs described in the previous section to hierarchical structures. Hierarchical structures are ubiquitous in nature and can be observed at many different scales in organic materials and biological systems (Aizenberg et al. 2005; Buehler 2006; Espinosa et al. 2011; Fratzl and Weinkamer 2007; Gibson et al. 2010; Lakes

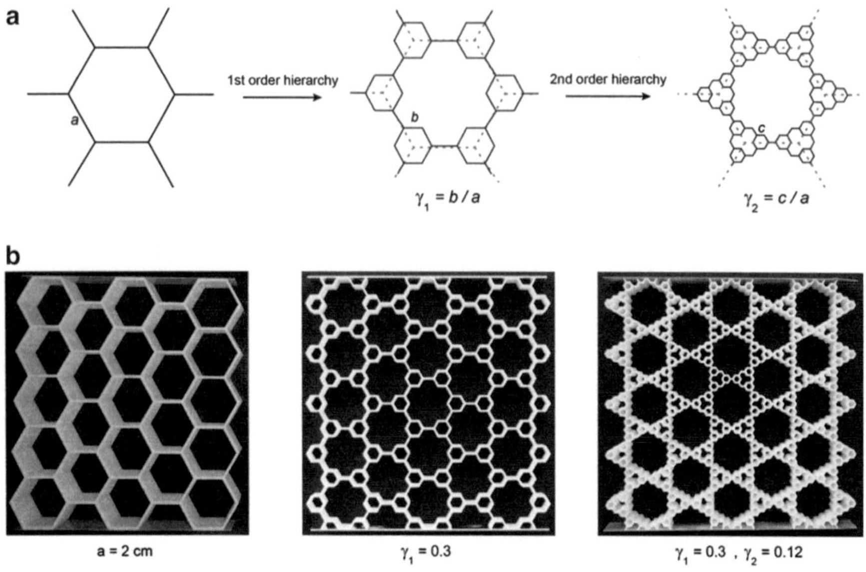

Fig. 3.11 Hierarchical honeycombs. (**a**) Unit cell of the hierarchical honeycombs with regular structure and with first- and second-order hierarchy. (**b**) Images of honeycombs with $a = 2$ cm fabricated using three-dimensional printing

1993; Ortiz and Boyce 2008; Qing and Mishnaevsky 2009). The hierarchical organization of these systems generally plays a key role in their properties, function, and survival (Fratzl and Weinkamer 2007; Gibson et al. 2010). Hierarchy is also important in engineering designs, materials, and architecture. Examples range from the Eiffel tower (Lakes 1993) and polymers with micro-level hierarchical structures (Lakes 1993) to sandwich panels with cores made of foams or composite lattice structures (Cote et al. 2009; Fan et al. 2008; Kazemahvazi et al. 2009; Kazemahvazi and Zenkert 2009; Kooistra et al. 2007). There, the hierarchical organization can lead to superior mechanical behavior and tailorable properties, as described recently for sandwich cores with hierarchical structure (Fan et al. 2008) and for hierarchical corrugated truss structures (Kooistra et al. 2007).

Ajdari et al. (2012) introduced a hierarchical family of honeycomb-based cellular structures by systematic introduction of successively smaller hexagonal cells wherever three cell walls meet. This replacement procedure for three-edge vertices can be repeated at smaller scales to achieve fractal-appearing honeycombs with higher orders of structural hierarchy. Figure 3.11a shows the evolution of a hexagonal honeycomb cell as structural hierarchy is increased. This class of hierarchical honeycombs was shown to be capable of attaining superior in-plane stiffness-to-mass ratio (i.e., specific Young's modulus). For experimentation, the hierarchical honeycomb samples were fabricated using 3D printing. The structural organization of the honeycomb at each level of hierarchy can be defined by the ratio of the introduced hexagonal edge length (*b* for first-order hierarchy and *c* for second-order hierarchy) to the original hexagon's edge length, *a*, as described in Fig. 3.11a (i.e., $\gamma_1 = b/a$ and $\gamma_2 = c/a$). For a honeycomb with first-order hierarchy, $0 \leq b \leq a/2$ and thus $0 \leq \gamma_1 \leq 0.5$, where $\gamma_1 = 0$, denote the regular honeycomb structure. For a honeycomb with second-order hierarchy, there are two geometrical constraints, $0 \leq c \leq b$ and $c \leq a/2 - b$. In terms of the ratio parameters, the constraints are $0 \leq \gamma_2 \leq \gamma_1$ if $\gamma_1 \leq 0.25$ and $0 \leq \gamma_2 \leq (0.5 - \gamma_1)$ if $0.25 \leq \gamma_1 \leq 0.5$. The dimensionless relative density (i.e., area fraction) can be given in terms of *t*/*a*:

$$\rho = 2/\sqrt{3} \bullet (1 + 2\gamma_1 + 6\gamma_2) \bullet t/a, \tag{3.1}$$

where *t* is the thickness of the cell walls, from which the special cases of $\gamma_2, \gamma_1 = 0$ can be read off immediately. (For regular honeycomb, $\rho = 2/\sqrt{3} \bullet t/a$; and for honeycomb with first-order hierarchy, $\rho = 2/\sqrt{3} \bullet (1 + 2\gamma_1) \bullet t/a$.) This relation clearly shows that *t*/*a* must decrease to maintain fixed relative density as γ_1, γ_2 are increased.

In the next subsections, we discuss analytical techniques to calculate the effective mechanical properties of hierarchical honeycombs. In particular we present the calculation of Young's modulus and plastic collapse strength of honeycombs.

3.4 Mechanical Properties of Hierarchical Honeycombs

The stiffness and strength of regular honeycombs (for example, when the illustrated geometry is compressed in the vertical direction) are governed by the bending deformation of cell walls which reaches a maximum at the honeycomb vertices (i.e., cell wall corners). Thus, moving the material from the middle part of each wall closer to the vertices can increase the stiffness and strength (Gibson and Ashby 1999; Rathbun et al. 2006; Chuang and Huang 2002).

Ajdari et al. (2012) explored a hierarchical pattern-based approach to this strengthening: they replaced every three-edge vertex with a smaller hexagon, adjusting all the wall thicknesses to maintain a fixed overall density. Repeating this process with potentially altered shrinkage ratio built a fractal-appearing structure. The structural organization of the honeycomb with two levels of hierarchy can be defined by size rations, γ_1 and γ_2, the ratio of the smallest honeycomb's edge length, *b* for first-order hierarchy and *c* for second-order hierarchy, to the original honeycomb's edge length, *a*. Figure 3.11b shows samples of regular honeycomb and honeycombs with one and two levels of hierarchy with 10 % relative density, fabricated using 3D printing, that were tested using an INSTRON machine.

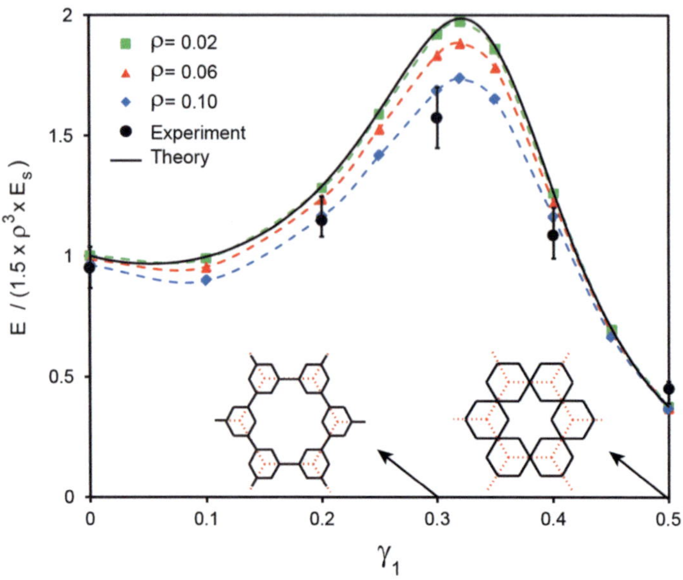

Fig. 3.12 Normalized stiffness for honeycombs with first-order hierarchy. The finite element results are shown for honeycombs with three different relative densities. Experimental results for structures with different hierarchy levels are also shown (*black circles*). The error bars show the result variation. Each experimental point is from three tested specimens

3.5 Elastic Properties of Hierarchical Honeycombs

It is well known that plane lattices with threefold symmetry will exhibit macroscopically isotropic in-plane elastic behavior (Christensen 1987). Thus, the macroscopic in-plane linear elastic behavior of hierarchical honeycomb can be characterized by just two constants, to be found by whatever loadings are most convenient. Ajdari et al. (2012) used Castigliano's second theorem (Boresi and Schmidt 2002) to determine the uniaxial in-plane deformation of hierarchical honeycombs made of an isotropic elastic material with elastic modulus, E_s.

To compute the effective properties, Ajdari et al. (2012) imposed a far-field *y*-direction stress, $\sigma_{yy} = -(2/3)F/a$, in a vertical direction (perpendicular to the horizontal hexagon edges in Fig. 3.11a) and subsequently transferred this remote load to a subassembly. Thereby they carried out force balance on this subassembly while using arguments of symmetry to calculate the energy of the structure. Thereafter a straightforward application of Castigliano's theorem yielded the following expression of Young's modulus of the structure:

$$E/E_s = (t/a)^3\ f(\gamma_1), \tag{3.2}$$

where $f(\gamma_1) = \sqrt{3}/(0.75 - 3.525\gamma_1 + 3.6\gamma_1^2 + 2.9\gamma_1^3.$

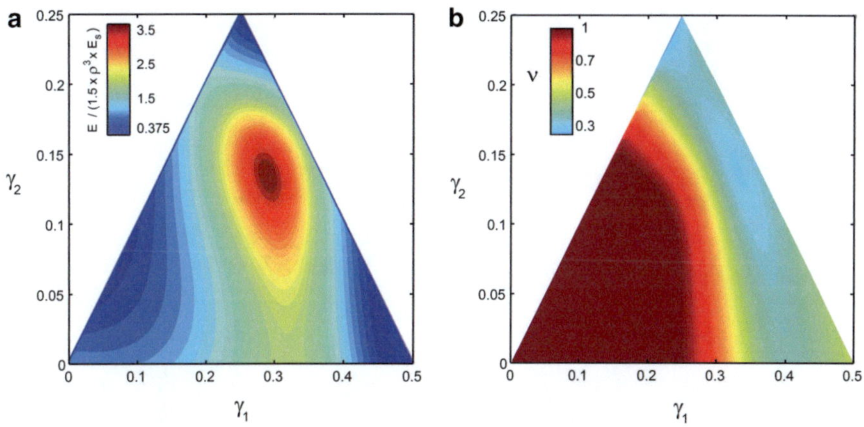

Fig. 3.13 Contour maps of the (**a**) effective elastic modulus and (**b**) Poisson's ratio of hierarchical honeycombs with second-order hierarchy for all possible geometries (i.e., admissible range of γ_1 and γ_2)

To find the maximum normalized elastic modulus for structures with first-level hierarchy and constant relative density, one can eliminate (*t*/*a*) from (3.3) by using the relative density expression of (3.2). The resulting expression for E/E_s was ρ^3 times a function of γ_1, and setting $(\delta(E/E_s)/\delta\gamma_1)_\rho = 0$ gave $\gamma_1 = 0.32$. Making this substitution led to $E/E_s = 2.97\rho^3$, a stiffness almost twice the stiffness of the regular honeycomb structure (Ajdari et al. 2012), for which $E_0/E_s = 1.5\rho^3$. (The regular honeycomb result could be found by letting $\gamma_1 = 0$ in (3.3), and using (3.2) to eliminate *t*/*a*.) The Poisson's ratios could be calculated using a similar procedure as above. For hierarchical honeycomb with one-order hierarchy, this yielded

$$\nu = 1 - \gamma_1^{\,3} / (2.9\gamma_1^{\,3} + 3.6\gamma_1^{\,2} - 3.525\gamma_1 + 0.75). \tag{3.3}$$

Figure 3.12 shows the effective elastic modulus of first-order hierarchical honeycombs for all possible values of γ_1. In this figure, the elastic modulus was normalized by the effective stiffness of the counterpart regular honeycomb with the same relative density, $1.5E_s\rho^3$, allowing us to present results for every density on a single curve. In the finite element simulations, structures with three different relative densities (2, 6, and 10 %) were analyzed. Results showed quite good agreement between numerical and theoretical approaches, even though the theoretical analysis ignored the axial and shear deformation of the beams (a good approximation only for low-density honeycombs with small beam thickness (Harders et al. 2005)). The authors speculated that the numerical incorporation of shear and stretching accounts for the FEA-determined modulus falling somewhat below the theory, particularly as density increases or beam lengths decreased. The FEA results confirmed the near-doubling of stiffness for $\gamma_1 = 0.32$. In this figure, experimental results are also plotted which show reasonable agreement with both theory and numerical results.

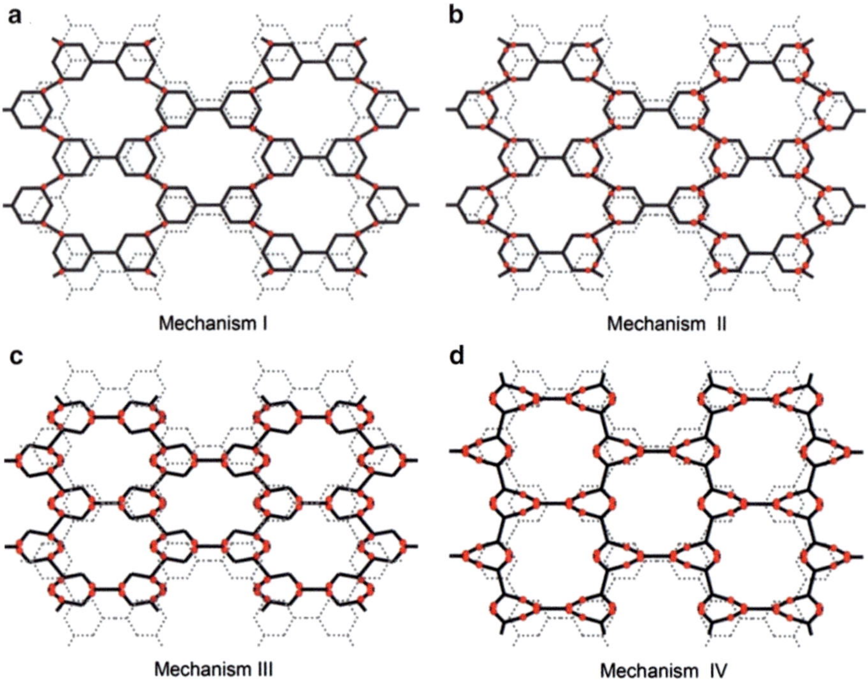

Fig. 3.14 (**a–d**) Deformed configurations of plastic collapse for hierarchical honeycomb according to the mechanisms I–IV, involving different plastic hinge locations marked by red bullets

The same analytical approach was used by the authors to evaluate the in-plane effective Young's modulus of honeycomb with two orders of hierarchy, as a function of hierarchy indices γ_1 and γ_2. In particular, they showed that the effective elastic modulus can be written as

$$E / E_s = (t / a)^3 f(\gamma_1, \xi), \tag{3.4}$$

where $\xi = \gamma_2 / \gamma_1$ and $f(\xi_1, \xi) = N_4(\xi) / (\xi_1^3 D_7(\xi) + \xi_1^2 D_6(\xi) + \xi_1 D_5(\xi) + D_4(\xi))$

$$N_4(\xi) = 29.62 - 54.26\xi + 31.75\xi^2 - 4.73\xi^3 - \xi^4$$

$$D_7(\xi) = 49.64 - 609.01\xi + 862.56\xi^2 - 195.50\xi^3 - 270.14\xi^4 + 159.95\xi^5 - 18.13\xi^6 - 2.20\xi^7$$

$$D_6(\xi) = 61.73 + 310.43\xi - 662.32\xi^2 + 334.12\xi^3 + 9.70\xi^4 - 29.38\xi^5 - 1.88\xi^6$$

$$D_5(\xi) = 60.43 + 12.80\xi + 123.22\xi^2 - 108.06\xi^3 + 20.50\xi^4 + 3.90\xi^5$$

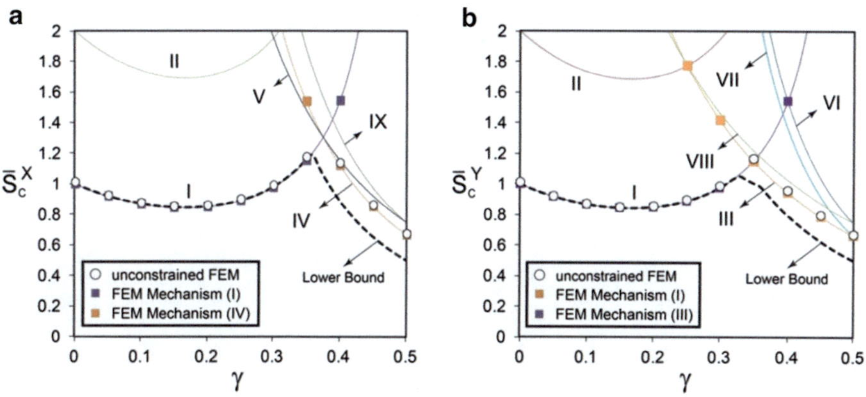

Fig. 3.15 Normalized plastic collapse strength of uniform thickness hierarchical honeycomb under uniaxial loading in (**a**) *x*- and (**b**) *y*-directions as a function of length ratio, γ. The plastic collapse load is normalized by that of regular honeycomb of the same density. The actual strength is on the curves marked by circles (where least *upper* bound happens to match unconstrained FEA)

$$D_4(\xi) = 12.80 - 23.46\xi + 13.74\xi^2 + 2.04\xi^3 - 0.43\xi^4.$$

For the second-order hierarchical structure, once again eliminating (*t*/*a*) in favor of density, and then differentiating at constant density, $(\delta(E/E_s)/\delta\gamma_1) = (\delta(E/E_s)/\delta\gamma_2)_\rho = 0$ gave $\gamma_1 = 0.32$, and $\gamma_2 = 0.135$, leading to a maximum normalized stiffness $E/E_s = 5.26\rho^3$, a stiffness almost 3.5 times that of the regular honeycomb. Figure 3.13a, b shows contour maps of the effective (normalized) elastic modulus and Poisson's ratio of hierarchical honeycombs with second-order hierarchy for all possible values of γ_1 and γ_2. The *x*-axis is γ_1 and ranges from 0 to 0.5, while γ_2 is limited by the two geometrical constraints, $\gamma_2 \le \gamma_1$ and $0 \le \gamma_2 \le (0.5 - \gamma_1)$.

3.6 Plastic Collapse Strength of Hierarchical Honeycombs

In this section, we discuss the in-plane plastic collapse behavior of hierarchical honeycomb at one order of hierarchy, for arbitrary biaxial in-plane loading in the principal structural directions (x, y). Figure 3.14 shows the deformed configuration of primary collapse mechanisms of a hierarchical honeycomb with first-order hierarchy. Haghopanah et al. (2012) provided analytical models and numerical simulations for estimation of the plastic collapse strength of first-order hierarchical honeycombs. In their analysis, the lower bound estimation was based on finding an equilibrium distribution of moments at or below the collapse moment, which balances the applied load. In contrast, the upper bound is based on finding the

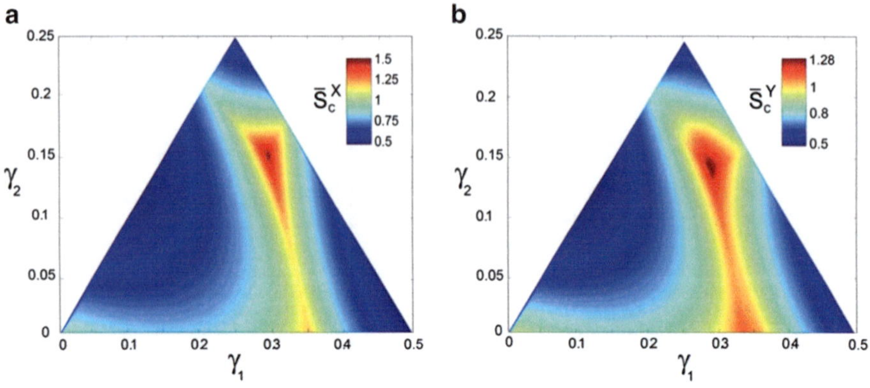

Fig. 3.16 Contours of plastic collapse strength of second-order honeycombs of uniform thickness ($\eta = 1$) under uniaxial loading in (**a**) x- and (**b**) y-directions for different values γ_1 and γ_2, respectively. The plastic collapse strength is normalized by that of regular honeycomb of the same density

minimum collapse load (as determined by virtual work) among various mechanisms of structural deformation involving different possible locations of the plastic hinges.

Next, to model each individual plastic collapse mechanism numerically using the finite element method, the authors permitted plastic behavior only at certain hinges specified for that mechanism, thus permitting direct comparison to the upper bound analytical result for that mechanism. The plastic collapse was simulated by the authors for both sequential uniaxial loading along two axes and biaxial loading. The values of collapse strength in each direction, after being normalized by the collapse strength of regular honeycomb of equal density, are denoted by $\bar{S}_c^x$ and $\bar{S}_c^y$. The normalized uniaxial collapse strength (in both x- and y-directions) of the uniform thickness first-order hierarchical honeycomb structure is shown as a function of γ in Fig. 3.15. These results show that the structure failed according to deformation mode I over the range $0 \le \gamma < 0.375$ for x-loading and $0 \le \gamma < 0.35$ for y-loading. For greater γ values, mode IV was controlling for uniaxial loading in the x-direction, and III for uniaxial loading in the y-direction.

Figure 3.16a, b shows the plastic collapse of second-order honeycombs with varying γ_1 and γ_2 values under uniaxial loading under $x0$ and $y0$ directions, respectively. The maximum value of normalized collapse strength was 1.5 and 1.28 in $x0$ and $y0$ directions, occurring at $(\gamma_1, \gamma_2) = (0.29, 0.15)$ and $(\gamma_1, \gamma_2) = (0.29, 0.14)$, respectively. The minimum value of normalized collapse strength along $x0$ and $y0$ directions is 0.5 which occurs at $\gamma_1 = 0.125$ and $\gamma_2 = 0.125$.

Figure 3.17a, b shows the maps of stiffness versus collapse strength for regular-, first-, second-, third-, and fourth-order honeycombs in $x0$ and $y0$ directions, respectively. The graphs show the large increase in the achievable stiffness and plastic collapse strength permitted by increasing the order of hierarchy. The maximum value of plastic collapse strength experienced an increase of [30, 50, and 68 %] in $x0$ and [15, 28, and 30 %] in $y0$ directions for [first, second, and third] orders of

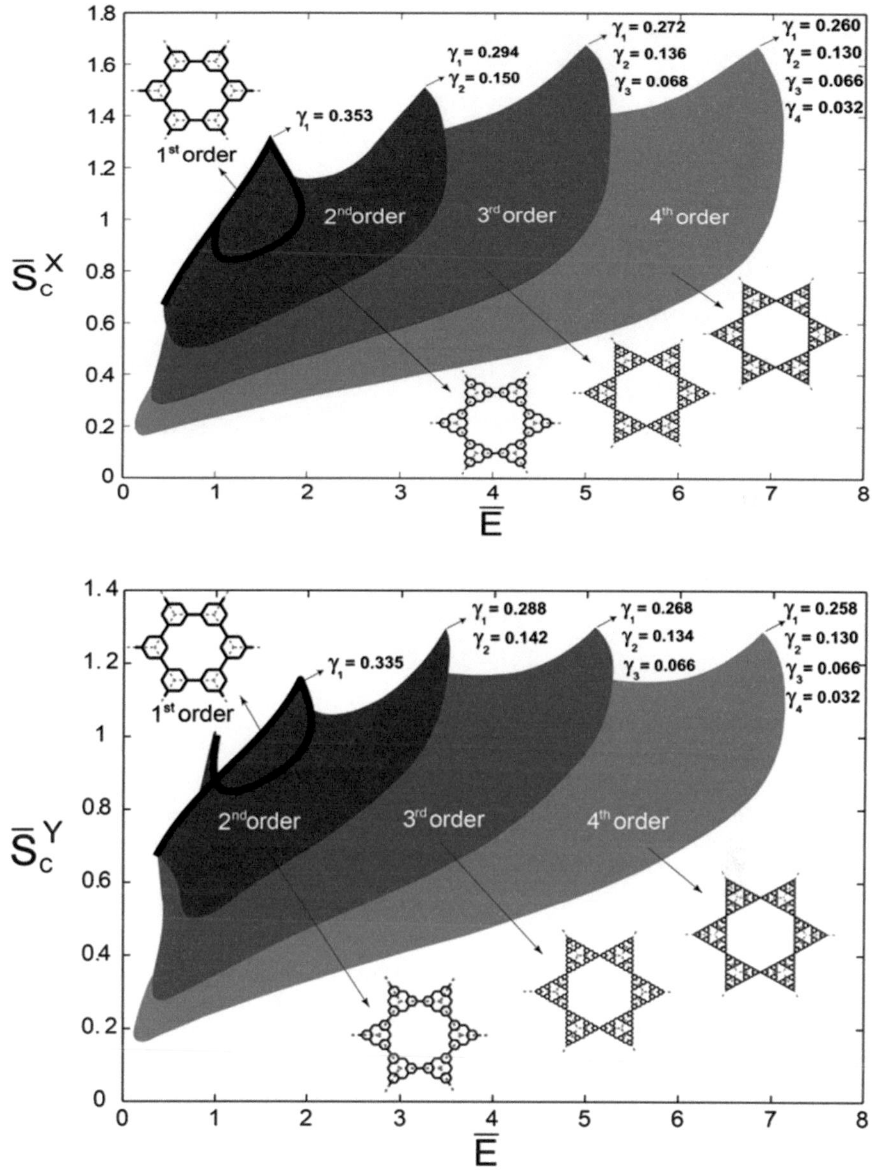

Fig. 3.17 Plastic collapse strength versus stiffness for first-, second-, third-, and fourth-order hierarchical honeycombs of uniform thickness ($\eta = 1$) under uniaxial loading in (**a**) x- and (**b**) y-directions, respectively. The plastic collapse strength and stiffness are normalized by those of regular honeycomb of same density

hierarchy in comparison with regular honeycomb, and remained almost constant afterwards. This can be attributed to the mechanisms of collapse remaining unchanged by introducing new orders of hierarchy. Depending on the substitution

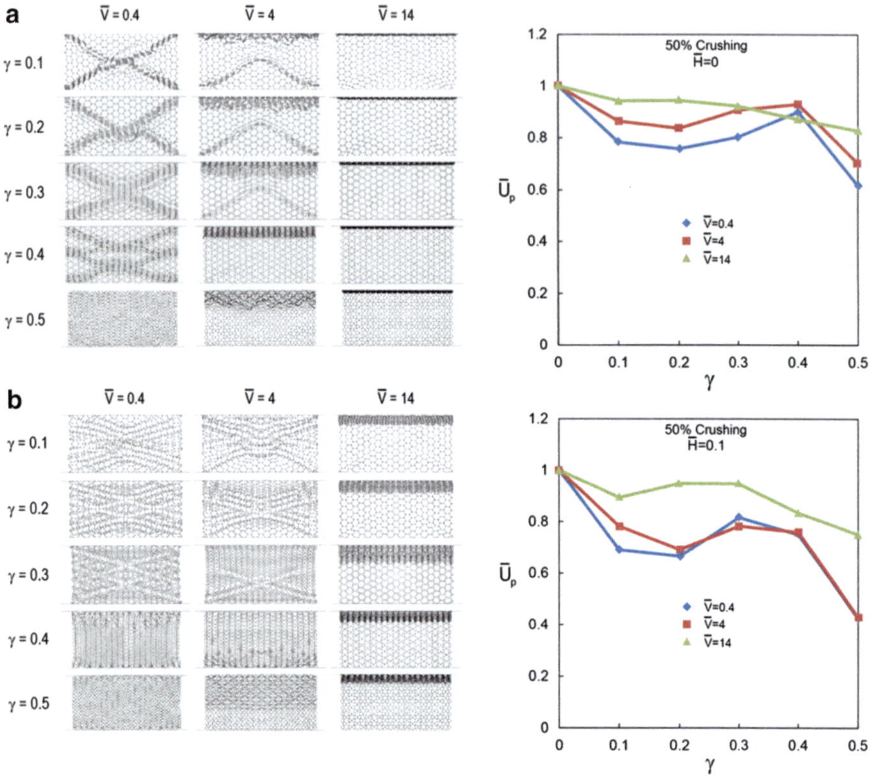

Fig. 3.18 Dynamic crushing response of first-order hierarchical honeycombs. Deformation modes, and the normalized plastic energy dissipation vs. the order of hierarchy, for three different normalized crushing velocities. (**a**) Honeycombs are made of a material with an elastic-perfectly plastic property. (**b**) Honeycombs are made of a material with 10 % strain hardening

ratios for length and thickness, the resulting structure could exhibit many possible stiffness–strength properties relative to regular honeycomb of same mass, including (stiffer/stronger), (stiffer/weaker), and (more compliant/weaker) properties. The significance of these results was most evident when examining the amount of improvement, relative to nonhierarchical honeycomb, in both uniaxial stiffness and uniaxial limit strength simultaneously. The maximum achievable specific stiffness increased from 2 to 3.5 for first- and second-order hierarchical honeycombs to 5.3 and 7.1 for third- and fourth-order hierarchical honeycombs.

3.6.1 Dynamic Crushing of Hierarchical Honeycombs

In this section, we discuss the numerical analysis on the dynamic crushing of hierarchical honeycombs which have shown superior mechanical behavior (Ajdari et al. 2012) under dynamic crushing. The authors studied the effect of cell wall material

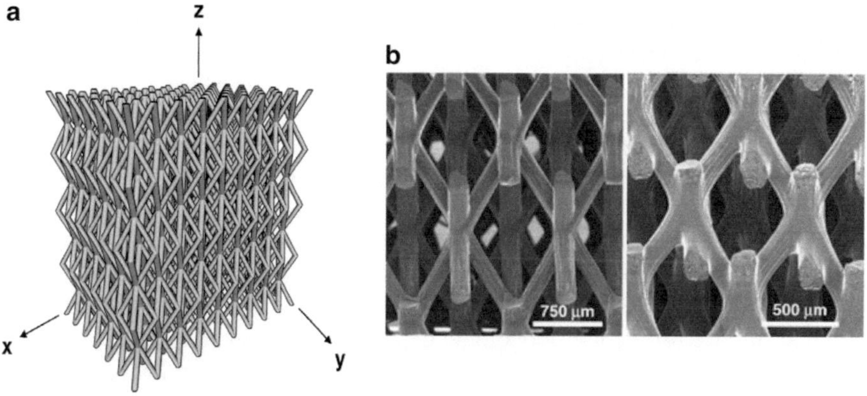

Fig. 3.19 Illustration of a micro-truss structure (Jacobsen et al. 2007a). (**b**) SEM images of the unit cell structures with different spacings (Jacobsen et al. 2007a)

hardening on the energy absorption of these structures under dynamic conditions. To this effect they have developed finite element simulations of first-order hierarchical honeycombs under dynamic in-plane loading. In Fig. 3.18 they show the deformation modes and the distribution of the normalized plastic energy dissipation, $\bar{U}_p$, versus the order of hierarchy, γ, for first-order hierarchical honeycombs with 0 and 10 % hardening of the cell wall material, respectively. The plastic energy dissipation was directly extracted from finite element results and then normalized by the plastic energy of a regular honeycomb ($\gamma = 0$) with 0 % (Fig. 3.18a) and 10 % (Fig. 3.18b) hardening of the cell wall material. The results were shown at 50 % crushing strain for three different normalized velocities and five different orders of hierarchy. The deformation modes observed in Fig. 3.18a show the distinctive X-shapes transitioning into V and then to localization with increasing velocities analogous to simpler honeycomb structures previously analyzed. However interestingly, the order of hierarchy played an important role in these deformation modes distinguishing them from simpler structures especially at lower and intermediate velocity regimes. For instance, Fig. 3.18a shows a hierarchical honeycomb made of a material with an elastic-perfectly plastic property at low crushing velocities ($\bar{V} = 0.4$). We can clearly see that as the order of hierarchy increased, the X-shape got thicker and for $\gamma = 0.4$ it transformed to a double-X-shape. For $\gamma = 0.5$ the thickness of the X-shape reached to a point in which we just see a uniform deformation for the honeycomb. For higher crushing rates ($\bar{V} = 4$), similar to regular honeycombs, V-shape was seen as the deformation mode of first-order hierarchical honeycombs. The deformation got thicker as the order of hierarchy increased. For the last two orders of hierarchy shown in this figure, the honeycomb was not in the transition mode anymore and they get the dynamic deformation mode. Finally at high crushing velocities ($\bar{V} = 14$), like regular honeycombs, we see dynamic mode for the deformation of hierarchical honeycombs. The deformation modes for hierarchical honeycombs made of a material with 10 % hardening (Fig. 3.18b) were much similar except exhibiting multi-branched X-shapes and multi-branched V-shapes instead of X-shapes and V-shapes, respectively.

To analyze how the variations in deformation mode translated to actual energy dissipation, they plotted normalized plastic energy dissipation with respect to variations of order of hierarchy. It was observed from these simulations that the plastic energy dissipation for hierarchical honeycombs was lower compared to regular honeycombs. As the crushing velocity increased the normalized plastic energy dissipation also increased, except for some values of γ. Thus, hierarchical honeycombs, even if made of a material with strain hardening property, had almost no advantages in energy absorption compared to regular honeycombs. In addition, in view of the non-monotonic dependence of energy dissipation on the order of hierarchy, the authors speculated a nontrivial and complex relationship between hierarchy order and energy dissipation.

3.7 Other Hierarchical Cellular Structures

In this section we discuss recent advances in the manufacture and characterization of other hierarchical cellular structures in brief, specifically 3D open cellular structures and hollow tube structures. Jacobsen et al. (2007a, b) highlighted the problem of meso-scale (micron level) 3D-microstructure fabrication and successfully executed a rapid 3D fabrication (see Fig. 3.19) using self-propagating photopolymer waveguides from a single 2D exposure surface. The orientation of these polymeric octahedral cell strut members and hence the geometry of the unit cell could be controlled conveniently by the waveguide angle. In initial compression tests, the authors found that axial loading as opposed to bending form of deformation caused the stiffness to scale linearly with the relative density as opposed to quadratic scaling exhibited by bending dominated cellular microstructures. The compression strength showed marked dependence on the strut angle (orientation) with a near doubling of maximum compressive stress when the angle was changed from *50* degrees to about *60* degrees. The axial loading transitioned into bending-dominated behavior once a significant percentage of struts buckled resulting in misalignments in the internal structure. This caused an extended plateau in the stress–strain curve leading to densification. Continuing on this effort, Jacobsen et al. (2007a, b) carried out a more detailed analysis of this structure and found that although effective elastic modulus was reliably predicted from the simple analytical models, the peak strength computed from inelastic buckling theory overpredicted the experimental results by a significant amount. This was attributed to a combination of imperfections as well as nonlinear compressive behavior of the parent polymeric material. It was also noted that although a general transition to bending behavior was observed after peak stress was reached, undulations in the characteristic "bending-plateau" were clearly observed due to the residual periodicity of the entire structure. The shear loading of these structures was further probed by Jacobsen et al. (2008a, b) who carried out an experimental investigation of these structures using a single lap shear fixture and compared with a simple model based on a unit cell. In their experiments, the authors found that the measured shear modulus deviated from the theoretical prediction in values in inverse proportion to the relative density. In addition, the authors pointed out that although in a continuum polymer sample the loading angle of the shear force would not affect the shear modulus, in these hierarchical structures, the distribution of load among the struts plays an important role in determining the shear modulus. Thus loading force direction is an important factor in these materials. The authors claimed that due to the asymmetry in the mechanical strength in the parent polymer with respect to compression and tension (compressive failure strength being far more than tensile), two competing mechanisms are at play—compressive buckling failure and tensile material failure. The ultimate plateau strain after initial member failure was determined by the distribution of buckling members through the thickness of the structure. When buckling was distributed evenly through the thickness, uniform plateau shear strains of up to 60 % were possible. However, when buckling was localized primarily in a single layer, this plateau strain was reduced twofold or more. Finally, when the micro-truss samples were post-cured in an oxidizing environment, the shear strength and modulus increased as expected, but the trusses were embrittled and complete fracture occurred in the structure at the peak load. In order to investigate the role of the geometry of the unit cell of the microstructure, Jacobsen et al. (2008a, b) fabricated unit cells with six- to threefold symmetry. The authors found that the measured compression modulus of the structures with sixfold symmetry was approximately 20–70 % greater than the modulus values for the structures produced with threefold symmetry. The discrepancy in the measured and predicted moduli of the threefold symmetric structures was attributed to the difference in the truss member slenderness ratios between the two architectures. In particular, the secondary nodes in the six-intersecting-waveguide structures reduced the slenderness ratio of the truss members without increasing the relative density. This reduced the structure's susceptibility to nonideal deformation. The smaller slenderness ratio was also found to increase the peak strength of the structures.

The energy absorption advantage of hollow microlattices was pointed out by Evans et al. (2010) through a combination of experiments and numerical simulation via impulsive loads on unit cells in the millimeter range. Experimental results showed that energy absorption per unit mass is even better than that predicted by the model developed. The authors reported that the energy absorption is governed primarily by the plastic deformations that occur subsequent to plastic buckling. In the simulations, conducted with walls of uniform thickness, the lattice members exhibit lateral buckling and kinking, with ensuing (relatively) rapid softening. Nevertheless, relative to other cellular media having isotropic response, such as foams, the hollow wall microlattices absorb more than twice the energy per unit mass.

In an interesting recent theoretical work Rayneau-Kirkhope et al. (2012) showed that the transition from solid to hollow struts (beams) changed the scaling in a manner analogous to increasing the hierarchical order by one. In this theoretical study, the authors used Euler buckling as a failure criteria for solid members whereas Euler buckling and Koiter or short wavelength buckling was assumed for the hollow tubes. The authors showed that if G is the order of hierarchy of the entire structure, then in terms of nondimensional loading (f) and volume (v) parameters, the scaling law can be computed as

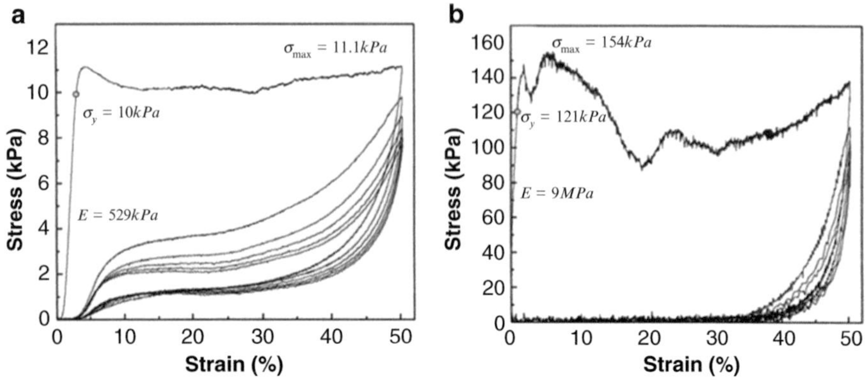

Fig. 3.20 Multicycle compression test results of nickel microlattices. (**a**) Stress–strain curves of a microlattice with 14 mg/cm³ exhibiting recoverable deformation. (**b**) Stress–strain curves of a microlattice with 43 mg/cm³ showing deformation more typical for metallic cellular materials showing no recovery (Schaedler et al. 2011)

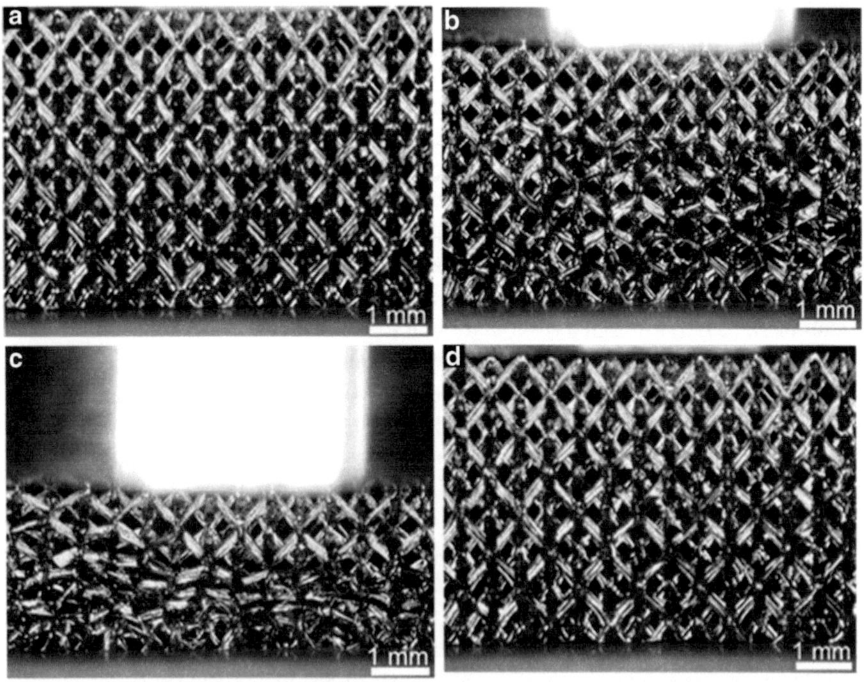

Fig. 3.21 Nickel microlattices exhibit recoverable deformation. (**a**) Before deformation. (**b**) 15 % compression. (**c**) 50 % compression. (**d**) Full recovery after removal of load (Schaedler et al. 2011)

$$v = K(G) f^{\left(\frac{G+2}{G+3}\right)},$$

where $\kappa(G)$ is a pre-multiplying parameter independent of loading. The authors showed using the above formula that the gain in scaling efficiency by hollowing a structure was similar to increasing the order of hierarchy (G) by one.

The above-mentioned advantage of hollow members is currently under increasing scrutiny and recently the potential of hollow thin-walled nanocrystalline cylinders to fabricate microlattices has been investigated. Schaedler et al. (2011) constructed an actual ultralight 3D microlattice using hollow nanometer thick tubes. The usual challenge in manufacturing these structures was overcome by the authors through coating of the sacrificial template with electroless nickel plating, and subsequently etching away the template during the self-propagating photopolymer waveguide fabrication process. The authors reported a quadratic scaling of elastic modulus as compared to cubic scaling observed in ultralight aerogels and carbon nanotube foams with stochastic architecture. Remarkable strain recovery even after compression exceeding 50 % strain—a range common for elastomers but not to typical metallic lattices which are not ultralight—was achieved as shown in Fig. 3.20. This feat was achieved due to the thin-shell buckling of the hollow struts which caused failure of the joints during compression causing relief cracks to form which allowed the rest of the structure to undergo fairly large compressive strains due to rotation of the failed joints but little straining of the parent material. In addition, a significant rise in stress was observed on further compression to about 40 % without the onset of densification. This was attributed to interaction between lattice members after localized compression at the nodes. In fact, densification was not observed until about 90 % strain. One full cycle of loading and unloading of this structure is shown in Fig. 3.21.

To explain this remarkable mechanical properties, two energy loss mechanisms have been proposed: (1) structural (kinking and buckling failure) and (2) frictional interaction between interacting members. The authors claimed that the peculiarity of this energy loss mechanism might explain the similarity of the overall stress–strain behavior of these structures to viscoelastic polymer foams even though the thin nickel film is known to be brittle. Further characterization was carried out on this structure (Torrents et al. 2012) to evaluate the effect of relative density and base metal properties. Two base metals electroless-Ni (EN Ni) film and electrodeposited nickel (ED Ni) film, which has a grain size several times that of EN Ni films, were utilized for the study. These nickel-based microlattice materials with structural hierarchy spanning three different length scales (nm, μm, mm) were prepared by plating a sacrificial template obtained by self-propagating photopolymer waveguide prototyping. Ni–P films with a thickness of 120 nm to 3 μm were deposited by electrolysis plating, whereas thicker films (5–26 μm) were obtained by subsequent electroplating of a pure Ni layer. This resulted in samples spanning three orders of magnitude in relative density, from 0.01 to 8.4 %. Uniaxial compression experiments revealed two distinct mechanical responses. At ultralow densities (<0.1 %), these lattices exhibited nearly full recovery after strains as large as 50 %, and damping coefficients an order of magnitude larger than for conventional Ni foams. This behavior is unprecedented for any metallic system. At higher densities (0.1–10 %), the compression behavior was fully plastic, similarly to traditional cellular metals. Optical and electron imaging of deformed lattices show that the deformation largely localized around the nodes. The pseudo-super-elastic behavior for ultralow-density lattices was attributed to nodal fracture events and local elastic buckling (kinking), which allowed extensive rotation around thin ligaments without the introduction of plastic strain. A simple mechanical model based on this observation clearly captured the order of magnitude of the critical thickness-to-diameter ratio for the transition, and may be used to design microlattices with desired recovery and damping characteristics. In the ultralight regime, the microlattice materials characterized in this work were approximately one order of magnitude stiffer and stronger than any existing alternative, and even at large densities (8.4 %) they were three times stronger than nickel foams of comparable weight. These substantial benefits stemmed from the hierarchical nature of the lattice topology, which allowed capitalization of plasticity size effects in a macroscale structural material.

In contrast to the structural level response of these structures made of thin-walled members, Lian et al. (2011) studied the behavior of single hollow wall nanocrystalline Ni cylinders. In particular the authors found a gradual brittle to ductile transition when the thickness of the hollow cylinder was decreased from 500 to 150 nm. It was reported that the physics of the thin-film plasticity significantly changed the compressive strength as compared to shell buckling theory. The size effects on plasticity on these micropillars were also clearly showed by Fan et al. (2012) using 3D-Discrete Dislocation Dynamics (3D-DDD). In their simulations, the authors confirmed that the compressive strength of these hollow micropillars strongly correlates to the wall thickness and very weakly on the inner or the outer diameter. The authors believed that this is due to the dominance of single-arm dislocation sources whose statistical lengths were determined by the wall thickness.

As previously discussed, engineered hierarchical structures with regular lattice exhibit more desirable characteristics due to suppression of bending modes of deformation. However, since their post-failure behavior is dominated by buckled struts, they may become less desirable for energy absorption application. J. Bernal Ostosa, et al. (2012), a hybrid composite-like structure was fabricated which comprised a low-density polymer lattice structure with strut lengths and void spaces between struts at the centimeter scale, and a lower density stochastic foam with porosity at the submillimeter scale in the space between the struts. The compressive stress–strain measurements clearly showed that the post-yielding flow stress can be elevated significantly with the addition of even weak foam: the strength exceeded the sum of the contributions from the lattice structure and the foam alone by a large margin. Using X-ray CT, the authors showed that the strength elevation was attributable to the stabilization of the strut members against buckling when surrounded by foam. This hypothesis was further bolstered by FE-based numerical simulations. However, it remains to be determined whether foams lead to improvements in the pertinent specific performance metrics (e.g., energy absorbed per unit mass).

In addition to solid and hollow struts, the use of composites to construct the truss members has also been investigated. The deployment of composites can, in addition to saving weight, enhance energy absorption significantly due to multiple damage modes. Zhao et al. (2012) constructed hierarchical composite honeycombs arranged in Kagome lattice structure where the usual metallic or polymeric struts were replaced with integrated woven textile sandwich composites made of glass fibers. The authors first characterized the individual composite members through three-point bending tests and determined that the skin strength of the sandwich strut (panel) will determine the peak load. However, once the peak load was reached and the skin fractured, the core was found to play a much bigger role in the residual strength. Experiments done on the lattice structure itself clearly showed four distinct loading stages: the elastic deformation, the stress-softening stage, the deformation plateau, and the densification. After the peak stress, failure of ribs led to stress softening. Strengths of the two Kagome HCHs have a disparity of 1.2 MPa, about 40 % of the strength of the un-strengthened HCH. The strength disparity was much larger than the difference of the skin thickness. Stresses of the deformation plateau have a disparity of 0.4 MPa, about 28 % of the smaller one. Strengthened HCHs had much greater peak loading than un-strengthened ones, but their stress levels of the deformation plateau did not have obvious advantages. The authors concluded that skins of the sandwich rib of the HCH determine the load capacity of the HCH, while the deformation plateau was determined jointly by the rib skin and the rib core. The three typical HCHs of the same relative density had similar mechanical properties, whether strengthened or un-strengthened. It was shown that the topology of the first-order lattice structures would be irrelevant to their out-of-plane compression properties. The main failure modes of the hierarchical lattice panel included (1) compression damage of the matrix of the rib, rendering by splays; (2) compression–bend coupling failure of ribs, which led to final break off of the rib; and (3) shear failure at the slit of the rib at the edge of the tested sample which could be looked as the edge effect of the limited dimensional samples. Up until densification, the panel maintains a long deformation plateau with relatively high stress level. The densification strain of the hierarchical structure varied from 0.7 to 0.8.

3.8 Concluding Remarks

In conclusion, in this book chapter, we have discussed how essential structural features like hierarchy and randomness in biological systems can potentially improve the mechanical performance of materials significantly. We have described a select set of results from the literature related to the behavior of biomimetic cellular materials under static and dynamic loading, including extension of hierarchical honeycombs to cellular structures with functional gradient and self-similar hierarchy. The results provide new insights into the relationship between material architecture and behavior and function, and also new opportunities and avenues for designing novel multifunctional cellular based materials.

Acknowledgments This work was supported in part by the US Air Force Office of Scientific Research under AFOSR YIP grant award, #FA 9550-10-1-0145 under the technical supervision of Dr. Joycelyn Harrison, and in part by the US Department of Homeland Security under Award

Number 2008-ST-061-ED0001. The views and conclusions contained in this document are those of the authors and should not be interpreted as necessarily representing the official policies, either expressed or implied, of the AFOSR or the US Department of Homeland Security.

References

Vincent JFV, Currey JD (1980) Mechanical properties of biological materials, society for experimental biology. UK, Cambridge

Vincent JFV (1990) Structural biomaterials. Princeton University Press, Princeton, NJ

Tai K, Dao M, Suresh S, Palazoglu A, Ortiz C (2007) Nanoscale heterogeneity promotes energy dissipation in bone. Nat Mater 6(6):454–462

Fabritius HO, Sachs C, Triguero PR, Raabe D (2009) Influence of structural principles on the mechanics of a biological fiber–based composite material with hierarchical organization: the exoskeleton of the lobster homarus americanus. Adv Mater 21(4):391–400

Chen C, Lu TJ, Fleck NA (1999) Effect of imperfections on the yielding of two dimensional foams. J Mech Phys Solids 47(11):2235–2272

Gibson L, Ashby M (1999) Cellular solids: structure and properties. CUP. doi:citeulike-article-id:7339572

Nonell JB, Levick M (2001) Antonio gaudi: master architect (tiny folio). Abbeville, New York

Lu TJ, Chen C (1999) Thermal transport and fire retardance properties of cellular aluminium alloys. Acta Mater 47(5):1469–1485

Hollister SJ (2005) Porous scaffold design for tissue engineering. Nat Mater 4(7):518–524

Hutmacher DW (2000) Scaffolds in tissue engineering bone and cartilage. Biomaterials 21(24):2529–2543

Dharmasena K, Queheillalt D, Wadley H, Chen Y, Knight D (2009) Dynamic response of a multilayer prismatic structure to impulsive loads incident from water, vol 36. 4. Elsevier, Kidlington, ROYAUME-UNI

Liang Y, Spuskanyuk AV, Hutchinson JW (2007) The response of metallic sandwich panels to water blast, vol 74. 1. ASME, New York, NY, ETATS-UNIS

Mori L, Lee S, Xue Z, Vaziri A, Queheillalt D, Wadley H, Hutchinson JW, Espinosa HD (2007) On the behavior of sandwich structures subjected to under water impulsive loads. J Mech Mater Struct 2:1981–2006

Mori L, Queheillalt D, Wadley H, Espinosa H (2009) Deformation and failure modes of I-Core sandwich structures subjected to underwater impulsive loads. Exp Mech 49(2):257–275. doi:10.1007/s11340-008-9166-9

Rathbun HJ, Radford DD, Xue Z, He MY, Yang J, Deshpande V, Fleck NA, Hutchinson JW, Zok FW, Evans AG (2006) Performance of metallic honeycomb-core sandwich beams under shock loading. Int J Solids Struct 43(6):1746–1763

Vaziri A, Hutchinson JW (2007) Metal sandwich plates subject to intense air shocks. Int J Solids Struct 44:2021–2035

Vaziri A, Xue Z, Hutchinson JW (2007) Performance and failure of metal sandwich plates subject to shock loading. J Mech Mat 2(10):1947–1964

Wadley H, Dharmasena K, Queheillalt D, Chen YC, Dudt P, Knight D, Xue Z, Vaziri A (2007) Dynamic crushing of square honeycomb structures during underwater impulsive loading. J Mech Mater Struct 2(10):2025–2048

Wadley H, Dharmasena K, Chen Y, Dudt P, Knight D, Charette R, Kiddy K (2008) Compressive response of multilayered pyramidal lattices during underwater shock loading. Int J Impact Eng 35(9):1102–1114

Wei Z, Deshpande VS, Evans AG, Dharmasena KP, Queheillalt DT, Wadley HNG, Murty YV, Elzey RK, Dudt P, Chen Y, Knight D, Kiddy K (2008) The resistance of metallic plates to localized impulse. J Mech Phys Solids 56(5):2074–2091

Xue Z, Hutchinson JW (2003) Preliminary assessment of sandwich plates subject to blast loads. Int J Mech Sci 45(4):687–705
Xue Z, Hutchinson JW (2004) A comparative study of impulse-resistant metal sandwich plates. Int J Impact Eng 30(10):1283–1305
Lu T, Valdevit L, Evans A (2005) Active cooling by metallic sandwich structures with periodic cores. Progr Mater Sci 50(7):789–815
Rakow J, Waas A (2004) Thermal buckling of metal foam sandwich panels for actively cooled thermal protection systems
Shuang L, Boming Z (2011) Effects of active cooling on the metal thermal protection systems. Aerosp Sci Technol 15(7):526–533
Ajdari A, Nayeb-Hashemi H, Vaziri A (2011) Dynamic crushing and energy absorption of regular, irregular and functionally graded cellular structures. Int J Solids Struct 48(3–4):506–516
Gibson LJ, Ashby MF (1997) Cellular solids: structures and properties, 2nd edn. Cambridge University Press, Cambridge
Ajdari A, Nayeb-Hashemi H, Canavan PK, Warner G (2008) Effect of defect on elastic-plastic behavior of cellular materials. J Mater Sci Eng A 487:558–567
Fazekas A, Dendievel R, Salvo L, Bre'chet Y (2002) Effect of microstructural topology upon the stiffness and strength of 2D cellular structures. Int J Mech Sci 44(10):2047–2066
Fortes MA, Ashby MF (1999) The effect of non-uniformity on the in-plane modulus of honeycombs. Acta Mater 47(12):3469–3473
Guo XE, Gibson LJ (1999) Behavior of intact and damaged honeycombs: a finite element study. Int J Mech Sci 41(1):85–105
Silva MJ, Gibson LJ (1997) The effects of non-periodic microstructure and defects on the compressive strength of two-dimensional cellular solids. Int J Mech Sci 39(5):549–563
Silva MJ, Hayes WC, Gibson LJ (1995) The effects of non-periodic microstructure on the elastic properties of two-dimensional cellular solids. Int J Mech Sci 37:1161–1177
Wang AJ, McDowell DL (2003) Effects of defects on in-plane properties of periodic metal honeycombs. Int J Mech Sci 45(11):1799–1813. doi:10.1016/j.ijmecsci.2003.12.007
Zhu HX, Hobdell JR, Windle AH (2001) Effects of cell irregularity on the elastic properties of 2D Voronoi honeycombs. J Mech phys Solid 49(4):857–870. doi:10.1016/S0022-5096(00)00046-6
Silva ECN, Walters MC, Paulino GH (2006) Modeling bamboo as a functionally graded material: lessons for the analysis of affordable materials. J Mat Sci 41:6991–7004
Chen PY, Lin AY, Lin Y-S, Seki Y, Stokes AG, Peyras J, Olevsky EA, Meyeres MA, McKittrick J (2008) Structure and mechanical properties of selected biological materials. J Mech Behav Biomed Mater 1:208–226
Cui L, Kiernan S, Gilchrist MD (2009) Designing the energy absorption capacity of functionally graded foam materials. Mat Sci Eng A 507:215–225
Ali M, Qamhiyah A, Flugrad D, Shakoor M (2008) Theoretical and finite element study of a compact energy absorber. Adv Eng Software 39:95–106
Aizenberg J, Weaver JC, Thanawala MS, Sundar VC, Morse DE, Fratzl P (2005) Skeleton of Euplectella sp.: structural Hierarchy from the nanoscale to the macroscale. Science 309(5732):275–278. doi:10.1126/science.1112255
Buehler MJ (2006) Nature designs tough collagen: explaining the nanostructure of collagen fibrils. Proc Natl Acad Sci 103(33):12285–12290. doi:10.1073/pnas.0603216103
Espinosa HD, Juster AL, Latourte FJ, Loh OY, Gregoire D, Zavattieri PD (2011) Tablet-level origin of toughening in abalone shells and translation to synthetic composite materials. Nat Commun 2:173, http://www.nature.com/ncomms/journal/v2/n2/suppinfo/ncomms1172_S1.html
Fratzl P, Weinkamer R (2007) Nature's hierarchical materials. Progr Mater Sci 52(8):1263–1334
Gibson LJ, Ashby MF, Harley BA (2010) Cellular materials in nature and medicine. University Press, Cambridge
Lakes R (1993) Materials with structural hierarchy. Nature 361(6412):511–515
Ortiz C, Boyce MC (2008) Bioinspired structural materials. Science 319(5866):1053–1054. doi:10.1126/science.1154295

Qing H, Mishnaevsky L Jr (2009) 3D hierarchical computational model of wood as a cellular material with fibril reinforced, heterogeneous multiple layers. Mech Mater 41(9):1034–1049
Cote F, Russell BP, Deshpande VS, Fleck NA (2009) The through-thickness compressive strength of a composite sandwich panel with a hierarchical square honeycomb sandwich core. J Appl Mech 76(6):061004–061008
Fan HL, Jin FN, Fang DN (2008) Mechanical properties of hierarchical cellular materials. Part I: analysis. Compos Sci Tech 68(15–16):3380–3387
Kazemahvazi S, Tanner D, Zenkert D (2009) Corrugated all-composite sandwich structures. Part 2: Failure mechanisms and experimental programme. Compos Sci Tech 69(7–8):920–925
Kazemahvazi S, Zenkert D (2009) Corrugated all-composite sandwich structures. Part 1: modeling. Compos Sci Tech 69(7–8):913–919
Kooistra GW, Deshpande V, Wadley HNG (2007) Hierarchical corrugated core sandwich panel concepts. J Appl Mech 74(2):259–268
Ajdari A, Jahromi BH, Papadopoulos J, Nayeb-Hashemi H, Vaziri A (2012) Hierarchical honeycombs with tailorable properties. Int J Solid Struct 49(11–12):1413–1419. doi:10.1016/j.ijsolstr.2012.02.029
Chuang C-H, Huang J-S (2002) Elastic moduli and plastic collapse strength of hexagonal honeycombs with plateau borders. Int J Mech Sci 44(9):1827–1844
Christensen RM (1987) Sufficient symmetry conditions for isotropy of the elastic moduli tensor. J Appl Mech 54(4):772–777
Boresi AP, Schmidt RJ (2002) Advanced mechanics of materials. Wiley, New York
Harders H, Hupfer K, Rösler J (2005) Influence of cell wall shape and density on the mechanical behaviour of 2D foam structures. Acta Mater 53(5):1335–1345
Haghopanah B, Papadoupolos J, Vaziri A (2012) Stiffness and plastic collapse of self-similar hierarchical honeycombs. Acta Mater (under review)
Jacobsen AJ, Barvosa-Carter W, Nutt S (2007a) Micro-scale truss structures formed from self-propagating photopolymer waveguides. Adv Mater 19(22):3892. doi:10.1002/adma.200700797
Jacobsen AJ, Barvosa-Carter W, Nutt S (2007b) Compression behavior of micro-scale truss structures formed from self-propagating polymer waveguides. Acta Mater 55(20):6724–6733. doi:10.1016/j.actamat.2007.08.036
Jacobsen AJ, Barvosa-Carter W, Nutt S (2008a) Shear behavior of polymer micro-scale truss structures formed from self-propagating polymer waveguides. Acta Mater 56(6):1209–1218. doi:10.1016/j.actamat.2007.11.018
Jacobsen AJ, Barvosa-Carter W, Nutt S (2008b) Micro-scale truss structures with three-fold and six-fold symmetry formed from self-propagating polymer waveguides. Acta Mater 56(11):2540–2548. doi:10.1016/j.actamat.2008.01.051
Evans AG, He MY, Deshpande VS, Hutchinson JW, Jacobsen AJ, Carter WB (2010) Concepts for enhanced energy absorption using hollow micro-lattices. Int J Impact Eng 37(9):947–959. doi:10.1016/j.ijimpeng.2010.03.007
Rayneau-Kirkhope D, Mao Y, Farr R (2012) Ultralight fractal structures from hollow tubes. Phys Rev Lett 109(204301):1–4. doi:10.1103/PhysRevLett.109.204301
Schaedler TA, Jacobsen AJ, Torrents A, Sorensen AE, Lian J, Greer JR, Valdevit L, Carter WB (2011) Ultralight metallic microlattices. Science 334(6058):962–965. doi:10.1126/science.1211649
Torrents A, Schaedler TA, Jacobsen AJ, Carter WB, Valdevit L (2012) Characterization of nickel-based microlattice materials with structural hierarchy from the nanometer to the millimeter scale. Acta Mater 60(8):3511–3523. doi:10.1016/j.actamat.2012.03.007
Lian J, Jang D, Valdevit L, Schaedler TA, Jacobsen AJ, Carter WB, Greer JR (2011) Catastrophic vs gradual collapse of thin-walled nanocrystalline Ni hollow cylinders as building blocks of microlattice structures. Nano Lett 11(10):4118–4125. doi:10.1021/Nl202475p
Fan HD, Li ZH, Huang MS (2012) Size effect on the compressive strength of hollow micropillars governed by wall thickness. Scripta Mater 67(3):225–228. doi:10.1016/j.scriptamat.2012.04.012
Zhao L, Zheng Q, Fan HL, Jin FN (2012) Hierarchical composite honeycombs. Mater Design 40:124–129. doi:10.1016/j.matdes.2012.03.009
J. Bernal Ostosa, R.G. Rinaldia, et al. (2012). "Deformation stabilization of lattice structures via foam addition." Acta Materialia 60(19): 6476–6485.

Xue Z, Hutchinson JW (2006) Crush dynamics of square honeycomb sandwich cores. Int J Num Meth Engrg 65:2221–2245
Liu Y, Zhang X (2009) The influence of cell micro-topology on the in-plane dynamic crushing of honeycombs. Int J Impact Eng 36(1):98–109
Zheng Z, Yu J, Li J (2005) Dynamic crushing of 2D cellular structures: A finite element study. Int J Impact Eng 32(1–4):650–664
Tan PJ, Reid SR, Harrigan JJ, Zou Z, Li S. (2005) Dynamic compressive strength properties of aluminum foams. Part II—'shock' theory and comparison with experimental data and numerical models. J Mech Phys Solids 53:2206–2230
Rakow JF, Waas AM. (2005) Thermal Buckling of Metal Foam Sandwich Panels for Convective Thermal Protection Systems. J Spacecraft Rockets 42(5):832–844

Chapter 4
Experimental and Theoretical Studies of Fiber-Reinforced Composite Panels Subjected to Underwater Blast Loading

Xiaoding Wei and Horacio D. Espinosa

Abstract Fluid–structure interaction (FSI) experiments on monolithic and sandwich composite panels were performed to identify key failure mechanisms resulting from underwater blast loading. Panel performance was compared in terms of impulse deflection. Various failure mechanisms such as delamination between laminas, matrix damage and fiber rupture in laminas, and foam crushing were identified. A 3-D rate-dependent numerical model was developed to understand the experimentally observed failure mechanisms. A new failure criterion that includes strain-rate effects was formulated and implemented to simulate different damage modes in unidirectional composite plies. This rate-dependent numerical model predicted the responses of composite panels subjected to underwater blast loading with more correlated material damage patterns with the experimental observation than previously developed models. The model also revealed the important role of the soft foam core in sandwich composite panels for improving panel performance by mitigating the transmitted impulse to the back-side face sheet while maintaining overall bending stiffness.

4.1 Introduction

Glass-reinforced plastic composite materials are of great interest in naval hull construction because they have strength comparable to steel yet are lightweight (Mouritz et al. 2001; Vinson and Sierakowski 2002). Two architectures are commonly used in marine hull design: monolithic structure and sandwich structure. Sandwich structures, in particular, have gained increasing attention as the composite material most

X. Wei • H.D. Espinosa (✉)
Department of Mechanical Engineering, Northwestern University,
2145 Sheridan Road, Tech B224, Evanston, IL 60208-3111, USA
e-mail: espinosa@northwestern.edu

A. Shukla et al. (eds.), *Blast Mitigation: Experimental and Numerical Studies*,
DOI 10.1007/978-1-4614-7267-4_4,

capable of mitigating high-intensity blast loadings (Tekalur et al. 2009; Andrews and Moussa 2009; Wang et al. 2009; Latourte et al. 2011). These sandwich structure composites contain a crushable core that is encased between front- and back-side face skins, which helps dissipate a substantial amount of energy by attenuating the impulse transmitted to the back-side face sheet and therefore protecting the whole structure from catastrophic failure.

Recently, a full-scale blast experiment has been developed by Arora et al. (2012) to assess the mechanical response of composite panels with dimensions comparable to real naval structures. However, the full-scale experiment is limited by its high cost and difficult implementation. Therefore, performing tests at the laboratory scale, extracting damage parameters, and properly extending the parameters to actual scale serves as a viable alternative (Porfiri and Gupta 2010), and many studies have shown the validity of this approach. Low-velocity impact behavior of composite panels was studied by Schubel et al. (2005) using a drop tower apparatus. Tagarielli et al. (2007) investigated the dynamic response of monolithic and sandwich composite beams impacted by metal foam projectiles. LeBlanc and Shukla (2010) examined the dynamic response and damage evolution in composite materials subjected to underwater explosive loading using a conical shock tube facility. One limitation of the experiments reported in the literature is that the impacted region was typically small compared to the panel or laminate dimension, resulting in localized damage. Localized damage features are not representative of the structural effects observed in large-scale blasts where material degradation spreads over a large section of the hull structure.

In this chapter, we present a study of the dynamic response of monolithic and sandwich composite panels assessed by the scaled down laboratory FSI apparatus initially developed by Espinosa and coworkers (2006). The advantage of this method is the accurate scaling to full-field loads, which enables testing of panels with dimensions small enough to be easily manufactured and handled (radius = 76.2 mm), yet with thicknesses that are sufficient to investigate layups consistent with full marine hulls in terms of stacking sequence and number of plies. Moreover, the setup is integrated with instrumentation that allows recording of the deflection profile histories over the entire span of the panels for a precisely defined applied impulse.

Another important aspect of assessing dynamic response is the development of theoretical models to predict the performance of composite materials under high strain-rate deformation based on simulations of FSI experiments. Among the limited literature on theoretical studies of the dynamic response of composite panels subjected to blast loading (LeBlanc and Shukla 2010; Librescu et al. 2006; Batra and Hassan 2007; Fatt and Palla 2009; Johnson et al. 2009; Fatt et al. 2010), a thorough numerical study has not yet been reported that investigates the strain-rate effect. We discuss in the second section of this chapter our development of a 3-D numerical model for predicting the performance of composite panels under high impulses that incorporates strain-rate effects.

4.2 FSI Experiments on Monolithic and Sandwich Composite Panels

4.2.1 *Scaled Down FSI Apparatus*

The FSI experimental setup developed by Espinosa and co-workers (2006) is shown in Fig. 4.1a where the shell of a naval hull is scaled down to a composite panel specimen, which is clamped at one end of an anvil tube filled with water. A water piston seals the other extremity of the chamber and the exponentially decaying pressure history is produced by impacting the water piston with a flyer plate launched by a gas gun (see Fig. 4.1b). A contact-pin type velocity sensor is used to measure the impact velocity of the flyer plate. The far-field momentum of the impulse is $I_0 = \int_0^\infty P\mathrm{d}t = \int_0^\infty P_0 \mathrm{e}^{-t/t_0}\mathrm{d}t = P_0 t_0$, where P_0 is the peak pressure and t_0 is the decay time. The peak pressure is a function of the acoustic impedances of the fluid and solid, and the decay time is a function of the dimensions of the water piston and flyer plate. The scaling is realized by setting the FSI to be the same as would be expected in the full scale application. In other words, the same fraction of I_0 should be transmitted to the specimen. Defining the scaling factor, K, as the ratio of the panel thickness in the full scale naval structure to the panel thickness in the experiment, $K = h_F/h_E$, analysis shows that the applied impulse should scale with K. In contrast, pressure does not scale down because it is an intensity quantity; therefore, impulse scaling is achieved by adjusting the decay time t_0 through modification of the dimensions of the projectile and water piston. When the impulsive wave propagates, the peak pressure is inversely proportional to the cross-sectional area of the anvil tube. In previous experiments carried out on solid stainless steel panels (Espinosa et al. 2006), pressure histories recorded with pressure transducers were consistent with the predicted exponential decay associated with blast loading. For each experiment, I_0, P_0, and t_0 were computed using solutions to the wave equation (Espinosa et al. 2006; Drumheller 2002). After the impulse hits the panel, the deflection history of the specimen is measured by shadow Moiré and high-speed photography.

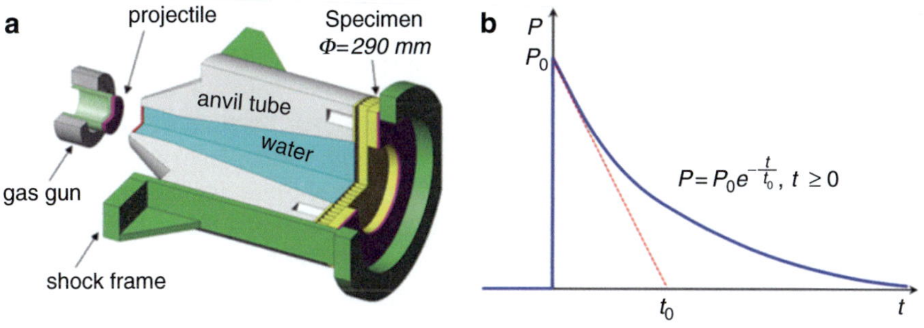

Fig. 4.1 (**a**) Schematic of the underwater blast experimental apparatus. (**b**) Pressure history of the impulse after impact

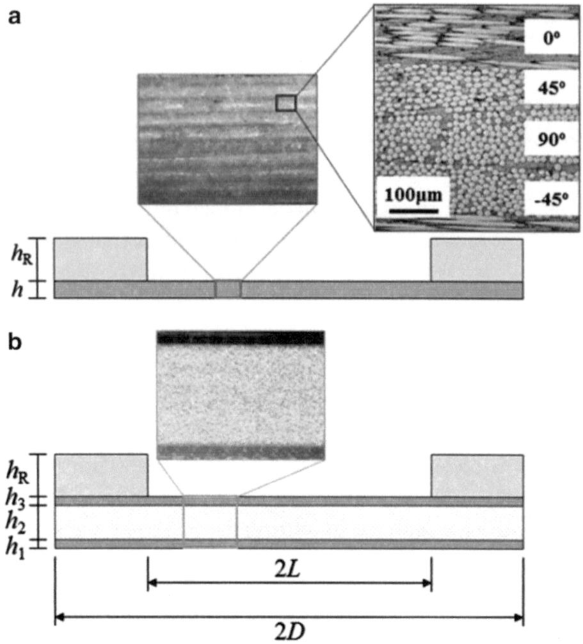

Fig. 4.2 Schematics and optical images of the composite panels. (**a**) Monolithic panel of thickness h bonded to a steel ring of thickness h_R. *Insets* are a photograph of the cross section, highlighting the different fabric layers, and a microscopy image of an E-glass/vinylester fabric showing four anisotropic laminates. (**b**) Sandwich panel with face sheet thicknesses h_1 and h_3 and core thickness h_2. *Inset* is a photograph of the cross section of the panel

4.2.2 Composite Panel Architectures

Three composite panel architectures—monolithic, symmetrical sandwich, and asymmetrical sandwich panels—were investigated. The composite face sheets are comprised of quasi-isotropic Devold DBLT850-E10 glass fiber non-crimp fabric infiltrated by vinylester Reichhold DION 9500. The sandwich panels incorporate a 15 mm Diab H250 divinycell PVC foam core. Each composite fabric is composed of four laminas comprised of unidirectional E-glass fibers that are assembled according to the following sequence: 0°/45°/90°/−45°. The fiber diameter, lamina thickness, and fabric thickness are 15 μm, 150 μm, and 600 μm, respectively. Figure 4.2 shows the geometries of the monolithic and sandwich composite panels and optical photographs of the cross section of each composite.

Composite monolithic and sandwich panels having approximately the same weight per unit area were manufactured at KTH-Royal Institute of Technology, Sweden. After manufacturing, the panels were cut with a water jet to obtain the final circular geometry with a diameter of $2D = 292.1$ mm. The circular panels were then bonded to a steel ring of thickness $h_R = 25.4$ mm (or $h_R = 19$ mm for monolithic panels tested at low impulse) with 3M Scotch-Weld™ DP460 epoxy adhesive. The steel ring has an inner diameter of $2L = 76.2$ mm, corresponding to the region where the impulsive wave hits the specimen. The three different specimen designs were the following: (1) Monolithic panels consisting of nine composite fabrics infiltrated by the resin as $(0/45/90/{-45})_4$–$(45/90/{-45}/0)_5$ for a total thickness of $h = 5.8$ mm;

Table 4.1 List of experimental impulse parameters and the normalized peak deflections recorded for each test

Geometry	Label	Flyer thickness (mm)	Flyer velocity (m/s)	P_0 (MPa)	t_0 (μs)	I_0 (Pa s)	$I_0 / \bar{M}$ (m/s)	δ_{max} / L
Monolithic	1-1	4.66	235	49.3	25	1,233	116	0.151
	1-2	4.67	340	70.6	25	1,766	166	0.188
	1-3	6.74	340	67.4	36	2,425	227	0.257
	1-4	8.74	352	76.3	43	3,283	314	0.350[a]
	1-5	10.82	368	76.3	57	4,349	416	(N/A)[b]
Symmetrical	2-1	6.74	343	68.5	36	2,465	217	0.198
	2-2	10.67	339	72.4	55	3,983	351	0.222
	2-3	13.42	360	101.5	55	5,581	495	0.268
	2-4	16.38	308	62.4	87	5,427	482	0.213[a]
Asymmetrical	3-1	10.87	333	68.1	58	3,948	350	0.222
	3-2	16.59	367	75.8	88	6,672	592	0.291

[a]Peak deflection preceding massive panel failure
[b]Peak deflection for 1-5 was not obtained because the grating used for the shadow Moiré was damaged by fly debris

(2) Symmetrical sandwich panels consisting of six composite fabrics separated by a foam core as $(0/45/90/-45)_3$–core–$(45/90/-45/0)_3$, yielding a total thickness of $h = 19$ mm ($h_1 = 2$ mm, $h_2 = 15$ mm, $h_3 = 2$ mm); (3) Asymmetrical sandwich panels consisting of six composite fabrics separated by a foam core as $(0/45/90/-45)_2$–core–$(45/90/-45/0)_4$, yielding a total thickness of $h = 19$ mm ($h_1 = 1.33$ mm, $h_2 = 15$ mm, $h_3 = 2.66$ mm). The areal masses ($\bar{M}$) of the monolithic, symmetrical sandwich, and asymmetrical sandwich panels were 10.7 kg m^{-2}, 11.3 kg m^{-2}, and 11.2 kg m^{-2}, respectively.

4.2.3 Experimental Results

Four monolithic panels (panel configuration 1), four symmetric sandwich panels (panel configuration 2), and two asymmetric sandwich panels (panel configuration 3) were tested at impulses ranging from $I_0 = 1{,}233$ to 6,672 Pa s. A notation of "panel configuration-test number" was used to distinguish individual tests of the panels. For example, test 1-3 refers to test #3 for the monolithic panel configuration. For all the experiments, except tests 2-4 and 3-2, the thickness of the water piston was 10.42 mm. Table 4.1 lists the panel configurations and corresponding impulse parameters for all tests. In addition, it provides the applied impulse per areal mass $I_0 / \bar{M}$ and the normalized peak deflection δ_{max} / L. These two quantities provide a direct means to compare the panel performance.

Figure 4.3 summarizes the performances of the three types of composite panels in terms of normalized maximum deflection vs. applied impulse per areal mass. Previous FSI experimental results on A304 stainless steel (A304SS) sandwich

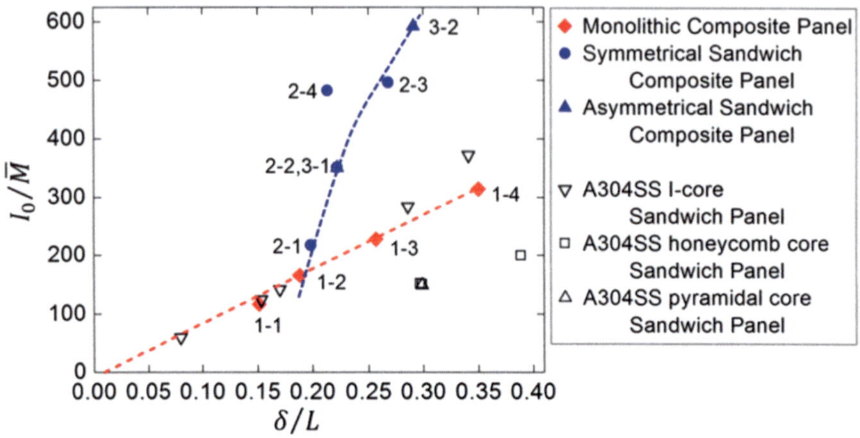

Fig. 4.3 Blast mitigation performances of monolithic and sandwich composite panels in terms of impulse per areal mass vs. normalized deflection. For comparison, the experimental results reported in Mori et al. (2007, 2009) for A304 stainless steel sandwich panels (*open symbols*) are overlaid

panels with various core topologies were included for comparison (Mori et al. 2007, 2009). The performances of monolithic composite panels were comparable to those observed for A304SS I-beam core sandwich panels, which were previously identified as the optimal core topology among various designs (Mori et al. 2007, 2009). This finding clearly shows the performance-to-weight advantage of monolithic composite panels. In addition, a linear relationship between the normalized deflection and impulse per areal mass was found for the monolithic composite panels where a least squares fit yielded $I_0/\bar{M} = (935 \times \delta_{\max}/L - 13)(\mathrm{ms}^{-1})$. In contrast, the symmetrical and asymmetrical composite panels showed a bilinear relationship described by $I_0/\bar{M} = (5561 \times \delta_{\max}/L - 884)(\mathrm{ms}^{-1})$ for impulses with $I_0/\bar{M} < 430\mathrm{ms}^{-1}$ and by $I_0/\bar{M} = (4211 \times \delta_{\max}/L - 633)(\mathrm{ms}^{-1})$ for high impulses of $I_0/\bar{M} > 430\mathrm{ms}^{-1}$. Both slopes for the sandwich panels are greater than those for the monolithic panels, effectively demonstrating the outstanding blast protection capability of the sandwich architectures.

In addition, comparing the results of tests 2-3 and 2-4, we found that the pressure history of the applied impulse plays an important role in panel response. Both panels in tests 2-3 and 2-4 were exposed to impulses of the same magnitude I_0; however, the impulse applied for test 2-4 had a lower peak pressure P_0 and longer decay time t_0. This difference in pressure profile resulted in 22 % less peak central deflection for test 2-4 compared to test 2-3. This suggests that there is a significant FSI effect in the experiments. According to Taylor's analysis (1963) and studies by Xue and Hutchinson (2004), the ratio of the impulse imparted to the plate to the free-field impulse can be approximated by $I/I_0 \sim 2(t_0/t_*)^{t_0/(t_*-t_0)}$, in which $t_* = (\rho h)/(\rho_f c_f)$ is the characteristic fluid–plate interaction time scale. ρ and h are the density and thickness of the plate, ρ_f is the density of the fluid, and c_f is the sound speed in the fluid. Thus, the impulse imparted to the plate decreases as the decay time t_0 becomes longer.

The plots in Fig. 4.4 display the central deflection histories for all tested monolithic, symmetrical sandwich, and asymmetrical sandwich panels. Note that for the

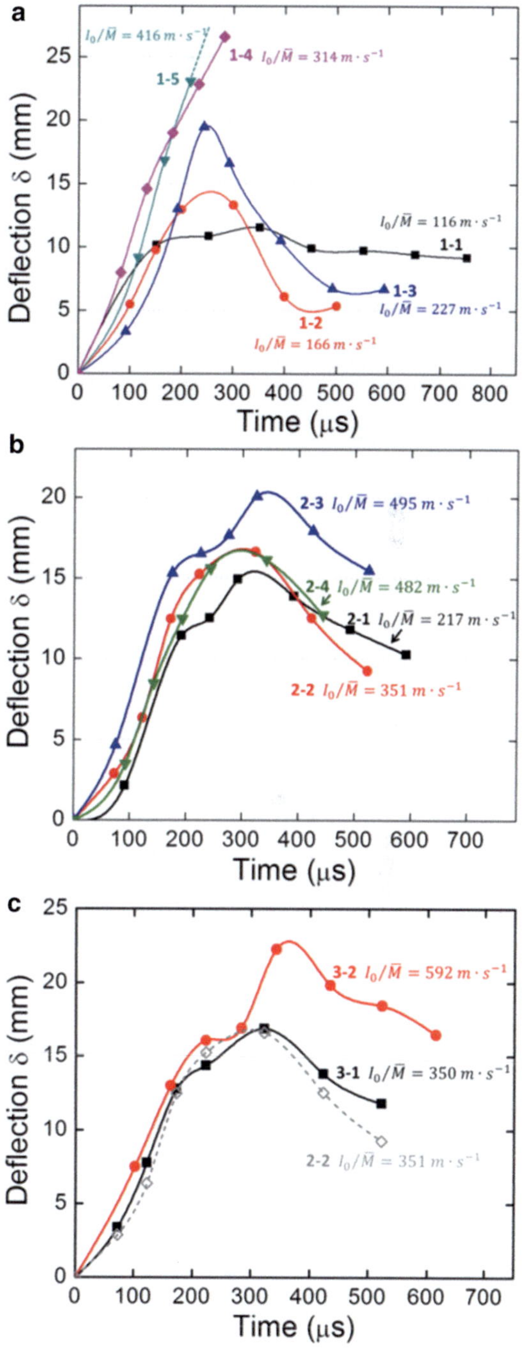

Fig. 4.4 Panel deflection histories for monolithic composite panels (**a**), symmetrical sandwich composite panels (**b**), and asymmetrical sandwich composite panels (**c**). Also plotted in (**c**) for comparison is the central deflection history for test 2-2, which was subject to an applied impulse that was approximately equivalent to that in test 3-1

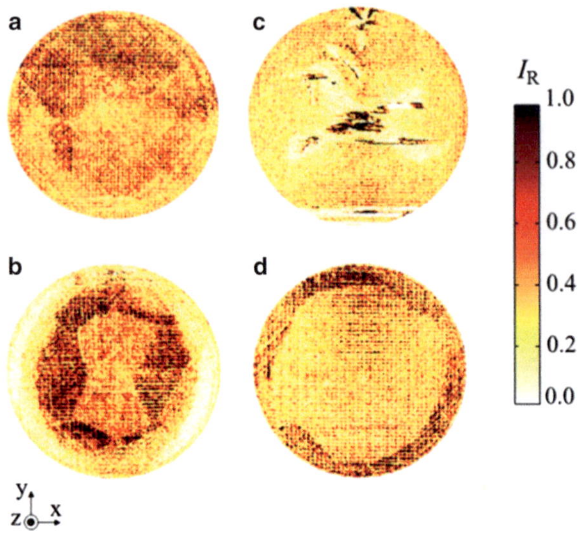

Fig. 4.5 C-scan results for typical monolithic and sandwich panels. (**a**) Water-side fabric in monolithic panel 1-1 subjected to an impulse of $I_0 = 1{,}233$ Pa s. (**b**) Water-side fabric in monolithic panel 1-3 subjected to an impulse of $I_0 = 2{,}425$ Pa s. (**c**) and (**d**) Water-side and air-side face sheets of sandwich panel 2-3, respectively. I_R is the magnitude of signal reflection from the material (0 corresponds to uniform medium and 1 corresponds to void)

monolithic panels the peak deflection was reached faster when exposed to greater impulses, whereas the peak deflection for sandwich panels was reached at approximately the same time for all impulses. In addition, sandwich panels reached the peak deflection later than monolithic panels due to compaction of the foam core. At the smallest impulse tested, the monolithic panel 1-1 showed a slow recovery rate after reaching the peak deflection; however, tests 1-2 and 1-3 for which higher impulses were applied showed a remarkable spring back after reaching the peak deflection. In tests 1-4 and 1-5, no spring back was observed because the panels suffered a catastrophic failure. Unlike the monolithic panels, all the tested sandwich panels consistently showed the spring back phenomenon with slower recovery rates compared to monolithic panels. Furthermore, we found that the symmetrical and asymmetrical sandwich panels subjected to the same impulse (tests 2-2 and 3-1) exhibited very similar responses.

4.2.4 Nondestructive Damage Identification

Ultrasonic C-scan was performed on the panels after FSI experiments to identify the damage in composite laminates. Figure 4.5 gives the C-scan results for monolithic panels 1-1 and 1-3, and sandwich panel 2-3. At low impulse (test 1-1), the mild signal contrast suggests that no appreciable localized damage feature occurred on the water-side fabric in the monolithic panel. At a higher impulse (test 1-3), a bright ring on the periphery of the water-side fabric in the panel was identified, which corresponds to bending-induced delamination near the clamped boundary. Sandwich panel 2-3, which was subjected to an impulse more than twice the magnitude of test

1-3, showed very limited delamination at the panel periphery on both the water-side and air-side face sheets. Instead, a clear X-shaped bright pattern was found on the water-side face sheet, which corresponds to the extensive fiber fracture identified later by optical microscope. In contrast, the air-side face sheet presented negligible material degradation, benefiting from the extraordinary blast protection capability of the sandwich configuration. Thus, it is worth pointing out that the damage feature identified for sandwich panels is usually not detrimental to the overall performance of the sandwich panels, as long as their integrity is preserved. For details in performing C-scan, the readers are referred to the report in Latourte et al. (2011).

4.3 Strain-Rate Dependent Numerical Model for Composite Panels

In this section, we will discuss the strain-rate dependent numerical model developed for composite panels under dynamic loadings. The model accounts for the rate dependence inherent to the mechanical properties of vinylester resin, PVA foam, and unidirectional fiber-reinforced composite plies.

4.3.1 Rate Dependence of Vinylester Resin and PVA Foam

To simulate the inter-lamina delamination in composite materials, the rate-dependent cohesive law was developed. As shown in Fig. 4.6, a user defined 3-D eight-node cohesive element was used to simulate the interface (vinylester resin) between two fiber-reinforced composite laminates. The cohesive elements relate the traction vectors $\mathbf{T} = \{T_n, T_{s1}, T_{s2}\}$ and the displacement jump vector $\delta = \{\delta_n, \delta_{s1}, \delta_{s2}\}$, where subscripts n, s_1 and s_2 denote the normal and two tangential shear modes. Under

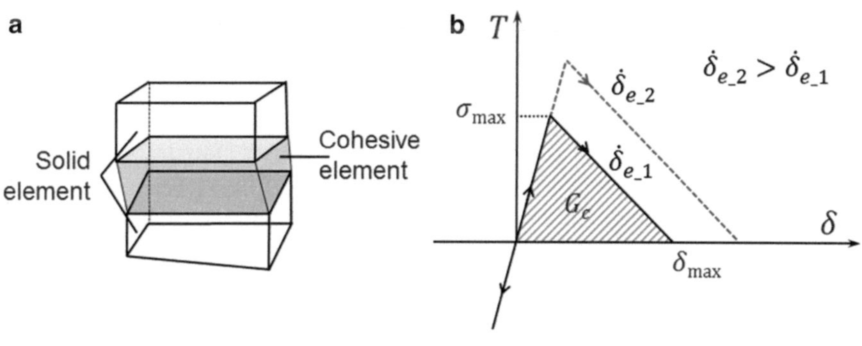

Fig. 4.6 (**a**) 3-D schematic of an eight-node cohesive element bonding to two eight-node solid elements. (**b**) Rate-dependent cohesive law with the bilinear damage evolution relationship

each mode, after reaching the maximum traction $\sigma_{\max}$, a linear strain-softening method is used to simulate the damage evolution of the cohesive element. To account for the rate dependence, $\sigma_{\max}$ and the maximum displacement jump $\delta_{\max}$ of the cohesive element are defined as functions of the opening rate $\dot{\delta}$:

$$\sigma_{\max} = \sigma^{\mathrm{ref}}_{\max}\left(1+n\ln\left(\frac{\dot{\delta}}{\dot{\delta}^{\mathrm{ref}}}\right)\right)$$

$$\delta_{\max} = \delta^{\mathrm{ref}}_{\max}\left(1+n\ln\left(\frac{\dot{\delta}}{\dot{\delta}^{\mathrm{ref}}}\right)\right)$$

i.e.

$$G_{\mathrm{c}} = \frac{1}{2}\sigma_{\max}\delta_{\max} = G_{\mathrm{c}}^{\mathrm{ref}}\left(1+n\ln\left(\frac{\dot{\delta}}{\dot{\delta}^{\mathrm{ref}}}\right)\right)^2 \tag{4.1}$$

where $\sigma^{\mathrm{ref}}_{\max}$, $\delta^{\mathrm{ref}}_{\max}$, $\dot{\delta}^{\mathrm{ref}}$, and n are the reference cohesive maximum traction, reference critical displacement jump, reference displacement jump rate, and rate sensitivity, respectively. G_{c} and $G_{\mathrm{c}}^{\mathrm{ref}}$ are the rate-dependent fracture toughness and reference fracture toughness, respectively. The damage variable d for the cohesive element is defined by

$$d = \max\left\{\sqrt{\left(\frac{\langle\delta_n\rangle}{\delta_{n,\max}}\right)^2+\left(\frac{\delta_{s1}}{\delta_{s1,\max}}\right)^2+\left(\frac{\delta_{s2}}{\delta_{s2,\max}}\right)^2},1\right\}, \tag{4.2}$$

where $\langle\ \rangle$ is the Macauley bracket operator, defined for any $x \in \mathbf{R}$ as $\langle x\rangle = (x+|x|)/2$. The bilinear damage-softening evolution law was achieved by using an approach similar to Espinosa and Zavattieri (2003) and Geubelle and Baylor (1998); namely,

$$T_i = \frac{1-d}{d}\frac{\sigma_{i,\max}}{1-d_{\mathrm{initial}}}\frac{\delta_i}{\delta_{i,\max}} \quad \text{where } i \in \{n, s1, s2\}, \tag{4.3}$$

in which d_{initial} is a positive value close to zero assigned to the cohesive elements before material degradation. During deformation, d starts from d_{initial} and increases monotonically to prevent material healing. The rate-dependent cohesive model was formulated and implemented in ABAQUS/explicit (Abaqus 2009) through a user-defined element subroutine (VUEL). The reference strain rate for vinylester was set to $\dot{\delta}_{\mathrm{e}}^{\mathrm{ref}} = 1\times10^{-3}\,\mathrm{s}^{-1}$, the reference strength was set to $\sigma^{\mathrm{ref}}_{\max} = 80\,\mathrm{MPa}$, and the reference fracture toughness values in normal and shear modes were both set to $G_{\mathrm{c}}^{\mathrm{ref}} = 1\times10^{3}\,\mathrm{Jm}^{-2}$ (Oguni and Ravichandran 2001; Compston et al. 1998).

The mechanical behavior of Divinycell H250 PVC foam in the composite sandwich panels was simulated by the crushable foam plasticity model first developed by Deshpande and Fleck (2000). The yield surface with volumetric hardening is described by

$$F = \sqrt{q^2 + \beta^2 (p - p_0)} - B = 0, \tag{4.4}$$

where p is the pressure stress and q is the von Mises stress. β is the shape factor of the elliptical yield surface, which is expressed as

$$\beta = \frac{3k}{\sqrt{(3k_t + k)(3 - k)}}, \quad \text{with } k = \frac{\sigma_c^0}{p_c^0} \quad \text{and } k_t = \frac{p_t}{p_c^0}, \tag{4.5}$$

where σ_c^0 is the initial yield stress for foam under uniaxial compression, p_c^0 is the initial yield stress in hydrostatic compression, and p_t is the yield strength in hydrostatic tension. Fitting the experimental data in Deshpande and Fleck (2001) and Tagarielli et al. (2008) for H250 PVC foam using (4.4) and (4.5) yields $k = k_t = 1.563$. The evolution of the yield surface was modeled by inputting the uniaxial compressive stress-strain response of the foam reported by Tagarielli et al. (2008). In addition, the rate sensitivity of the foam strength was also taken into account by Deshpande and Fleck's model as implemented in ABAQUS (Abaqus 2009). The dynamic experiments performed in Tagarielli et al. (2008) suggest that the yield strength of H250 PVC foam follows a power law relationship with the strain rate, i.e.

$$\frac{\sigma_y}{\sigma_y^0} = \left(\frac{\dot{\varepsilon}}{\dot{\varepsilon}_0} \right)^m, \tag{4.6}$$

with the reference strain rate $\dot{\varepsilon}_0 = 1\text{s}^{-1}$, the reference yield stress $\sigma_y^0 = 7.44\text{MPa}$, and the power law exponent $m = 0.048$. An elastic modulus of 170 MPa and a Poisson's ratio of 0.3 were also identified for H250 PVC foam.

Thus far, the numerical model has incorporated the rate dependence of the mechanical behavior of the vinylester resin and the PVA foam. In the following section, a rate-dependent model developed for unidirectional fiber-reinforced composite plies will be discussed.

4.3.2 *Rate-Dependent Model for Unidirectional Fiber-Reinforced Composite Plies*

4.3.2.1 Mechanical Response of Damaged Materials

As originally proposed by Kachanov (1986), in continuum damage mechanics the effect of damage is modeled as a reduction in material stiffness constants. For instance, Matzenmiller et al. (1995) used five damage variables to model the reduction of the material's elastic constants in the damage model for unidirectional fiber-reinforced composite plies. To simulate their elastic-brittle behavior, Lapczyk and Hurtado (2007) combined the model suggested by Matzenmiller et al. (1995) with Hashin's failure criteria (Hashin and Rotem 1973; Hashin 1980). This model differentiates four failure modes: fiber tension, fiber compression, matrix tension, and matrix compression. Here, we used an approach similar to that in Lapczyk and Hurtado (2007). Furthermore, as suggested by Chaboche (1981), the effective stress tensor $\hat{\sigma}$ of a material is defined as the material's response when no damage occurs. That is, for an orthotropic linearly elastic unidirectional ply, the effective stresses in 2D form are defined as

$$\hat{\sigma} = \mathbf{C}_0 \varepsilon \quad \text{i.e.}, \quad \begin{bmatrix} \hat{\sigma}_{11} \\ \hat{\sigma}_{22} \\ \hat{\sigma}_{12} \end{bmatrix} = \frac{1}{1-\nu_{12}\nu_{21}} \begin{bmatrix} E_{11} & \nu_{21}E_{11} & 0 \\ \nu_{12}E_{11} & E_{22} & 0 \\ 0 & 0 & (1-\nu_{12}\nu_{21})G_{12} \end{bmatrix} \begin{bmatrix} \varepsilon_{11} \\ \varepsilon_{22} \\ \varepsilon_{12} \end{bmatrix} \tag{4.7}$$

where $\mathbf{C_0}$ is the stiffness matrix of the undamaged material. The notation used in (4.7) refers to the coordinate system in which the *1*-axis and *2*-axis are parallel and transverse to the fiber direction, respectively. After damage initiation, the components in the stiffness matrix decrease to account for the material's degradation. The nominal stress tensor of the damaged material takes the form

$$\sigma = \mathbf{C}\varepsilon \tag{4.8}$$

where $\mathbf{C}$ is the stiffness matrix of the damaged material. It can be expressed as

$$\mathbf{C} = \mathbf{M}\mathbf{C}_0\mathbf{M}^{\mathrm{T}}, \tag{4.9}$$

where $\mathbf{M}$ is the damage operator. Here we propose that it takes the form

$$M = \begin{bmatrix} \sqrt{1-d_{\mathrm{f}}} & 0 & 0 \\ 0 & \sqrt{1-d_{\mathrm{m}}} & 0 \\ 0 & 0 & \sqrt{1-d_{\mathrm{s}}} \end{bmatrix} \tag{4.10}$$

in which d_{f}, d_{m}, and d_{s} are the damage variables for fiber, matrix, and shear failure modes, respectively. Furthermore, similar to the failure modes proposed by Hashin (Hashin and Rotem 1973; Hashin 1980), the fiber and matrix damage variables are defined as

$$d_{\mathrm{f}} = \begin{cases} d_{\mathrm{ft}} & for\ \hat{\sigma}_{11} \geq 0 \\ d_{\mathrm{fc}} & for\ \hat{\sigma}_{11} < 0 \end{cases}$$

$$d_{\mathrm{m}} = \begin{cases} d_{\mathrm{mt}} & for\ \hat{\sigma}_{22} \geq 0 \\ d_{\mathrm{mc}} & for\ \hat{\sigma}_{22} < 0 \end{cases} \tag{4.11}$$

where d_{ft} and d_{fc} are the damage variable for fiber tension and compression modes, respectively; d_{mt} and d_{mc} are the damage variable for matrix tension and compression modes, respectively. The shear damage variable is not independent and takes the form (Lapczyk and Hurtado 2007)

$$d_s = 1-(1-d_{ft})(1-d_{fc})(1-d_{mf})(1-d_{mc}). \tag{4.12}$$

All the damage variables take values from zero (no damage) to one (total damage), and are constrained to increase monotonically in order to prevent material healing.

4.3.2.2 Failure Criteria

Degradation of unidirectional fiber-reinforced composite materials involves different failure modes such as matrix cracks, fiber-matrix debonding, fiber rupture, and fiber buckling and kinking. Over the past three decades, numerous failure criteria have been developed with the aim to accurately predict the onset of composite material failure. In summary, there are four forms of failure criteria proposed: limit theories (e.g., maximum stress and maximum strain criteria Rowlands 1985); polynomial theories (Tsai–Wu failure criterion (Tsai and Wu 1971) in which a polynomial of stress terms was used to describe a failure surface); strain energy theories (Azzi and Tsai 1965 failure criterion which used a nonlinear energy based criterion to define material failure); and direct mode theories (Hashin failure criterion (Hashin and Rotem 1973; Hashin 1980) in which separate equations are used to describe the onset of each failure mode). Because Hashin failure criteria are able to distinguish damage initiation in various failure modes, it has been widely used in commercial codes. The failure surface functions that determine the onset of damage in Hashin failure criterion are given by,

Fiber tension failure$(\hat{\sigma}_{11} \geq 0)$:

$$F_{ft} = \left(\frac{\hat{\sigma}_{11}}{X^T}\right)^2 + \alpha\left(\frac{\sigma_{12}}{S^L}\right) = 1$$

Fiber compression failure$(\hat{\sigma}_{11} < 0)$:

$$F_{fc} = \left(\frac{\hat{\sigma}_{11}}{X^C}\right)^2 = 1$$

Matrix tension failure$(\hat{\sigma}_{22} \geq 0)$:

$$F_{mt} = \left(\frac{\hat{\sigma}_{22}}{Y^T}\right)^2 + \left(\frac{\sigma_{12}}{S^L}\right)^2 = 1$$

Matrix compression failure$(\hat{\sigma}_{22} < 0)$:

$$F_{mc} = \left(\frac{\hat{\sigma}_{22}}{2X^T}\right)^2 + \left[\left(\frac{Y^C}{2S^T}\right)^2 - 1\right]\frac{\hat{\sigma}_{22}}{Y^C} + \left(\frac{\hat{\sigma}_{12}}{S^L}\right)^2 = 1 \tag{4.13}$$

where X^T and X^C are the tension and compression strengths in the fiber direction; Y^T and Y^C are the tension and compression strengths in the matrix (transverse) direction; S^L and S^T are the longitudinal and transversal shear strengths. The coefficient

α denotes the contribution of the shear effect on fiber rupture. Usually, $\alpha = 1$ as suggested by Hashin (1980).

In a round-robin study, it was found that the failure prediction using Hashin criterion deviated from experimental results in matrix and fiber compression modes (Lapczyk and Hurtado 2007). Daniel et al. (2009a, b) proposed a new interfiber failure criterion based on multi-axial tests on unidirectional E-glass/vinylester and carbon/epoxy composite laminates. The onset of matrix failure is determined by three subcriteria:

Tension dominated matrix failure:

$$\left(\frac{\hat{\sigma}_{22}}{Y^T}\right)+\left(\frac{E_{22}}{2G_{12}}\right)^2\left(\frac{\hat{\sigma}_{12}}{Y^T}\right)^2=1$$

Shear dominated matrix failure:

$$\left(\frac{\hat{\sigma}_{12}}{S^L}\right)^2+\frac{2G_{12}}{E_{22}}\frac{\hat{\sigma}_{22}}{S^L}=1$$

Compression dominated matrix failure:

$$\left(\frac{\hat{\sigma}_{22}}{Y^C}\right)+\left(\frac{E_{22}}{G_{12}}\right)^2\left(\frac{\hat{\sigma}_{12}}{Y^C}\right)^2=1 \tag{4.14}$$

The matrix failure surface in the $\hat{\sigma}_{22}-\hat{\sigma}_{12}$ plane described by (4.14) is not only dependent on a material's strength, but also elastic modulus and shear modulus, E_{22} and G_{12}. Furthermore, the criterion acknowledges the increase in the shear strength when the matrix in under compression. However, continuity of the matrix failure surface at the $\hat{\sigma}_{12}$-axis is not guaranteed since the tension- and shear-dominate failure criteria [the first and second functions in (4.14)] do not necessarily intersect at the same point on the $\hat{\sigma}_{12}$-axis for all materials. This limits the generality of the criterion proposed by Daniel et al. To overcome this shortcoming, we reexamined Tsai–Wu failure criterion (Tsai and Wu 1971), in which the failure surface in the matrix plane is given by

$$F_{\mathrm{m}}=\frac{1}{Y^{\mathrm{T}}Y^{\mathrm{C}}}\hat{\sigma}_{22}{}^2+\left(\frac{1}{Y^{\mathrm{T}}}-\frac{1}{Y^{\mathrm{C}}}\right)\hat{\sigma}_{22}+\left(\frac{\hat{\sigma}_{12}}{S^{\mathrm{L}}}\right)^2=1. \tag{4.15}$$

Since (4.15) has a linear term of $\hat{\sigma}_{22}$, the contribution from the matrix compression to the failure surface is taken into account. To augment Tsai–Wu criterion to better describe the matrix failure for $\hat{\sigma}_{22}<0$, the modification approach proposed by Sun (2000), consisting of the inclusion of an internal friction coefficient for the shear failure of matrix materials under compression, was adopted here. We proposed a new failure criterion in the matrix plane expressed as

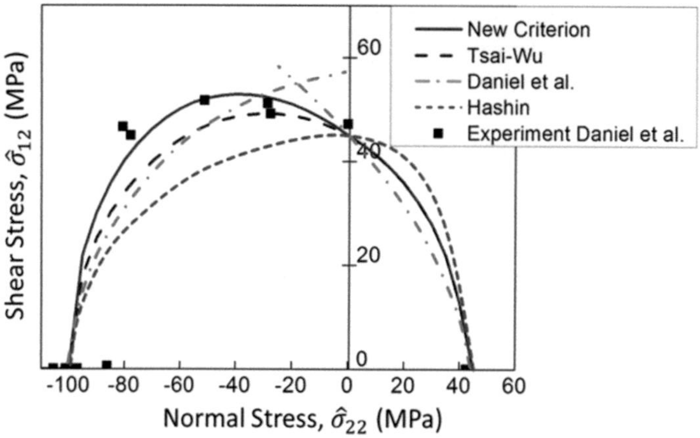

Fig. 4.7 Comparison of the matrix failure envelopes predicted by different criteria to the experimental results for E-glass/vinylester composite laminates reported by Daniel et al. (2009b)

Matrix tension failure $\left(\hat{\sigma}_{22} \geq 0\right)$:

$$F_{mt} = \frac{1}{Y^T Y^C}\hat{\sigma}_{22}{}^2 + \left(\frac{1}{Y^T} - \frac{1}{Y^C}\right)\hat{\sigma}_{22} + \left(\frac{\hat{\sigma}_{12}}{S^L}\right)^2 = 1$$

Matrix compression failure $\left(\hat{\sigma}_{22} < 0\right)$:

$$F_{mc} = \frac{1}{Y^T Y^C}\hat{\sigma}_{22}{}^2 + \left(\frac{1}{Y^T} - \frac{1}{Y^C}\right)\hat{\sigma}_{22} + \left(\frac{\hat{\sigma}_{12}}{S^L - \eta\hat{\sigma}_{22}}\right)^2 = 1 \qquad (4.16)$$

where the internal friction coefficient η is a material property and can be identified by off-axis tests. The experimental results by Daniel et al. (2009b) on unidirectional E-glass/vinylester plies were used to verify our new failure criterion. Figure 4.7 compares the resulting failure envelopes predicted by various previously reported criteria and our new criterion for unidirectional E-glass/vinylester plies in the matrix plane. Inputting $Y^{\mathrm{T}} = 45\mathrm{MPa}$, $Y^{\mathrm{C}} = -100\mathrm{MPa}$, $S^{\mathrm{L}} = 45\mathrm{MPa}$, and η=0.1, the new criterion predicted a failure surface that compares very well with the experimental results. In addition, the new criterion given by (4.16) assures the continuity of the failure surface at $\hat{\sigma}_{12} = 0$, in contrast to the criterion by Daniel et al. (2009a, b).

In summary, the new criterion for predicting the damage onset for unidirectional fiber-reinforced composite laminates includes Hashin criterion for fiber damage and a modified Tsai–Wu criterion for matrix damage. The formulas are expressed as

$$\begin{aligned}
&\text{Fiber tension failure}\left(\hat{\sigma}_{11} \geq 0\right):\\
&F_{ft} = \left(\frac{\hat{\sigma}_{11}}{X^T}\right)^2 + \alpha\left(\frac{\sigma_{12}}{S^L}\right) = 1\\
&\text{Fiber compression failure}\left(\hat{\sigma}_{11} < 0\right):\\
&F_{fc} = \left(\frac{\hat{\sigma}_{11}}{X^C}\right)^2 = 1\\
&\text{Matrix tension failure}\left(\hat{\sigma}_{22} \geq 0\right):\\
&F_{mt} = \frac{1}{Y^T Y^C}\hat{\sigma}_{22}{}^2 + \left(\frac{1}{Y^T} - \frac{1}{Y^C}\right)\hat{\sigma}_{22} + \left(\frac{\hat{\sigma}_{12}}{S^L}\right)^2 = 1\\
&\text{Matrix compression failure}\left(\hat{\sigma}_{22} < 0\right):\\
&F_{mc} = \frac{1}{Y^T Y^C}\hat{\sigma}_{22}{}^2 + \left(\frac{1}{Y^T} - \frac{1}{Y^C}\right)\hat{\sigma}_{22} + \left(\frac{\hat{\sigma}_{12}}{S^L - \eta\hat{\sigma}_{22}}\right)^2 = 1
\end{aligned} \tag{4.17}$$

4.3.2.3 Damage Evolution

A complete description of the model requires defining the damage evolution law. After the damage initiation criterion is satisfied, further loading will cause degradation of the components in the material stiffness matrix. As shown in (4.9) and (4.10), reduction of the stiffness constants are realized by the damage variables from zero (no damage) to one (full damage). Camanho and Dávila (2002) proposed that the evolution of the damage variable after damage initiation should be based on the energy release rate G_C during crack propagation. Linear strain-softening (Lapczyk and Hurtado 2007; Iannucci and Willows 2006; Donadon et al. 2008) and exponential strain-softening relationships (Maimí et al. 2006; Tijssens et al. 2000) are the two most common damage evolution laws used to simulate damage degradation. In our model, the linear strain-softening evolution law is similar to that implemented by Lapczyk and Hurtado (2007), as shown in Fig. 4.8. This evolution law yields

$$G_C = \frac{1}{2}\sigma_{eq}^0 \delta_{eq}^f, \tag{4.18}$$

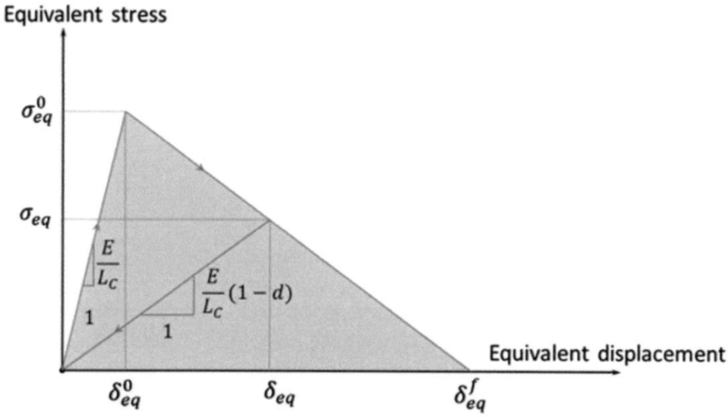

Fig. 4.8 Damage evolution law for a linear softening material. The reduced slope in the equivalent plane is given by the reduced modulus $E(1-d)$ divided by the element characteristic length L_C

where σ_{eq}^0 is the equivalent strength (i.e., the equivalent stress level at which the failure initiation criterion is satisfied); δ_{eq}^0 denotes the equivalent displacement when the equivalent stress equals σ_{eq}^0; and $\delta_{\text{eq}}^{\text{f}}$ is the equivalent displacement limit at which the material is fully damaged. The damage variable for each damage mode i is given by

$$d_i = \begin{cases} 0\,, & \text{for } \delta_{i,\text{eq}} < \delta_{i,\text{eq}}^0 \\ \dfrac{\delta_{i,\text{eq}}^{\text{f}}\left(\delta_{i,\text{eq}} - \delta_{i,\text{eq}}^0\right)}{\delta_{i,\text{eq}}\left(\delta_{i,\text{eq}}^{\text{f}} - \delta_{i,\text{eq}}^0\right)}\,, & \text{for } \delta_{i,\text{eq}}^0 \le \delta_{i,\text{eq}} \le \delta_{i,\text{eq}}^{\text{f}}\,; \\ 1\,, & \text{for } \delta_{i,\text{eq}} > \delta_{i,\text{eq}}^{\text{f}} \end{cases} \tag{4.19}$$

where $i \in \{\text{ft}, \text{fc}, \text{mt}, \text{mc}\}$

Similar to the approach in Lapczyk and Hurtado (2007), the mesh dependency in the finite element model is alleviated by setting

$$\delta_{\text{eq}}^{\text{f}} = \varepsilon_{\text{eq}}^{\text{f}} L_{\text{C}} = \frac{2G_{\text{C}}}{\sigma_{\text{eq}}^0}, \tag{4.20}$$

where $\varepsilon_{\text{eq}}^{\text{f}}$ is the equivalent strain at which the material totally fails and L_{C} is the characteristic length of the element (defined as the square root of the area associated with an integration point). Formulas for calculating the equivalent displacement and stress for each failure mode are listed in Table 4.2.

Besides modification of the failure criterion, we also reexamined the damage evolution law for fiber damage under tension. In contrast to fiber damage in compression, which is mainly through fiber kinking and buckling, damage to fibers under tension is dominated by modes that cause fiber rupture or pull-out. The latter is primarily caused by in-plane shear deformation. This shear contribution to failure

Table 4.2 Formulas for equivalent displacements and equivalent stresses in all damage modes

Failure mode	δ_{eq}	σ_{eq}
Fiber tension $(\hat{\sigma}_{11} \geq 0)$	$L_C\sqrt{\langle\varepsilon_{11}\rangle^2 + \alpha\varepsilon_{12}^2}$	$\sqrt{\langle\hat{\sigma}_{11}\rangle^2 + \alpha\hat{\sigma}_{12}^2}$
Fiber compression $(\hat{\sigma}_{11} < 0)$	$L_C\langle-\varepsilon_{11}\rangle$	$\langle-\hat{\sigma}_{11}\rangle$
Matrix tension $(\hat{\sigma}_{22} \geq 0)$	$L_C\sqrt{\langle-\varepsilon_{22}\rangle^2 + \varepsilon_{12}^2}$	$\sqrt{\langle\hat{\sigma}_{22}\rangle^2 + \hat{\sigma}_{12}^2}$
Matrix compression $(\hat{\sigma}_{22} < 0)$	$L_C\sqrt{\langle-\varepsilon_{22}\rangle^2 + \varepsilon_{12}^2}$	$\sqrt{\langle-\hat{\sigma}_{22}\rangle^2 + \hat{\sigma}_{12}^2}$

initiation for a fiber under tension is accounted for in Hashin's failure criteria. However, the shear contribution to damage evolution in fiber tension mode was omitted in the previous implementation in Lapczyk and Hurtado (2007). The same fracture toughness G_C^{ft} is used no matter if the fiber fails in rupture or pull-out. This violates common sense because resistance to fiber pull-out caused by shear deformation is usually much lower than resistance caused by fiber rupture. Therefore, we proposed a modification in the fracture toughness for the fiber tension failure mode that accounts for the shear contribution. When the fibers rupture due to pure normal tension, the fracture toughness is identified by G_C^{ft}; however, when the fibers fail by pull-out due to fiber-matrix interface debonding, the fracture toughness should decrease to G_C^{mt}. Therefore, similar to the approach proposed by Benzaggagh and Kenane (1996) for modeling mixed mode delamination, the fracture toughness for fiber tension damage mode is defined as a function of the ratio between longitudinal normal stress and effective shear stress:

$$\tilde{G}_C^{ft} = \left(G_C^{ft} - G_C^{mt}\right)\sqrt{\frac{\hat{\sigma}_{11}^2}{\hat{\sigma}_{11}^2 + \alpha\hat{\sigma}_{12}^2}} + G_C^{mt}. \tag{4.21}$$

In this way, the modified fracture toughness $\tilde{G}_C^{ft}$ accounts for the shear contribution to damage evolution for the fiber tension damage mode.

4.3.2.4 Strain-Rate Dependence of Strength and Toughness

One important characteristic of the finite element model described here is that it incorporates the strain-rate dependence of the strength of the fiber-reinforced composites. Experiments have shown that the strain rate affects the modulus, strength, and toughness of glass-fiber reinforced composites (Oguni and Ravichandran 2001; Daniel et al. 2009a, b; Akil et al. 2003; Shokrieh and Omidi 2009a, b, c; Naik et al. 2010). To preserve implementation simplicity, the strain-rate effect on modulus was neglected here. The rate dependence of the strength of a unidirectional E-glass/vinylester ply in each damage mode is modeled by

$$
\begin{aligned}
X^T &= X_0^T\left[m_f \log\left(\frac{|\dot{\varepsilon}_{11}|}{\dot{\varepsilon}_{0,11}}\right)+1\right] \\
X^C &= X_0^C\left[m_f \log\left(\frac{|\dot{\varepsilon}_{11}|}{\dot{\varepsilon}_{0,11}}\right)+1\right] \\
Y^T &= Y_0^T\left[m_m \log\left(\frac{|\dot{\varepsilon}_{22}|}{\dot{\varepsilon}_{0,22}}\right)+1\right] \\
Y^C &= Y_0^C\left[m_m \log\left(\frac{|\dot{\varepsilon}_{22}|}{\dot{\varepsilon}_{0,22}}\right)+1\right] \\
S^L &= S_0^L\left[m_m \log\left(\frac{|\dot{\varepsilon}_{12}|}{\dot{\varepsilon}_{0,12}}\right)+1\right]
\end{aligned} \tag{4.22}
$$

where parameters with subscript 0 denote the reference values; m_f and m_m are the strain rate sensitivities for the fiber and matrix modes, respectively. These values can be determined by performing multi-axial dynamic experiments on the unidirectional plies. To ensure stability of the linear softening damage evolution model, we assumed that the failure strain of the composite has the same rate dependence as the strength. Therefore, the toughness of the composite, under dynamic loading, for each damage mode is expressed as:

$$
\begin{aligned}
G_C^{ft} &= G_{0,C}^{ft}\left[m_f \log\left(\frac{|\dot{\varepsilon}_{11}|}{\dot{\varepsilon}_{0,11}}\right)+1\right]^2 \\
G_C^{fc} &= G_{0,C}^{fc}\left[m_f \log\left(\frac{|\dot{\varepsilon}_{11}|}{\dot{\varepsilon}_{0,11}}\right)+1\right]^2 \\
G_C^{mt} &= G_{0,C}^{mt}\left[m_m \log\left(\frac{|\dot{\varepsilon}_{22}|}{\dot{\varepsilon}_{0,22}}\right)+1\right]^2 \\
G_C^{mc} &= G_{0,C}^{mc}\left[m_m \log\left(\frac{|\dot{\varepsilon}_{22}|}{\dot{\varepsilon}_{0,22}}\right)+1\right]^2
\end{aligned} \tag{4.23}
$$

For unidirectional E-glass/vinylester composite plies, the rate dependence of the strengths for the fiber failure modes was adapted from the dynamic tests by Oguni and Ravichandran (2001). The rate dependence of the strengths for the matrix failure modes was adapted from the dynamic tests by Daniel et al. (2009b). The reference strengths were adapted from quasi-static results (Daniel et al. 1994). The parameters used in modeling the dynamic behavior of E-glass/vinylester composite panels are listed in Table 4.3.

Table 4.3 Material properties of dynamic behavior for E-glass/vinylester unidirectional composite

Material property	Value	Material property	Value
Density, ρ	1,850 kg/m^3	X_0^T	1.2 GPa (Daniel et al. 1994)
E_{11}	39 GPa (Daniel et al. 1994)	X_0^C	620 MPa (Daniel et al. 1994)
E_{22}	11.5 GPa (Daniel et al. 1994)	Y_0^T	45 MPa (Daniel et al. 2009b)
G_{12}	3.5 GPa (Daniel et al. 1994)	Y_0^C	100 MPa (Daniel et al. 2009b)
Poisson's ratio, ν	0.28 (Daniel et al. 1994)	S_0^L	45 MPa (Daniel et al. 2009b)
m_f	0.177 (Oguni and Ravichandran 2001)	$G_{0,C}^{ft} = G_{0,C}^{fc}$	35×10^3 N/m (Roy et al. 2001)
m_m	0.2 (Daniel et al. 2009b)	$G_{0,C}^{mt} = G_{0,C}^{mc}$	2×10^3 N/m (Compston et al. 1998)
$\dot{\varepsilon}_{0,11}$	1×10^{-3}s^{-1} (Oguni and Ravichandran 2001)	$\dot{\varepsilon}_{0,22}$	1×10^{-3}s^{-1} (Daniel et al. 2009b)

The rate-dependent progressive damage model discussed in the previous sections was then implemented using VUMAT. In the implementation, at the beginning of a strain step, information such as the previous strain tensor, the strain increment tensor in the current step, and the time increment for the current step are passed into the VUMAT subroutine. The strain-rate tensor is calculated and then used to update the strength and toughness of the composite for each damage mode. Concurrently, the effective stress tensor is updated and input, along with the dynamic strength, into the failure criteria equations (4.17) to check whether the element is damaged. If the failure criteria are satisfied, the linear softening damage evolution law is applied with the updated toughness. The damage variables for all damage modes are updated (and monotonic growth is constrained) and saved as state variables. At the end of each step, the nominal stress tensor is updated and passed back into ABAQUS/explicit.

4.3.2.5 Numerical Simulation

In order to apply the finite element model introduced in the previous section to predict the panel performance subjected to underwater blast loading, the interaction between fluid and structure domains needs to be simulated properly. The FSI domains were simulated with an Acoustic-Lagrangian (A-L) formulation. The fluid body was modeled as an acoustic medium with assumptions of inviscid, linear, and compressible mechanical behavior. The fluid (water) properties input in ABAQUS are material density $\rho_f = 982.27$ kg/m^3 and bulk modulus $K_f = 2.19$ GPa. In addition, because water cannot withstand significant tensile stress and is likely to form cavities under tension pressure, a fluid cavitation limit of $p_c = 3{,}270$ Pa the saturation vapor pressure at room temperature, was used to simulate water cavitation. The interaction between the fluid and solid domains was realized by tying the solid structure surfaces (set as master surface) to the acoustic medium surfaces (set as slave surfaces).

After the 3-D finite element model was developed, rate-sensitive models for the vinylester resin, PVC foam, and composite plies were implemented in ABAQUS to simulate the FSI experiments. In this section, simulation results on monolithic and sandwich composite panels will be discussed.

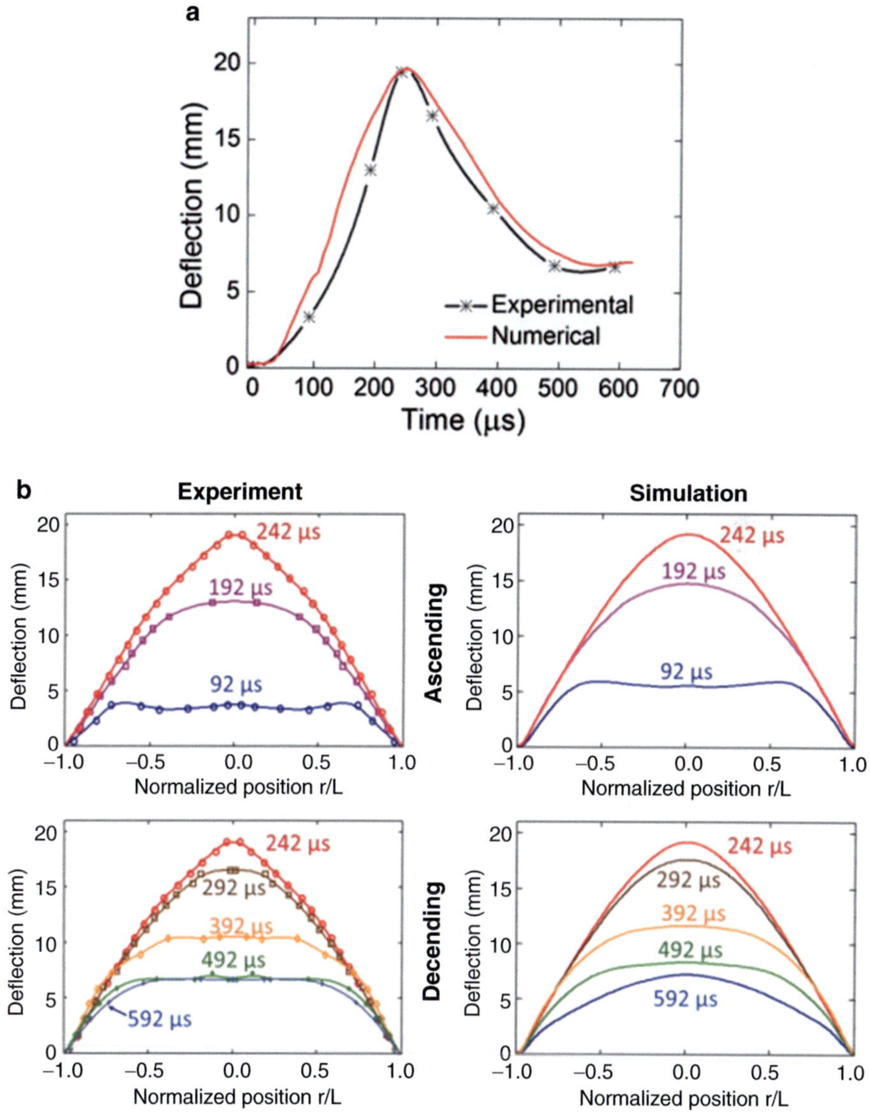

Fig. 4.9 (**a**) Experimental central deflection history of test 1-3 compared with that predicted by numerical simulations. (**b**) Ascending and descending deformation histories across the panel as obtained from experimental measurements (*left*) and simulations (*right*)

4.3.2.6 Monolithic Composite Panel

To assess the performance of the numerical model, test 1-3 was simulated. This test of a monolithic panel impacted by an impulse of 2,425 Pa s presented multiple interesting failure mechanisms, such as complicated inter-lamina delamination and matrix damage patterns. First, the central deflection history predicted by the numerical model was compared with the experiment measurement (Fig. 4.9a).

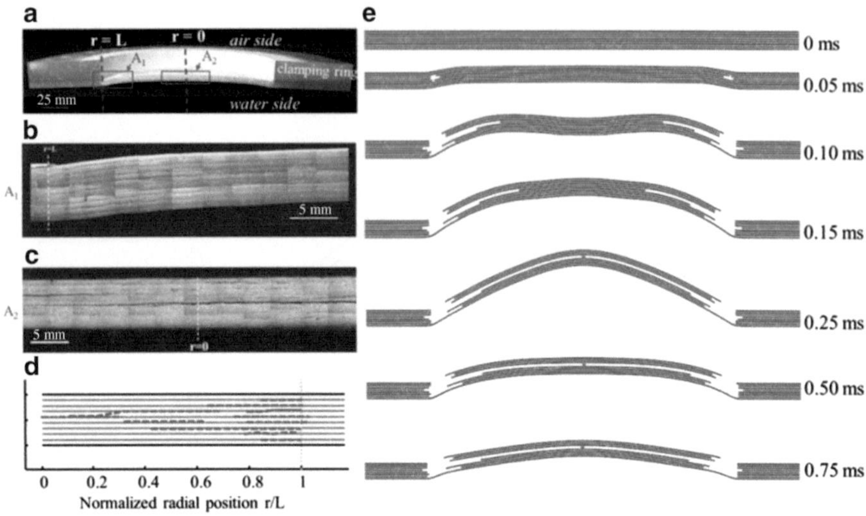

Fig. 4.10 (**a**) Cross-sectional view of the monolithic panel 1-3 after the test. (**b**) and (**c**) Optical micrographs of inter-lamina delamination close to the clamped edge (region A_1 in **a**) and at the center (region A_2 in **a**), respectively. (**d**) Schematic of the interface delamination pattern along the radial position. (**e**) Numerically predicted interface delamination history

A maximum deflection of ~19.6 mm and a final deflection of ~7.0 mm is predicted, which agrees well with the experimental values of 19.5 mm and 6.7 mm, respectively. The deflection rate in the ascending segment (from 0 to 250 μs after the pulse impact) compares favorably with the experimental measurement. Figure 4.9b compares the deflection profiles across the panel at various time intervals with those measured by means of shadow Moiré. Clearly, the simulated deformations in the bulging and spring-back regimes agree well with the experimental profiles.

The material failure predicted by the model was compared with the experimental records, and a parametric study was performed to determine the effect of the strain-rate sensitivity coefficient η on the delamination pattern. By comparing the simulated interface failure distribution to that observed in the test, we found that $\eta = 0.25$ yields the best correlation with the experimentally measured delamination patterns. The inter-lamina delamination pattern predicted by the numerical model correlates well with the experimental measurement. Micrographs of the cross section (Fig. 4.10a–c) reveal characteristic delamination patterns for the monolithic composite panel. At the clamped edge, delamination at all eight interfaces developed due to shearing, with partial propagation to the center of the panel (Fig. 4.10b). As deformation of the panel increased, a delamination front propagates along the central interface (Fig. 4.10c). Figure 4.10d depicts the delamination pattern obtained from the specimen after the test, and the numerically predicted delamination history is shown in Fig. 4.10e. Delamination initiates at the clamped boundaries and propagates toward the center. Both the panel deflection and delamination reach their maximum at approximately 250 μs, after which, spring back of the panel is observed and no further delamination is predicted. The final delamination pattern agrees well with the one observed in the experiment (Fig. 4.10a–d).

Fiber and matrix damage patterns were extracted from the simulation results. As shown in Fig. 4.11a, the fibers on the water-side and air-side laminates begin to show damage at the clamped edge, yet the whole panel shows no evidence of fiber catastrophic failure, which agrees with the experimental observation. In the simulation, similar anisotropic damage patterns were found on the matrices in laminates on both sides, with a more pronounced damage level presenting on the air side. Moreover, the matrix damage d_m along the radius was extracted from the simulation and compared with the experimental measurements (Fig. 4.11b). The crack density and spacing measured in the post mortem microscopy images of the panel cross section were translated into the matrix damage level using the approach discussed in Latourte et al. (2011). The experiment and simulation follow a similar trend—the matrix damage is maximized at the center due to high bi-axial tension, which becomes more pronounced when the water cavity collapses due to the FSI effect. The matrix damage is reduced toward the clamped edge, but is still present due to bending effects. Overall, our rate-dependent model successfully predicts the matrix damage level that is measured experimentally.

4.3.2.7 Sandwich Composite Panel

The second application of the numerical model we discuss here is the predicted deformation and failure mechanisms for sandwich composite panels in which a soft core is present between the water-side and air-side face sheets. At the same areal weight, the sandwich composite outperforms the monolithic panels by sustaining higher impulses and showing suppressed panel deflections. Among the six sandwich composite panel tests, 2-3 (impacted at an impulse of 5,581 Pa s) is of greatest interest because it exhibited a complex synergy of failure mechanisms such as foam crushing, initiation of fiber rupture failure, inter-lamina delamination, and signs of shear off at the clamped boundary. First, the numerically predicted central deflection history of the air-side face sheet is compared with the experimentally measured result (Fig. 4.12a). The ascending slope of central deflection predicted by the numerical model compares accurately with the experiment. A maximum deflection of ~20.2 mm and a final deflection of ~14.0 mm are predicted, in good agreement with the experimental values of 20.1 mm and 15.5 mm. The spring-back rate predicted by the model after the panel deflection reaches the maximum value correlates favorably with the test. Unlike the monolithic panel, which deforms elastically until a brittle failure, the plastic deformation of the foam core resulted in a slower spring-back rate and a larger residual deformation for the sandwich composite panel.

To better understand the extraordinary energy absorption capability of the sandwich composite panel, we investigated the role of each component during the deformation. The deformation history of the foam core is represented by plotting contours of the Lagrangian normal strain in the panel thickness direction in Fig. 4.12b. This demonstrates the unique energy dissipation mechanism of foam core crushing that is present in sandwich panels. Examining both the foam deformation history and the center-point velocity histories of the air-side and water-side face sheets in Fig. 4.12c, we found slightly more complex deformation phases than those proposed in Fleck and Deshpande (2004) and Liang et al. (2007). Stage I (0–20 μs) is the initial FSI

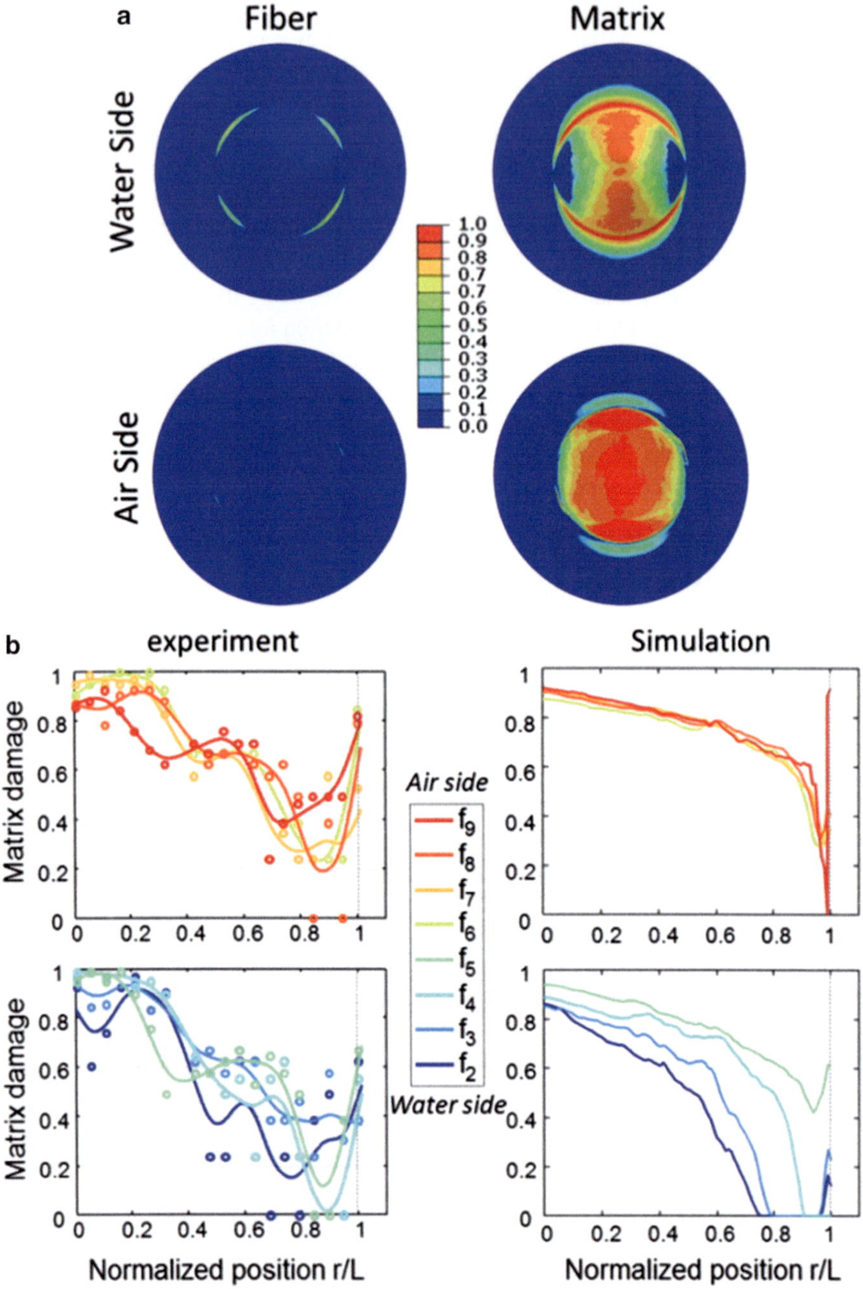

Fig. 4.11 (**a**) Fiber and matrix damage patterns in the water-side and air-side laminas for the monolithic composite panel where the damage variable from 0 to 1 represents the damage level from low to high. (**b**) Comparison of matrix damage variables in individual lamina as a function of panel radial position from experiments and simulations. In this plot, $f_i \left(i = 1, 2, \ldots 9\right)$ represents the fabrics from the water-side to the air-side (the result for f_1 was not given due to extensive noise and uncertainties in the measurements)

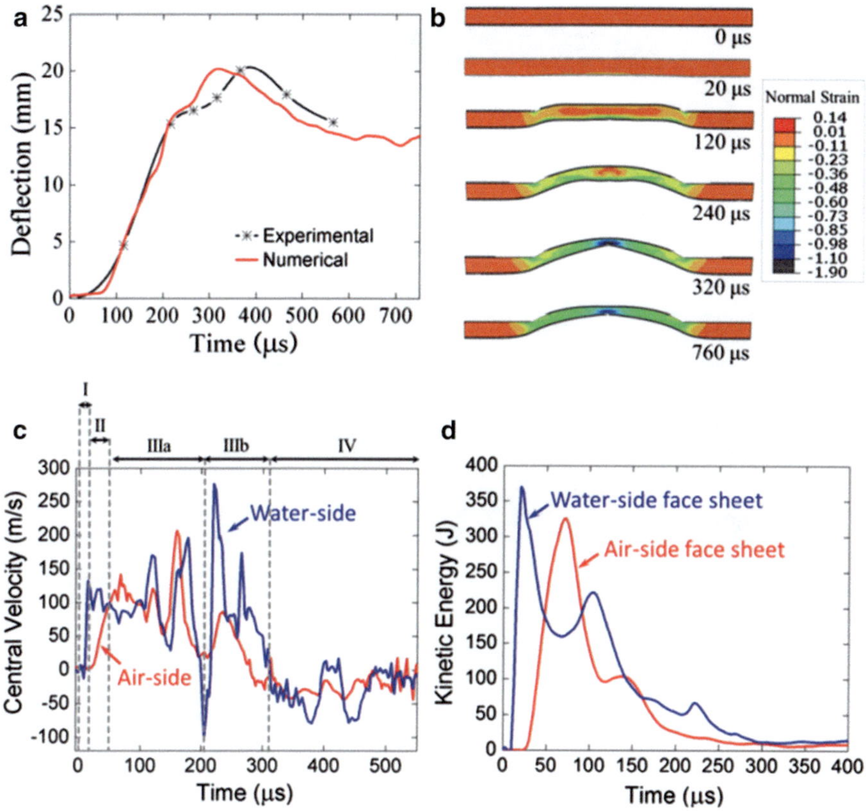

Fig. 4.12 (**a**) Comparison between the central deflection histories measured experimentally and predicted by numerical simulation for the sandwich composite panel at an impulse $I_0 = 5{,}581$ Pa s. (**b**) Simulated deformation of the foam core represented as contour plots of Lagrangian normal strain in the panel thickness direction. (**c**) Plots of center-point velocity for the water-side and air-side face sheets vs. time. Stages I, II, IIIa, IIIb, and IV are labeled to differentiate the dynamic responses of foam core, air-side and water-side face sheets. (**d**) Plots of kinetic energy of the water-side and air-side face sheets vs. time

phase when the panel behaves similarly to a free-standing plate. In this phase, the impulse impinges on the water-side face sheet and crushes a thin layer of the foam core behind it. Meanwhile, the air-side face sheet remains static. At the end of stage I, water cavitation begins. Stage II (20–40 μs) is the core compression phase in which the wave transmits through the whole panel and accelerates the air-side face sheet. Stage IIIa (40–220 μs) is the panel bending and stretching phase when the air-side face sheet moves at a velocity similar to that of the water-side face sheet with lateral waves traveling across the panel. Stage IIIb (220–310 μs) is the cavitation collapse and foam crushing phase in which the water bubble collapses and the foam core crushes. A large amount of energy is dissipated during this phase. Stage IV (after 310 μs) is the panel bending and stretching "recovery" phase when the foam and both the air-side and water-side face sheets behave similar to a solid plate. Spring-back of the panel is observed in this phase.

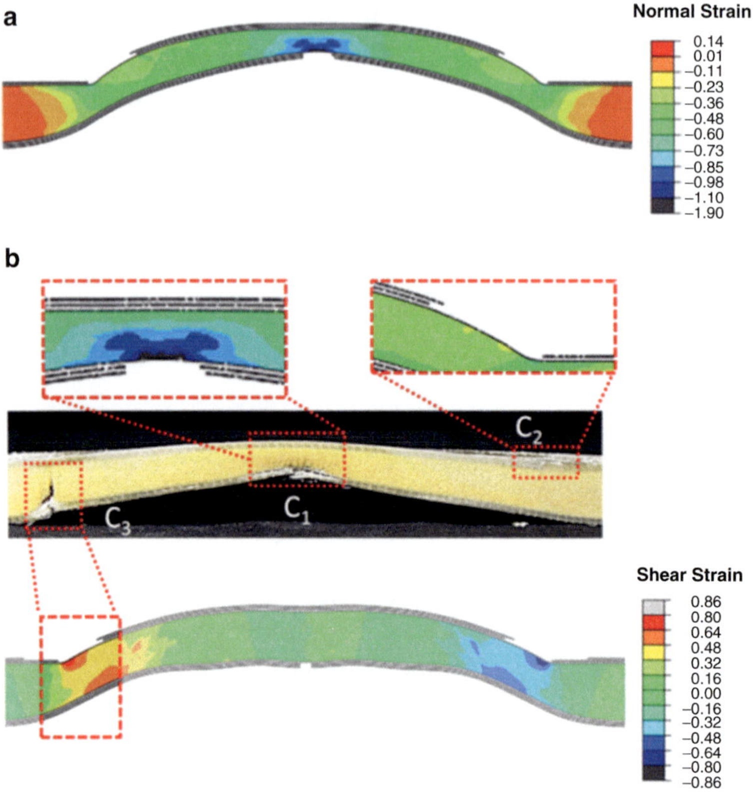

Fig. 4.13 (**a**) Cross-sectional view of the sandwich panel from the numerical simulation with contour plots of the Lagrangian normal strain in the foam in the panel thickness direction. (**b**) Cross-sectional view of the sandwich panel after a test showing three prominent damage regions, C_1, C_2, and C_3. Magnified views of the numerical simulation results for the corresponding regions are provided. For regions C_1 and C_2, contour plots of the Lagrangian normal strain for the foam in the panel thickness direction are given. For region C_3, the contour plot of the shear strain in the foam at 180 μs is shown

As expected, numerical simulation aid identification of the sequence of various damage mechanisms in the sandwich panel. Figure 4.13a shows the cross-sectional view of the sandwich panel from the simulation of the interface delamination and the contours of Lagrangian normal strain in the foam along the panel thickness direction. The panel deformation and foam compression contour correlate closely with the cross-sectional view of the specimen after the experiment. Furthermore, magnified views from the numerical simulation of the three major damage regions (C_1, C_2, and C_3 in Fig. 4.13b) provide valuable insight into the damage processes.

For instance, the compressive strain in the central region of the panel (C_1 in Fig. 4.13b) indicates a severe core crushing due to the cavitation collapse. A V-shaped compressive strain localization pattern in the foam matches two cracks observed experimentally that extend from the water-side face sheet surface at ~45°. The water-side face sheet in region C_1 displays extensive delamination, matrix damage, and fiber rupture associated with the impulse induced by the water cavitation collapse. All these features are captured by the model. Moreover, by examining the air-side face sheet near the clamped boundaries (region C_2 in Fig. 4.13b), inter-lamina delamination is found in both experiments and simulations. In addition, another important failure mechanism is identified: shear-off at the boundaries. As shown in Fig. 4.13b, one side of the sandwich panel at the clamped region exhibits failure in the foam beginning on the water-side face sheet (region C_3). The crack initiated from the bottom and propagated nearly vertically to the air-side face sheet. This is a typical shear-off failure feature, which is indicative of a failure mode in sandwich panels as suggested by Hutchinson and Xue (2005) where shear-off occurs very early in the history of the panel response prior to core crush and overall bending and stretching. This behavior is also observed in the present simulation, as is evident from the Lagrangian shear strain contour plot at 180 μs after impact, just before major foam core crushing in the boundary.

We next examine the model-predicted fiber and matrix damage patterns and compare them with the experimental measurements. In Fig. 4.14b, an X-shaped fiber failure pattern is clearly present in the middle of the water-side face sheet. With our model, we obtained a highly correlated X-shaped fiber failure pattern. More importantly, the new rate-dependent composite damage model introduced in Sect. 3.2 can differentiate between two fiber tension failure modes; at the intersection of the "X," the fiber failed primarily as a result of rupture, whereas toward the end of the pattern, the fiber failed primarily by pull-out accompanied with matrix failure due to shear. We also notice that fiber damage is predicted on the air-side face sheet along the boundary due to the constraint imposed by the steel ring. In the experiment, evidence of fiber damage in those regions was found; however, the damage level is slightly less than the model prediction. This discrepancy probably occurs because accurate damage modeling at the contact between the face sheet and the steel clamp ring would require a finer mesh to better describe the rounded edge of the steel ring.

Another important feature revealed by this numerical model is the matrix damage distribution in the air-side and water-side face sheets. As shown in Fig. 4.14a, a much more uniformly distributed matrix damage pattern is predicted for the air-side face sheet compared to the water-side face sheet because the foam core crushing dissipates a significant amount of energy and mitigates the impulse transmitted to the air-side face sheet. More importantly, in stage IIIa and IIIb, the foam core helps to homogenize the pressure on the air-side face sheet, which results in a uniform matrix damage pattern.

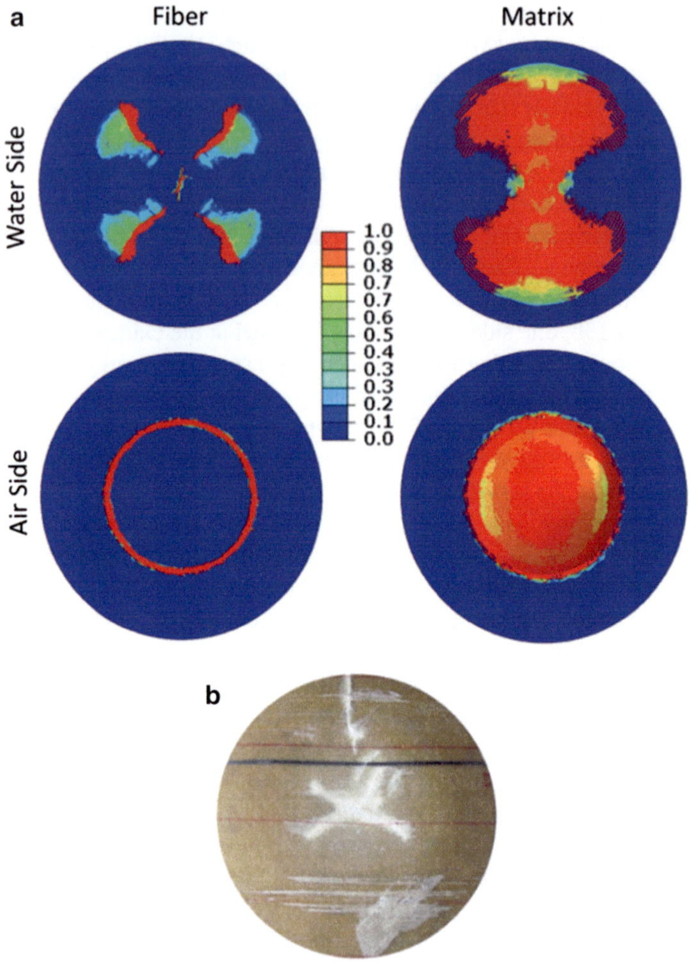

Fig. 4.14 (**a**) Model-predicted fiber and matrix damage patterns in the water-side and air-side face sheets (the damage variable from 0 to 1 represents the damage level from low to high). (**b**) Photograph of the fiber failure pattern in the water-side face sheet of the sandwich composite panel (the *black line* denotes the fiber orientation in the first laminate on the water-side face sheet)

4.4 Summary and Outlook

The current chapter focused on FSI experimental and numerical studies on the performance of monolithic and sandwich fiber-reinforced composite panels subjected to underwater blast loading. A scaled-down FSI laboratory apparatus was used to assess the performance of monolithic and sandwich architecture designs. At the same areal weight, the sandwich panels outperform the monolithic panels by sustaining higher impulses and showing suppressed panel deflections, benefiting from their higher bending stiffness and the extraordinary energy absorption capability of the core material.

Limitations exhibited by existing unidirectional composite models prompted us to develop a new rate-dependent unidirectional laminate model to simulate scaled FSI laboratory experiments. The simulations used a coupled acoustic-solid technique capable of accurately describing the interaction between water and composite panels. The numerical model takes into account the strain-rate effect for the inter-laminate interfaces, the foam core, and the fiber-reinforced composite laminates. The new laminate model uses Hashin's fiber failure criterion and a modified Tsai–Wu matrix failure criterion. The composite moduli are degraded using five damage variables, which are updated in the post-failure regime by means of a linear softening law governed by an energy release criterion. A key feature in the formulation is the distinction between fiber rupture and pull-out by introducing fracture toughness, which varies from a fiber tensile toughness to a matrix tensile toughness as a function of the ratio of longitudinal normal stress to effective shear stress. The experimentally validated failure criteria for unidirectional E-glass/vinylester composite plies were formulated and implemented into the FSI model. Rate-dependent damage initiation and evolution was accounted for in the model, providing a capability for differentiating between fiber rupture and pull-out failure mechanisms.

The numerical model presented here is able to depict the various material deformation and damage mechanisms observed in monolithic and sandwich composite panels such as inter-lamina delamination, fiber and matrix damage, and foam crushing. The details given by the numerical simulations provide valuable insights into the material degradation process, as well as the key energy dissipation mechanisms present in composite panels. Specifically, the numerical model reveals several key features of the sandwich panels subjected to severe impulses. First, water cavitation leads to heterogeneous core crushing. The panel response can be divided into four well-defined phases including one associated with water cavity collapse. Second, core crushing helps homogenize the impulses imparted on both skins. Last, shear off is another failure mechanism observed under high impulses, which tends to happen before the core crushes and the laminates undergo stretching.

Validated by the underwater blast experiments, this numerical model can serve as a powerful tool to efficiently and accurately predict the performance of composite structures. Furthermore, with proper parameterization, this rate-dependent unidirectional composite model can be applied to a variety of fiber and matrix combinations.

Acknowledgments This research was carried out under the financial support of the Office of Naval Research (ONR) under grant number N00014-08-1-1055. We greatly appreciate the support and encouragement provided by Dr. Rajapakse throughout the study.

References

Abaqus, Simulia (2009) Abaqus 6.9-ef online documentation.

Akil Ö, Yıldırım U, Güden M, Hall IW (2003) Effect of strain rate on the compression behaviour of a woven fabric s2-glass fiber reinforced vinyl ester composite. Polymer Test 22(8): 883–887

Andrews EW, Moussa NA (2009) Failure mode maps for composite sandwich panels subjected to air blast loading. Int J Impact Eng 36(3):418–425

Arora H, Hooper PA, Dear JP (2012) The effects of air and underwater blast on composite sandwich panels and tubular laminate structures. Exp Mech 52:1–23

Azzi VD, Tsai SW (1965) Anisotropic strength of composites. Exp Mech 5(9):283–288

Batra RC, Hassan NM (2007) Response of fiber reinforced composites to underwater explosive loads. Compos B Eng 38(4):448–468

Benzeggagh ML, Kenane M (1996) Measurement of mixed-mode delamination fracture toughness of unidirectional glass/epoxy composites with mixed-mode bending apparatus. Compos Sci Tech 56(4):439–449

Camanho PP, Dávila CG (2002) Mixed-mode decohesion finite elements for the simulation of delamination in composite materials. NASA-Technical paper, 211737

Chaboche JL (1981) Continuous damage mechanics–a tool to describe phenomena before crack initiation. Nucl Eng Des 64(2):233–247

Compston P, Jar PYB, Davies P (1998) Matrix effect on the static and dynamic interlaminar fracture toughness of glass-fibre marine composites. Compos B Eng 29(4):505–516

Daniel IM, Ishai O, Daniel I (1994) Engineering mechanics of composite materials, vol 22. Oxford University Press, New York, NY

Daniel IM, Luo JJ, Schubel PM, Werner BT (2009a) Interfiber/interlaminar failure of composites under multi-axial states of stress. Compos Sci Tech 69(6):764–771

Daniel IM, Cho JM, Werner BT, Fenner JS (2009) Mechanical behaviour and failure criteria of composite materials under static and dynamic loading. Proceeding of 17th international conference on composite materials (ICCM17), Edinburgh, UK, 27–31 July 2009

Deshpande VS, Fleck NA (2000) Isotropic constitutive models for metallic foams. J Mech Phys Solid 48(6–7):1253–1283

Deshpande VS, Fleck NA (2001) Multi-axial yield behaviour of polymer foams. Acta Mater 49(10):1859–1866

Donadon MV, Iannucci L, Falzon BG, Hodgkinson JM, De Almeida SFM (2008) A progressive failure model for composite laminates subjected to low velocity impact damage. Comput Struct 86(11–12):1232–1252

Drumheller DS (2002) Introduction to wave propagation in nonlinear fluids and solids. J Acoust Soc Am 111(3):1142

Espinosa HD, Zavattieri PD (2003) A grain level model for the study of failure initiation and evolution in polycrystalline brittle materials. Part I: theory and numerical implementation. Mech Mater 35(3–6):333–364

Espinosa HD, Lee S, Moldovan N (2006) A novel fluid structure interaction experiment to investigate deformation of structural elements subjected to impulsive loading. Exp Mech 46(6):805–824

Fatt MSH, Palla L (2009) Analytical modeling of composite sandwich panels under blast loads. J Sandw Struct Mater 11(4):357–380

Fatt MSH, Palla L, Sirivolu D (2010) Modeling blast and high-velocity impact of composite sandwich panels. Major accomplishments in composite materials and sandwich structures. Springer, Dordrecht, The Netherlands, pp 661–690

Fleck NA, Deshpande VS (2004) The resistance of clamped sandwich beams to shock loading. J Appl Mech 71(3):386–401

Geubelle PH, Baylor JS (1998) Impact-induced delamination of composites: a 2d simulation. Compos B Eng 29(5):589–602

Hashin Z (1980) Failure criteria for unidirectional fiber composites. J Appl Mech 47:329

Hashin Z, Rotem A (1973) A fatigue failure criterion for fiber reinforced materials. J Compos Mater 7(4):448–464

Hutchinson JW, Xue ZY (2005) Metal sandwich plates optimized for pressure impulses. Int J Mech Sci 47:545–569

Iannucci L, Willows ML (2006) An energy based damage mechanics approach to modelling impact onto woven composite materials—part I: numerical models. Compos Appl Sci Manuf 37(11):2041–2056

Johnson HE, Louca LA, Mouring S, Fallah AS (2009) Modelling impact damage in marine composite panels. Int J Impact Eng 36(1):25–39

Kachanov LM (1986) Introduction to continuum damage mechanics, vol 10. Martinus Nijhoff Publishers, Dordrecht, The Netherlands

Lapczyk I, Hurtado JA (2007) Progressive damage modeling in fiber-reinforced materials. Compos Appl Sci Manuf 38(11):2333–2341

Latourte F, Grégoire D, Zenkert D, Wei X, Espinosa HD (2011) Failure mechanisms in composite panels subjected to underwater impulsive loads. J Mech Phys Solid 59:1623–1646

LeBlanc J, Shukla A (2010) Dynamic response and damage evolution in composite materials subjected to underwater explosive loading: an experimental and computational study. Compos Struct 92(10):2421–2430

Liang Y, Spuskanyuk AV, Flores SE, Hayhurst DR, Hutchinson JW, McMeeking RM, Evans AG (2007) The response of metallic sandwich panels to water blast. J Appl Mech 74(1):81–99

Librescu L, Oh SY, Hohe J (2006) Dynamic response of anisotropic sandwich flat panels to underwater and in-air explosions. Int J Solid Struct 43(13):3794–3816

Maimí P, Camanho PP, Mayugo JA, Davila CG (2006) A thermodynamically consistent damage model for advanced composites. National Aeronautics and Space Administration, Langley Research Center, Hampton, VA

Matzenmiller A, Lubliner J, Taylor RL (1995) A constitutive model for anisotropic damage in fiber-composites. Mech Mater 20(2):125–152

Mori LF, Lee S, Xue ZY, Vaziri A, Queheillalt DT, Dharmasena KP, Wadley HNG, Hutchinson JW, Espinosa HD (2007) Deformation and fracture modes of sandwich structures subjected to underwater impulsive loads. J Mech Mater Struct 2(10):1981–2006

Mori LF, Queheillalt DT, Wadley HNG, Espinosa HD (2009) Deformation and failure modes of i-core sandwich structures subjected to underwater impulsive loads. Exp Mech 49(2): 257–275

Mouritz AP, Gellert E, Burchill P, Challis K (2001) Review of advanced composite structures for naval ships and submarines. Compos Struct 53(1):21–42

Naik NK, Yernamma P, Thoram NM, Gadipatri R, Kavala VR (2010) High strain rate tensile behavior of woven fabric e-glass/epoxy composite. Polymer Test 29(1):14–22

Oguni K, Ravichandran G (2001) Dynamic compressive behavior of unidirectional e-glass/vinylester composites. J Mater Sci 36(4):831–838

Porfiri M, Gupta N (2010) A review of research on impulsive loading of marine composites. Major accomplishments in composite materials and sandwich structures. Springer, Dordrecht, The Netherlands, pp 169–194

Rowlands RE (1985) Strength(failure) theories and their experimental correlation. Handbook of Composites, vol 3. Elsevier Science Publishers B.V., Amsterdam, pp 71–125

Roy R, Sarkar BK, Bose NR (2001) Impact fatigue of glass fibre-vinylester resin composites. Compos Appl Sci Manuf 32(6):871–876

Schubel PM, Luo JJ, Daniel IM (2005) Low velocity impact behavior of composite sandwich panels. Compos Appl Sci Manuf 36(10):1389–1396

Shokrieh MM, Omidi MJ (2009a) Tension behavior of unidirectional glass/epoxy composites under different strain rates. Compos Struct 88(4):595–601

Shokrieh MM, Omidi MJ (2009b) Compressive response of glass-fiber reinforced polymeric composites to increasing compressive strain rates. Compos Struct 89(4):517–523

Shokrieh MM, Omidi MJ (2009c) Investigation of strain rate effects on in-plane shear properties of glass/epoxy composites. Compos Struct 91(1):95–102

Sun CT (2000) Strength analysis of unidirectional composites and laminates. Comprehensive composite materials, vol 1. Elsevier Science, Ltd., Oxford, pp 641–666

Tagarielli VL, Deshpande VS, Fleck NA (2007) The dynamic response of composite sandwich beams to transverse impact. Int J Solid Struct 44(7):2442–2457

Tagarielli VL, Deshpande VS, Fleck NA (2008) The high strain rate response of pvc foams and end-grain balsa wood. Compos B Eng 39(1):83–91

Taylor GI (1963) The pressure and impulse of submarine explosion waves on plates. The scientific papers of GI Taylor, vol 3. Cambridge University Press, Cambridge, UK, pp 287–303

Tekalur SA, Bogdanovich AE, Shukla A (2009) Shock loading response of sandwich panels with 3-d woven e-glass composite skins and stitched foam core. Compos Sci Tech 69(6):736–753

Tijssens MGA, Sluys BLJ, van der Giessen E (2000) Numerical simulation of quasi-brittle fracture using damaging cohesive surfaces. Eur J Mech A Solid 19(5):761–780

Tsai SW, Wu EM (1971) A general theory of strength for anisotropic materials. J Compos Mater 5(1):58–80

Vinson JR, Sierakowski RL (2002) The behavior of structures composed of composite materials, vol 105. Springer, Dordrecht, The Netherlands

Wang E, Gardner N, Shukla A (2009) The blast resistance of sandwich composites with stepwise graded cores. Int J Solid Struct 46(18):3492–3502

Xue Z, Hutchinson JW (2004) A comparative study of impulse-resistant metal sandwich plates. Int J Impact Eng 30(10):1283–1305

Chapter 5
Underwater Explosive Response of Submerged, Air-backed Composite Materials: Experimental and Computational Studies

James LeBlanc and Arun Shukla

Abstract The response of composite plates subjected to underwater explosive (UNDEX) loading has been investigated through laboratory experiments and corresponding computational simulations. The focus of the work is the transient plate response and the evolution of material damage. The study is comprised of three main sections: (1) Response of flat E-Glass/Epoxy plates, (2) Response of curved E-Glass/Vinyl Ester plates, and (3) Effects of polyurea surface coatings. The first two sections present comparisons between experimental results and computational simulations, while the third section contains experimental observations. The UNDEX experiments are carried out using a water-filled conical shock tube (CST) which imparts pressure loading representative of the far field, underwater detonation of a spherical charge. During the experiments, the transient response of the plates is captured through either strain gages (flat plates) or the Digital Image Correlation Method (curved Plates). The computational models utilize the commercial finite element code LS-DYNA and are shown to accurately simulate the dynamic response of the plates as well as the fluid structure interaction (FSI). Polyurea coatings are shown to have both positive and adverse effects on the shock response of the composite plates depending on coating thickness and more importantly, coating location.

J. LeBlanc (✉)
Naval Undersea Warfare Center (Division Newport), Building 1346/407F,
1176 Howell Street, Newport, RI 02841, USA
e-mail: james.m.leblanc@navy.mil

A. Shukla
Dynamic Photomechanics Laboratory, Department of Mechanical,
Industrial and Systems Engineering, University of Rhode Island,
92 Upper College Road, Kingston, RI 02881, USA

A. Shukla et al. (eds.), *Blast Mitigation: Experimental and Numerical Studies*,
DOI 10.1007/978-1-4614-7267-4_5,

5.1 Introduction

Composite materials are increasingly being employed in a wide range of industries and applications due to numerous advantages over traditional construction materials such as wood and metallics. The marine industry in particular has utilized composite materials in the construction of leisure and commercial watercraft with civilian applications. Recently, there has been a widespread desire to implement these materials in naval applications such as advanced ship hull designs, ship decks, unmanned underwater vehicles, and submarine components owing to their high strength to weight ratios, improved corrosion resistance, and overall reduced maintenance costs. Utilization of these materials in a military environment, however, introduces unique service requirements, particularly the ability to withstand severe dynamic loads including those associated with underwater explosions (UNDEX) and high speed impacts. The response of these materials under quasi-static loading rates associated with normal operating conditions is well understood. However, their behavior under high loading rates that are imparted due to shock loading is not as well understood. In particular, the evolution of damage, the transient response, and the load carrying capacity in these materials after a shock event needs to be better understood. Historically, this reduced knowledge base has resulted in conservative designs with large safety factors in order to ensure damage will not occur. Ultimately, these large safety factors result in structural over-designs that do not afford the full weight saving benefits that composite materials can provide.

The United States Navy (USN), as well as navies throughout the world, has recently shown a more pronounced interest in the utilization of composites as suitable materials for construction. The use of fiber reinforced plastics (FRP) for marine vessels primarily dates to the end of the Second World War when the USN contracted the construction of the first fiberglass/polyester personnel crafts. The use of these materials has continued to expand and recent USN applications include the MHC 51 mine hunter, Virginia Class submarine bow domes, DDG 1000 deckhouse, and various unmanned underwater vehicles/autonomous underwater vehicles/remotely operated vehicles (UUV/AUV/ROV). UUV/AUVs which employ composite technology include the Deepglider, Talisman, and Manta. Worldwide, composite materials have been utilized in the construction of the Norwegian Skjold Class and the Swedish Visby Class vessels. An extensive review of the use of composite materials in naval ships and submarines has been presented by Mouritz et al. (2001).

The use of composite materials in marine applications offers distinct advantages over metallic materials, including high strength to weight ratios, reduced maintenance costs, and higher corrosion resistance. There are however disadvantages to the use of these materials including lower resistance to impact damage and potential aging effects when subjected to long term submergence in marine environments. When these types of materials are employed in naval vessels there is an increased risk of shock and/or explosive loading within their service life. This results in a need to understand the behavior of these materials not only at strain rates associated with static load levels (10^{-4}–10^{-3}) but also at loading rates many orders of magnitude higher (10^{-1}–10^{3}).

When composite materials are subjected to severe loading conditions they may experience damage in the form of several distinct mechanisms occurring in the in-plane and through-thickness directions. The in-plane mechanisms consist of fiber breakage and matrix cracking, while the through-thickness damage is dominated by delamination of the plies. Composite materials typically have high in-plane strengths due to their construction using oriented layers of continuous fibers. They are weak in the through-thickness direction. In this direction the material strength is governed by the resin as there are no fibers oriented in this direction. The resin has lower tensile and shear strengths as compared to the fibers. This through-thickness weakness typically leads to delamination when the plates are subjected to through-thickness loading or flexural conditions. Recently, weaving processes have evolved to the point at which it is possible to make structural cloth materials that include through-thickness fibers and show the ability to improve through-thickness weakness (LeBlanc et al. 2007; Tekalur et al. 2009; Bogdanovich et al. 2001). The through-thickness damage in the form of delamination is increasingly found in composite materials subjected to shock or impact loading. Under shock loading a compression wave passes through the thickness of the material and reflects off the back face as a tensile wave. Simultaneously a tensile unloading wave passes from the front face of the material toward the back. This unloading wave (also referred to as a "release wave") corresponds to the decay phase (negative slope) of the incident loading (Wang 2004). In reality, the incident wave is composed of a loading front followed by an unloading front. If the unloading wave and the reflected wave meet in a ply interface, de-bonding may occur if the waves combine to exceed the tensile strength of the resin. These waves have a greater chance of meeting in composite materials due to the lower wave speed compared to metallic parts (Dyka and Badaliance 1998). The interaction of the tensile release and reflected waves is the same phenomenon that is often responsible for the formation of surface spalling in metallic materials.

Past experimental studies on composite materials have evaluated the material response over a range of loading rates. Work by Zaretsky et al. (2004) and Yuan et al. (2007) has studied the damage characteristics of composite materials when subjected to low velocity impacts. Franz et al. (2002) and Mouritz (1995) studied the effects of an underwater explosion at different standoff distances on a glass composite laminate. Tekalur et al. (2008a) investigated the effects of shock loading on both E-Glass and Carbon fiber based laminates. Mouritz (2001) studied the effectiveness of adding a light weight, through-thickness stitching material to increase the damage resistance of composites subjected to UNDEX loading, while LeBlanc et al. (2007) studied three-dimensional woven composite materials. The results of these experimental studies are limited to evaluating post mortem properties of the materials. In particular, the residual strength or fatigue properties were determined from the damaged samples. Recently, there has been an increased interest in the study of the effect of shock loading on sandwich structures. These studies include the effects of shock and impact loading conditions (Jackson and Shukla 2011; Schubel et al. 2007; Arora et al. 2010).

Analytical damage models for composites have been widely developed and are continually being refined and updated. These models typically assign an internal damage variable to each of the types of damage of interest (i.e., matrix cracking, fiber rupture) which, in simple form, are ratios of the stress state to a failure criteria (Matzenmiller et al. 1995; Zako et al. 2003; Dyka and Badaliance 1998). Matzenmiller et al. (1995) have presented a mathematical model for damage of composite materials that develops a relationship between the level of material damage and the effective elastic properties of the material. For each of the significant damage mechanisms (fiber rupture, fiber buckling, matrix crushing, and matrix cracking) an internal variable describes the evolution of the damage as a function of loading. Based upon the expression representing each damage variable, the effective elastic properties can be degraded when the variable reaches a critical value. As the mechanical properties must be continually updated to account for the damage degradation, this methodology lends itself well to implementation in finite element codes.

The finite element modeling of damage in composites has been performed primarily on models simulating strain rates up to those representing drop test experiments, with some work performed at the high strain rate regimes expected in shock loading. Material models are currently being implemented in existing commercial finite element codes (O'Daniel et al. 2005; McGregor et al. 2007; Williams and Vaziri 2001); however, the validation work with these models has been limited to the low strain rate regime not experienced under blast/shock loading conditions. Recent publications involving computational modeling of damage progression in composites have utilized LS-DYNA and the Mat_162 (Mat_Composite_OPTION) material models which simulate fiber breakage, matrix cracking and delamination damage. This material model combines the progressive failure theory of Hashin and the damage mechanics approach of Matzenmiller et al. (1995). Gama et al. (2004) and Xiao et al. (2007) have published results from quasi-static punch shear loading experiments which correlate well with simulations utilizing the Mat_162 material model. Simulations of low velocity impact experiments have been documented in the work by Donadon et al. (2008) and Hosseinzadeh et al. (2006). Furthermore, Batra and Hassan (2007) have studied the response of composites to UNDEX loading through numerical simulations; however, there are no comparisons to experimental results. Although the dynamic experiments have been simulated, the results taken from these models have been limited to the prediction of damage area and final deformations rather than comparisons to transient response. Historically, the material inputs have been determined from quasi-static test data (Chan et al. 2007), an assumption which has been shown to be reasonable in simulations of composite materials when the in-plane strain rates are low. Low in-plane strain rates typically occur under loading conditions in which the primary deformation mode is out-of-plane flexure. It is acknowledged that the through-thickness strain rates for blast and ballistic impact events are likely significantly higher and the derivation of material inputs from quasi-static test data is not a valid assumption.

In recent years, the use of polyurea materials to enhance the failure resistance of materials subjected to explosive loading has become a topic of interest. Polyurea is a synthetic, high strength/high elongation coating that is typically spray cast onto existing structures to increase their resistance to shock and ballistic/shrapnel loading events such as those of an improvised explosive device (IED). The armed forces have begun to investigate the suitability of these materials for use on military and naval vehicles such as Humvees, troop carriers and ship hulls, Hodge (2004). Research efforts have focused on the use of polyurea coatings on composite plates, steel plates and as inner layers of sandwich composite structures. Tekalur et al. (2008b) studied polyurea based layered composite materials subjected to blast loading and showed that sandwich materials prepared by sandwiching the polyurea between two composite skins had the best blast resistance compared to the EVE composite and polyurea layered plates. Amini et al. (2010a, b, c) have studied the effects of monolithic and polyurea coated steel plates subjected to impulsive loads and showed that polyurea has a positive damage mitigation effect when applied to the back face of the material. Similar results were observed by Gardner et al. (2012) when investigating the effect of polyurea in sandwich composites. Results indicated that when a layer of polyurea is placed between the foam core and the back face of the sandwich the blast resistance is improved, while conversely if the polyurea is placed between the front face and the foam core the performance is degraded.

The widespread use of composite materials for structural applications in which shock and high velocity impact loading is a concern is typically impeded by a lack of information in regards to the material behavior under these loading rates. Specifically, the ability to accurately predict the material response in terms of damage mechanisms and post mortem structural integrity through the use of design tools is lacking. There are very few analysis approaches suitable for predicting the failure modes that evolve in composites under shock loading, and those that do exist typically involve computational material models that are complex and require very specific material characterization to obtain all of the required material parameters for accurate results. Furthermore, there is very limited experimental data involving the shock loading of composites against which finite element model results can be validated. The goal of a fully validated modeling method is the development of a reliable, simple to use modeling tool for optimization of structural designs. Without a validated modeling approach that can be applied early in the design phase, the development of composite structures is limited to the design/test/redesign cycle in which structural deficiencies under shock loading are only identified through testing. As a result of this lack of experimental data and modeling capabilities, composite structures have historically been designed with large safety factors which negate the weight savings afforded by composite materials.

5.2 Experimental Methods

The experiments that are the basis for the work described in the following sections are designed to subject composite panels to shock loading conditions representative of far field underwater explosions. A conical shock tube facility was utilized to impart the desired loading to the panels in a controlled laboratory environment. During the experiments several methods were employed to capture the transient response of the panels. These measurements form the basis for correlation between the experiments and corresponding finite element simulations.

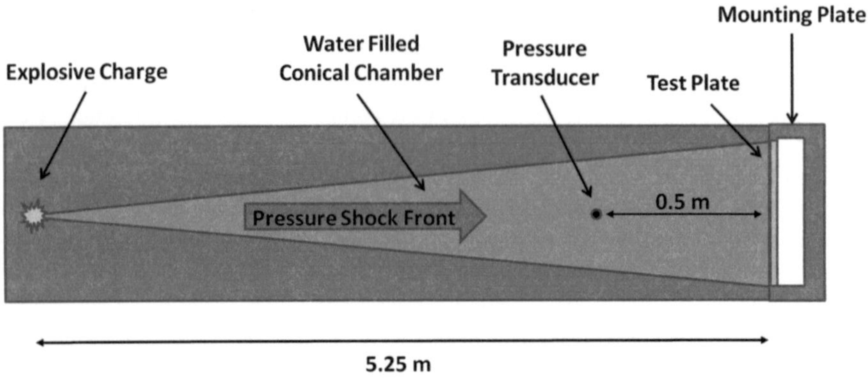

Fig. 5.1 Conical shock tube schematic (not to scale)

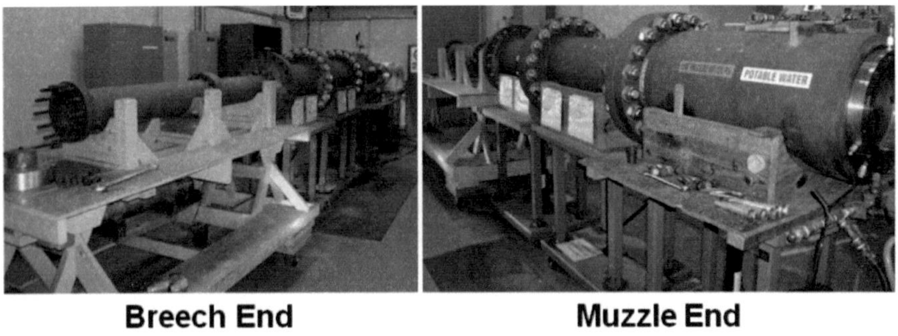

Fig. 5.2 Conical shock tube facility

5.2.1 *Conical Shock Tube*

A conical shock tube (CST) facility located at the Naval Undersea Warfare Center Division Newport (NUWCDIVNPT), was utilized in the shock loading of the composite materials. The shock tube is a horizontally mounted, water filled tube with a conical internal shape, Figs. 5.1 and 5.2. The tube geometry represents a solid angle segment of the pressure field which results from the detonation of a spherical, explosive charge, Fig. 5.3. In an open water environment (free field), the pressure wave expands from the charge location as a spherically propagating wave. In the shock tube, the rigid wall acts to confine the expansion of the pressure wave in a manner that simulates a conical sector of the pressure field. The guiding principle is that if the walls of the tube are sufficiently rigid then the radial distribution of the free field pressure wave can be replicated (Coombs and Thornhill 1967). In order to compare free field and shock tube pressure values, it is useful to define an amplification

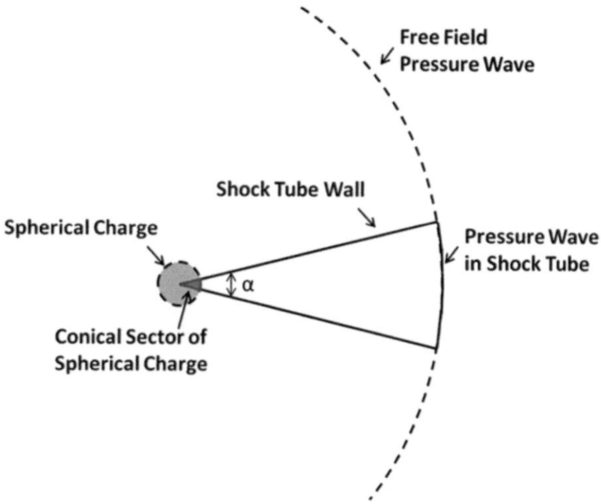

Fig. 5.3 Conical shock tube approximation (Matzenmiller et al. 1995)

factor, which is the ratio between the volume of a spherical charge to the volume of the conical sector charge and is defined by Poche and Zalesak (1992) as:

$$\mathrm{AF} = \frac{1}{\sin^2\left(\frac{\alpha}{4}\right)}$$

Where α is the cone angle.

This equation assumes perfectly rigid wall boundary conditions which are not fully realized. Therefore, the actual amplification factor is less than the calculated value and is typically reported as an effective weight amplification factor. This is defined by Poche and Zalesak (1992) as the ratio between the weight of a spherical charge, W, required to produce the same peak pressure at a given standoff distance as that produced in the shock tube by a segment of charge weight, w:

$$\mathrm{AF} = \frac{W}{w}$$

The reduction in the amplification factor (AF) is typically attributed to the elastic deformation of the shock tube walls due to the pressure wave acting normal to the walls. Further discussion on the development and history of the water filled conical shock tube is provided by Coombs and Thornhill (1967) and Filler (1964).

The internal cone angle of the shock tube used in the current work is 2.6°. The tube is 5.25 m (207 in.) long from the charge location to the location of the test specimen and internally contains 98.4 L (26 gal.) of water at atmospheric pressure. The pressure shock wave is initiated by the detonation of an explosive charge at the breech end of the tube (left side of Fig. 5.1) which then proceeds down the length of

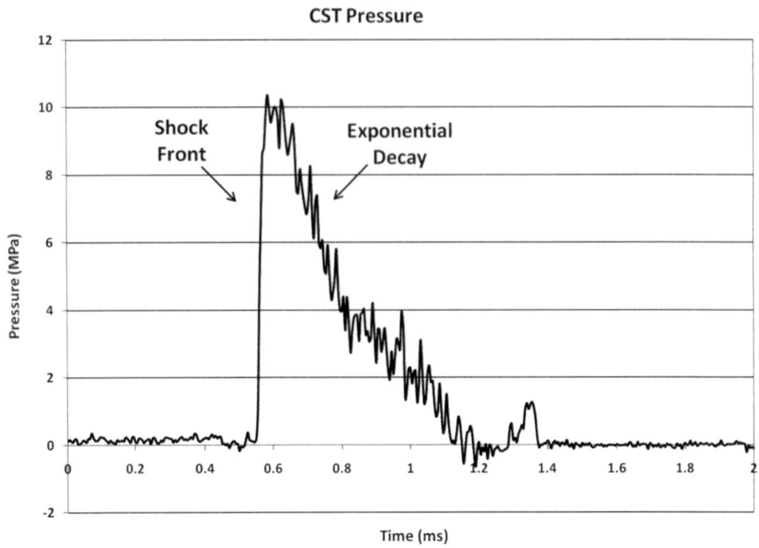

Fig. 5.4 Typical pressure profile generated in the conical shock tube

the tube. Peak shock pressures from 10.3 MPa (1,500 lb/in.^{2}) to 20.6 MPa (3,000 lb/in.^{2}) can be obtained depending on the amount of explosive charge used. A typical pressure profile, as obtained from the pressure transducers, is shown in Fig. 5.4. This figure illustrates the rapid pressure increase associated with the shock front followed by the exponential decay of the wave. This profile was obtained using a M6 blasting cap containing 1.32 g (0.00292 lb) TNT equivalence and is measured 0.508 m (20 in.) from the impact face of the plate. The length of the tube is considered sufficient so that plane wave conditions are nearly established at the specimen.

A mounting fixture was designed in such a way that the test specimens were air backed with fully clamped edges. The specimens had an overall diameter of 26.54 cm (10.5 in.) with a 22.86 cm (9.0 in.) unsupported middle section. The mounting arrangement for the curved plates is shown in Fig. 5.5, and is identical to that of the flat plates. For the case of the curved plates, the specimens were mounted with the convex surface towards the incoming shock fronts. This was chosen so that the experiments would represent geometries commonly used in underwater applications with curved surfaces typically facing into the fluid (i.e., submersible vehicle hull forms).

5.2.2 Transient Plate Response Measurements

As part of the experimental methods utilized in the study, two measurement techniques were employed to capture the transient response of the composite plates. During the testing of the flat panels, strain gages were applied to the back (dry) face to record the material strains during loading. However, this method presented

Fig. 5.5 Shock tube mounting configuration

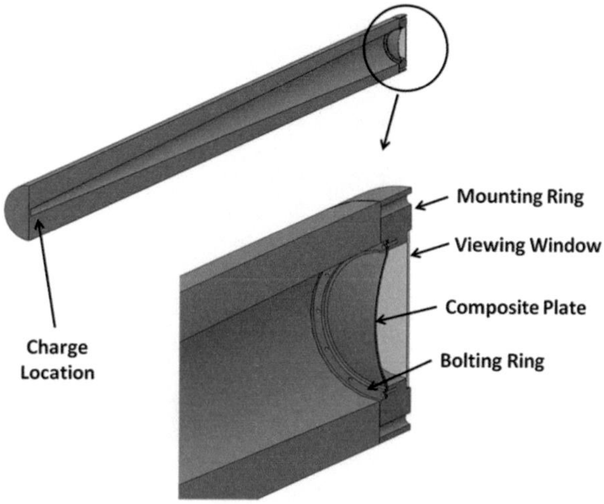

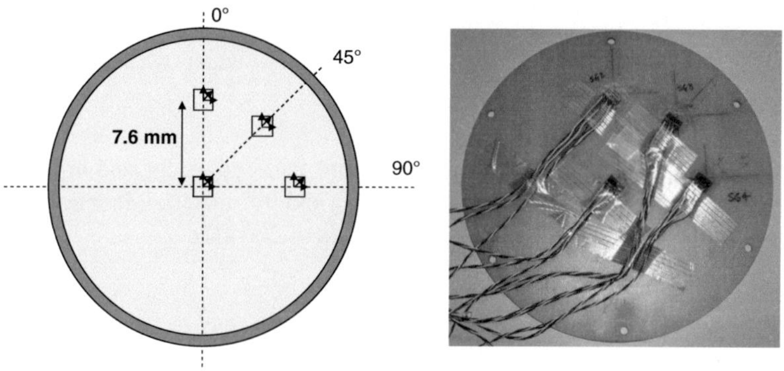

Fig. 5.6 Strain gage mounting arrangement

difficulties in that there is a tendency for gage de-bonding as the loading becomes more severe, and only limited data was obtained during the experiments. During the subsequent experiments involving the curved and polyurea-coated plates, the Digital Image Correlation (DIC) method was utilized to capture the transient response of the back face. An overview of the application of each method is presented in the following sections.

5.2.2.1 Strain Gages

For each of the flat plates that were tested, the dry (non-impact) face was instrumented with four (4) tri-axial strain gages. Each of the gages measures strains in the 0°, 45°, and 90° gage directions. The gages were mounted as shown in Fig. 5.6, with

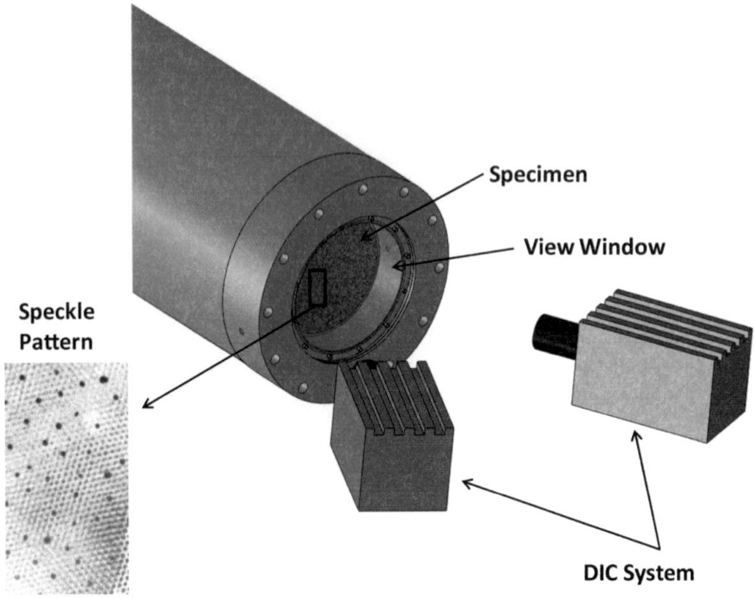

Fig. 5.7 Schematic of the digital image correlation

one gage located at the center of the panel and three gages located at a 7.62 cm (3 in.) radius from the center along the 0°, 45°, and 90° material directions. For all tests the strains were recorded at a sampling rate of 200 kHz.

5.2.2.2 Digital Image Correlation Method

Digital Image Correlation is a nonintrusive optical technique for capturing the full-field transient response of the panels through the use of high-speed photography and specialized software. Capturing the three-dimensional response of the plates requires that two cameras be used in a stereo configuration. If a single camera is utilized the data would be limited to the in-plane results. To record the transient response with this system, the cameras must be calibrated and have synchronized image recording throughout the event. The calibration of the cameras is performed by placing a grid containing a known pattern of points in the test space where the composite sample would be located during the experiment. This grid is then translated and rotated in- and out-of-plane while manually recording a series of images. Since this grid pattern is predetermined, the coordinates of the center of each point (dot) are extracted from each image. The coordinate locations of each dot are extracted uniquely for each camera, thus allowing for a correspondence of the coordinate system between both cameras (Tiwari et al. 2007). The DIC is then performed on the image pairs that are recorded during the shock event. Prior to the experiment, the back face of the sample is painted white and then coated with a randomized speckle pattern, as shown in Fig. 5.7. The post processing is performed

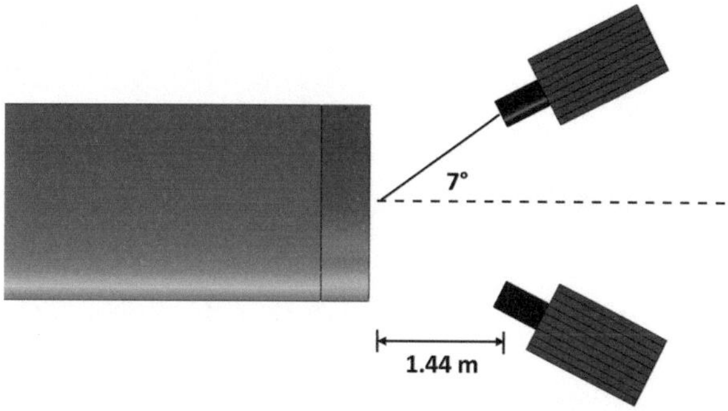

Fig. 5.8 Digital image correlation setup (not to scale)

with the VIC-3D software package (Correlated Solutions) which matches common pixel subsets of the random speckle pattern between the deformed and un-deformed images. The matching of pixel subsets is used to calculate the three-dimensional location of distinct points on the face of the panel throughout time. This technique has been applied as a full-field measurement technique in many applications including shock loading (Tiwari et al. 2009)

Two high-speed digital cameras (Photron SA1) are positioned behind the shock tube (Fig. 5.8). The use of two cameras allows for the out-of-plane behavior to be captured. In this particular case, the distance from the lens of the camera to the specimen was 1.44 m (57 in.) and both cameras were angled at approximately 7° with respect to the symmetry plane. A frame rate of 20,000 was used, which corresponded to an inter-frame time of 50 μs.

5.3 Finite Element Modeling

Finite element modeling of the experiments was performed utilizing the LS-DYNA software code available from the Livermore Software Technology Corporation (LSTC). The complete finite element model of the CST test setup is shown in Fig. 5.9. This specific model corresponds to the simulations of the curved composite plates as indicated by the plate geometry at the right side of the figure. The model for the simulations of the flat plates only differs by the plate geometry. The model consists of the internal fluid of the shock tube and the composite test sample. The fluid within the tube is included in the simulations so as to capture the fluid structure interaction (FSI) at the interface of the fluid and test plate. As will be shown later, this is a critical interaction to consider as the pressure loading on the plate is not uniform across its face. Only the first 1.01 m (40 in.) of the fluid extending from the test sample towards the charge location is modeled. This was deemed to be acceptable for two reasons: (1) the fluid is loaded with the pressure profile measured 0.508 m (20 in.) from the test sample, Fig. 5.1, and (2) a nonreflecting boundary layer is applied at the charge side boundary of the fluid domain. The nonreflecting boundary allows the wave that is reflected from the plate to leave the fluid domain but not

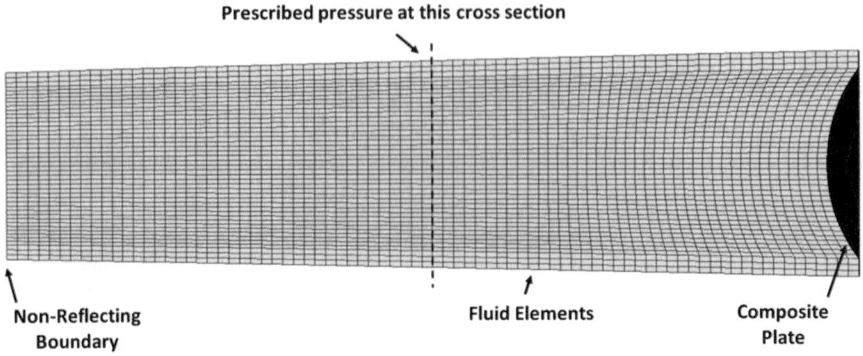

Fig. 5.9 Finite element model of CST

reenter. This assumption holds as long as the duration of the plate deformation is small enough such that the wave which is reflected off the plate cannot reach the end of the tube and reflect back to the plate location during this time. The fluid is modeled with solid elements and a null material definition. The use of the null material allows for the material to be defined with an equation of state (EOS) definition. A linear polynomial EOS is utilized for this model for which only the bulk modulus and density of the water is defined. This allows for an accurate propagation of the pressure wave in the water in a computationally efficient manner. The pressure load is applied as a plane wave at the location of the test pressure transducer (assuming near planar wave fronts at large distances from the charge detonation location). The pressure profile that was measured for the respective experiments is applied to the model. The FSI is handled through the use of a tied-surface-to-surface (Ls-Dyna keyword *Contact_Tied_Surface_To_Surface) contact definition between the water and composite plate. In this method, two contact surfaces (*Set_Segment) are defined for which the nodes are tied together. For the coupling of the fluid and composite plate, the two surfaces are (1) the fluid face where it contacts the plate, and (2) the plate face where it contacts the fluid. In the model, the element sizing at this interface is not continuous in that the elements comprising the composite plates are much more refined than the fluid mesh. The meshes however are geometrically conformal at the plate-water boundary as shown in Fig. 5.9.

The composite plates in the simulations have been modeled utilizing both solid and shell element formulations. The flat plates are represented by solid elements as shown in Fig. 5.10 and the curved plates utilize a shell element formulation, Fig. 5.11. Each layer (solid or shell) represents a 0° and 90° combined ply of the biaxial laminate construction (later discussion) with a thickness equal to twice that of the individual ply. For the models which utilize shell elements, the mid-surface of each ply is meshed and the individual shell layers are offset by the ply thickness. The choice of element formulations is a result of the response of the plate geometries to shock loading. The flat plates responded to the loading in a flexural mode with small angles of rotation at the supported plate edges, while the curved plates

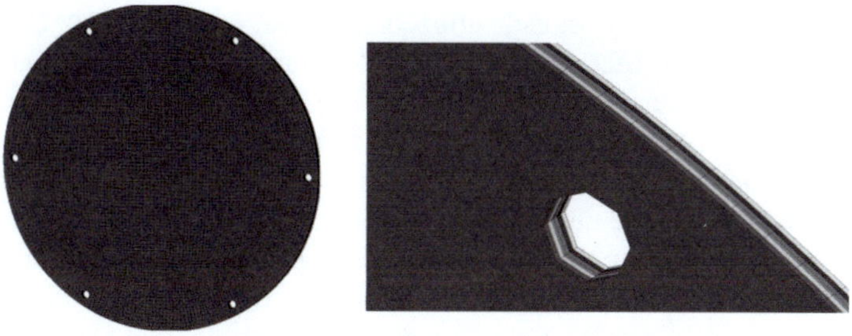

Fig. 5.10 Finite element model of composite plate: solid elements

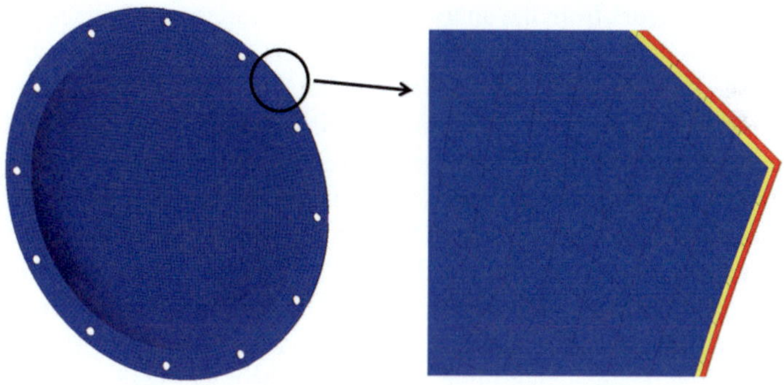

Fig. 5.11 Finite element model of composite plate: shell elements

underwent a complete inversion in which the elements at the plate boundary underwent significant angles of rotation. Each element formulation has its unique advantages and disadvantages relating to the current models. When using solid elements it is possible to uniquely define each ply interface since each element has a top and bottom face. Conversely when using shell elements each element is associated with two ply interface definitions which add to the complexity of the contact definitions. The shell elements are advantageous for the modeling of bending behavior in which large rotations occur in that a single element, with sufficient through-thickness integration points, accurately captures the associated response, whereas multiple through-thickness solid elements are required. The holes along the outer edges of the plates represent the through bolt holes present in the test samples used for mounting the plate to the fixture. These holes were found to be the initiation point for much of the material damage during the experiments and are included in the finite element models to capture this damage evolution. The influence of the bolt holes is primarily due to the pressure loading being applied over the entire surface of the composite plates, over which the plate dimensions are not large enough for boundary effects to be ignored.

The Ls-Dyna material model utilized in this work is Mat_Composite_ Failure_*Option*_Model (Mat_059) where the "Option" parameter is equal to either "Shell" or "Solid" depending on the element type selected for the plate. This is an orthotropic material definition capable of modeling the progressive failure of the material due to any of several failure criterion including tension/compression in the longitudinal and transverse directions, compression in the through-thickness direction, and through-thickness shear. Published descriptions of how each failure mode is handled are scarce; however, there is some informal documentation available from LSTC. For each possible failure mode, there is an internal variable which is checked throughout the analysis to determine if failure in that mode is present. Once failure due to one mode is triggered, the load carrying ability of the material in that direction is reduced to zero after a finite number of analysis time steps. It is important to note that failure in one direction does not cause the element to be deleted. An element is only deleted from the analysis after it has failed in all directions and can no longer carry any load. The material model inputs are derived from static tensile testing with no modifications for strain rate effects. It was seen that the static properties provide reasonable results for shock loading conditions. Similar observations were made by Chan et al. (2007) for the ballistic impact problem.

Delamination damage is considered and is taken into account through the use of a surface-to-surface tiebreak contact definition. For the models which contain solid elements each ply is modeled as a solid layer of elements but the nodes between plies are not equilibrated. Similarly for the shell element approach, each layer of elements represents the mid-surface of each ply and the shell layers are offset by the ply thickness. The tie break definition initially ties the nodes between plies together to inhibit sliding motion. The force at each node is then monitored by the software and the corresponding normal and shear stresses are computed. Failure is defined by the following formula:

$$\left(\frac{\sigma_N}{\mathrm{NFLS}}\right)^2 + \left(\frac{\sigma_S}{\mathrm{SFLS}}\right)^2 \geq 1$$

where: σ_N and σ_S are the computed normal and shear stresses, respectively

NFLS and SFLS are the failure normal and shear stresses, respectively.

At any point in time during the analysis when the stress state at a node in the contact definition exceeds the failure criteria then the tie definition for that node is deleted and the node is free to slide. It is important to note that once the tie component of the contact definition is deleted, the contact for that node transitions to a standard definition. This allows the slave node to slide and separate from the master surface but not pass through it. Therefore, individual plies can separate but not pass thorough one another.

5.4 Correlation of FE Results to Experiments: Russell Error Measure

The transient data (Strain, Deformation) that is captured during the experiments is used as a basis to correlate and validate the finite element model results. The quality of the correlation between the test data and numerical results is quantified using the Russell Comprehensive Error measurement. The Russell error technique is a method which evaluates the differences in two transient data sets by quantifying the variation in magnitude and phase. The magnitude and phase error are then combined into a single error measure, the comprehensive error factor. The full derivation of the error measure is provided by Russell (1997) with the phase (RP), magnitude (RM), and comprehensive (RC) error measures respectively given as:

$$\mathrm{RP} = \frac{1}{\pi}\cos^{-1}\left(\frac{\sum c_i m_i}{\sqrt{\sum c_i^2 \sum m_i^2}}\right)$$

$$\mathrm{RM} = \mathrm{sign}(m)\log_{10}\left(1+|m|\right)$$

$$\mathrm{RC} = \sqrt{\frac{\pi}{4}\left(\mathrm{RM}^2 + \mathrm{RP}^2\right)}$$

In the above equations c_i and m_i represent the calculated (simulated) and measured responses, respectively. Excellent, acceptable, and poor correlation using the Russell error measure are defined as: Excellent—$\mathrm{RC} \le 0.15$, Acceptable—$0.15 < \mathrm{RC} \le 0.28$, and Poor $\mathrm{RC} > 0.28$. The selection of these criteria levels are the result of a study that was undertaken to determine the correlation opinions of a team in support of a ship shock trial. A summary of the process used to determine the criteria is presented by Russell (1998).

5.5 Response of Flat Composite Plates

The response of flat E-Glass/Epoxy plates subjected to far field explosive loading is presented in the following section. The results consist of: (1) a discussion of the fluid structure interaction at the water/composite interface, (2) comparisons between strain gage data obtained through CST experiments and the corresponding simulations, and (3) the post mortem material damage types/extents in the plates resulting from the testing and predicted by the simulations. Additional details involving this work are found in LeBlanc and Shukla (2010).

Table 5.1 Cyply 1002 crossply—mechanical properties

	N/m^2 ($lb/in.^2$)
Tensile modulus (0°)	23.4×10^9 (3.4×10^6)
Tensile modulus (90°)	23.4×10^9 (3.4×10^6)
Tensile strength (0°)	482×10^6 (70×10^3)
Tensile strength (90°)	482×10^6 (70×10^3)
Compressive strength (0°)	689×10^6 (100×10^3)
Compressive strength (90°)	689×10^6 (100×10^3)

5.5.1 Composite Panel Construction

The composite material studied in this investigation is Cyply® 1002, a reinforced plastic manufactured by Cytec Engineered Materials. The material is a cured epoxy composite which utilizes a nonwoven, parallel fiber construction of continuous E-Glass filaments. The laminate is a cross-ply construction with alternating plies of 0° and 90° fibers. Each ply has a thickness of 0.254 mm (0.01 in.), an areal weight of 0.46 kg/m^2 (0.85 lb/yd^2), and a specific gravity of 1.85. The resin content is 36 ± 3 %. Total plate thicknesses of 3.3 mm (0.130 in.) and 4.82 mm (0.190 in.) have been chosen with each thickness having 13 and 19 plies, respectively. Due to the odd number of plies there is an additional ply in the 0° direction. The laminate schedule for the 3.3 mm (0.130 in.) material is as follows [0/90/0/90/0/90/0/90/0/90/0/90/0] with a similar schedule for the 4.82 mm (0.190 in.) material, only with three more additional 0/90 layers. The thick plates have been used to reduce the overall panel deflection during the shock event in order to allow the strain gages to remain attached. The thinner plates allow increased material damage to develop while holding the peak shock pressure consistent among all experiments. The specimens are round plates; 26.54 cm (10.45 in.) in overall diameter, with a 22.8 cm (9 in.) unsupported middle section. The material properties for the material are provided in Table 5.1.

5.5.2 Fluid Structure Interaction

The finite element simulation of the shock tube experiments allows for a full field visual representation of the interaction between the pressure wave and the composite plate, whereas the pressure profile obtained during the experiment from the pressure transducer gives only a single point history. The pressure field in the fluid as it interacts with and loads the plate is shown in the left side of Fig. 5.12. The associated plate response is shown in right side of the figure. The time as shown in these figures, is the analysis time with zero taken at the initiation of the pressure field 50.8 cm (20 in.) from the test sample. Figure 5.12 illustrates several key points. First, although the pressure wave is uniform (planar) prior to its arrival at the test plate, the pressure distribution becomes both complex and nonuniform when it

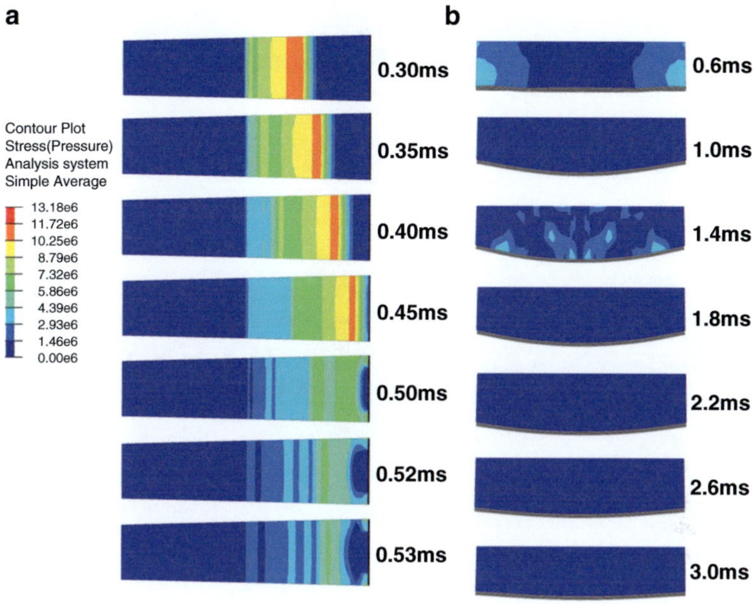

Fig. 5.12 (**a**) Fluid structure interaction, (**b**) Plate response (units of Pa)

interacts with and loads the plate itself. It is evident that there is a low pressure area that develops in the center of the plate while the clamped edge is loaded with high pressure. This can be attributed to the relatively low stiffness of the unsupported area of the plate as compared to the clamped edge of the plate where it is supported by the mounting ring. The arrival and full reflection of the pressure wave take place over approximately 0.2 ms. A second observation is that the loading of the plate and the resulting response can be separated into two distinct time regimes. While the pressure wave is fully reflected by 0.53 ms, the plate does not start to deform until 0.6 ms. The plate reaches full deformation at 1.8 ms and returns to its initial shape at 3 ms. Therefore, there is clearly a time lag from when the plate is loaded due to the pressure wave, to when plate structurally responds.

5.5.3 Strain Time History Correlation

The strain gage data comparisons for the shock experiment involving the thick (4.8 mm) plate and corresponding simulation is shown in Figs. 5.13 and 5.14 for the 0° and 90° gage directions. The Russell error for each of these correlations is provided below the figure. From these graphical comparisons and error summary it is seen that there is a high level of correlation between the experiment and corresponding simulation. Three of the four compared strain profiles fall within the acceptable regime. The gage located along the 90° plate direction and measuring strain in the

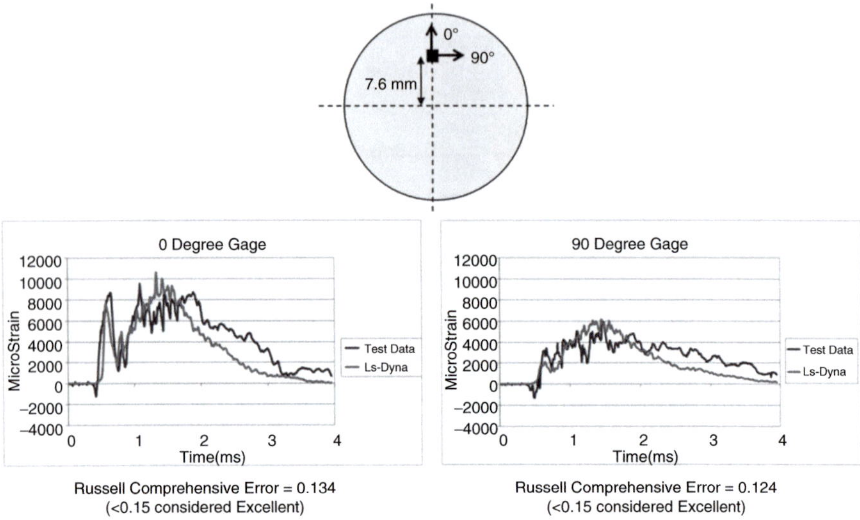

Fig. 5.13 0° strain gage correlation

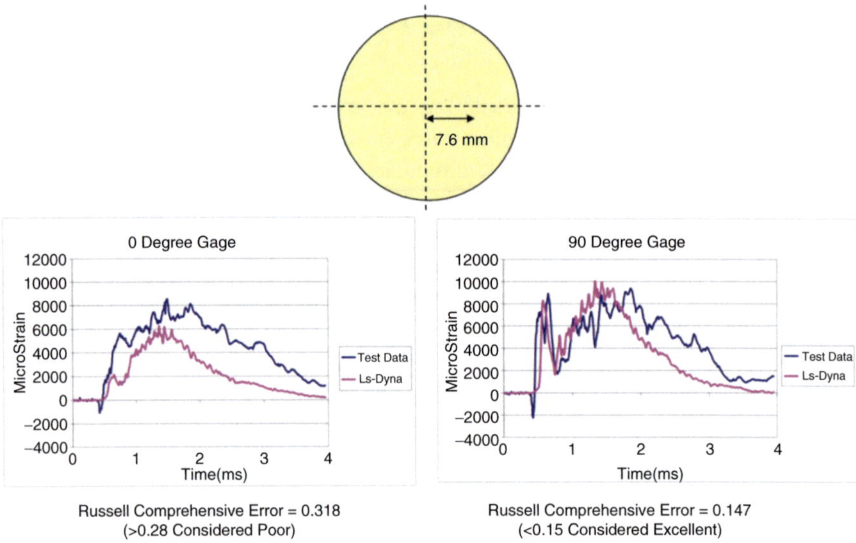

Fig. 5.14 90° strain gage correlation

0° material direction does indicate poor correlation. It is noted the trends in these two signals are comparable and it is the magnitude that differs. Furthermore, this gage is measuring what would be considered to be strain in the circumferential direction at that point, whereas the radial strains are expected to be more dominant due to the plate flexure. This level of agreement between the experiment and finite

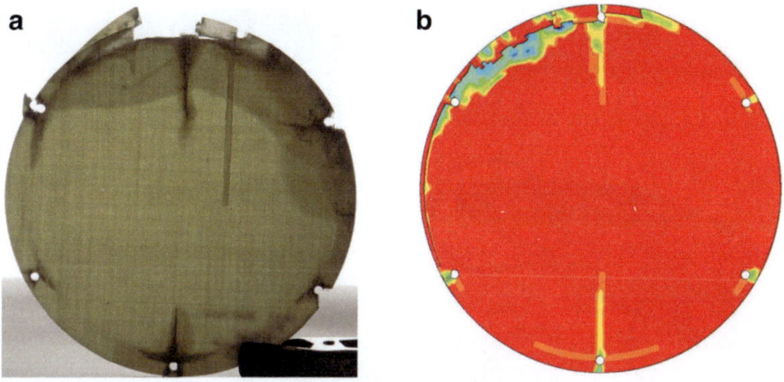

Fig. 5.15 (**a**) Material damage during test, (**b**) Material damage from simulation

element data is encouraging since strain gage data is notoriously difficult to correlate to and match with simulations.

5.5.4 *Material Damage Correlation*

The damage that was imparted to the plate during a typical test is shown in the left image of Fig. 5.15. This figure is from the 3.3 mm (0.130 in.) thick plate that was tested at a shock pressure of 11.7 MPa (1,700 lb/in.^{2}). The corresponding finite element model result is shown in the right side image. The image of the test sample has been backlit to highlight the internal delamination that has occurred.

From these two images several qualitative observations can be made. First, both the experimental and computational results show that there are two cracks that initiate from the through holes located at the top and bottom (0° material direction) of the sample. These cracks propagate to a final length of approximately 63.5 mm (2.5 in.) in the experiments and approximately 76.2 mm (3 in.) in the computational result. In both results the cracks run along the 0° material direction. The second observation is that there is material damage located between each of the holes and the edge of the sample, which was also predicted by the simulation. It is important to note that an initial finite element model of the plate was made in which the holes were omitted. This model developed neither the localized damage near the hole locations nor the crack along the 0° direction. This highlights two key aspects of this type of experimental and computational work. The first is that the damage that is observed is initiated due to the stress concentrations induced by the interaction of the mounting bolts and the plate as the material flexes and pulls towards the center of the plate during deformation. The second is that when undertaking small scale testing, where edge effects and geometric discontinuities can play a key part in the

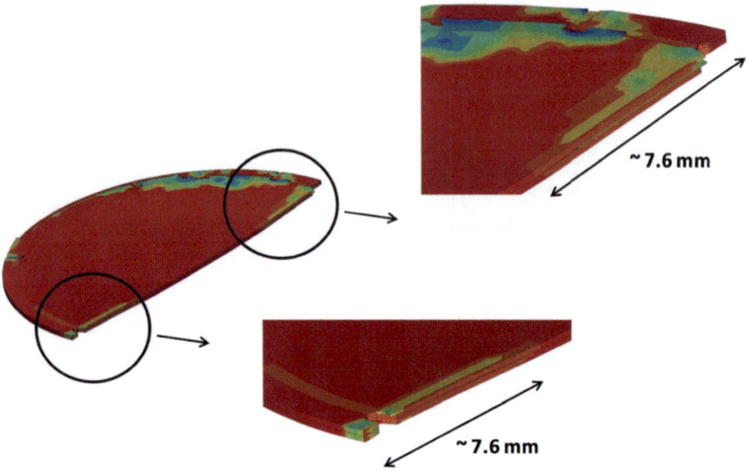

Fig. 5.16 Finite element model delamination

material response, it is important to include these features in the computational model. Otherwise the amount of damage predicted by the simulation will be less than that seen in the experimental component.

In the left image of Fig. 5.15 there is a region of delamination damage that developed along the top edge of the test specimen. This delamination zone extends from the edge of the plate inward to a radial distance of ~50.8 mm (2 in.) between the 10 and 3 o'clock positions. In the computational model this delamination zone also develops, Fig. 5.16, but it occurs both along the top and bottom edges. The delamination in the finite element model extends from the edge inward to a radial distance of 7.62 cm (3 in.). Although the amount of delamination is somewhat larger in the computational model than is observed in the test it is encouraging that the model is able to predict the onset of the delamination itself and propagate it to a comparable distance.

In the current model the choice of a delamination criterion was taken to be 34.4 MPa (5,000 lb/in.^{2}) for both tensile and shear stresses. The choice of this value was based on discussions of the developer of the material model (Materials Sciences Corporation). Based on these discussions and past experience it was determined that an appropriate knock down factor for the delamination criteria is approximately one-half of the tensile strength of the pure epoxy. The degradation by ½ of the tensile strength accounts for voids and interfacial defects/flaws between the layers of fibers during the manufacturing of the material. The exact epoxy resin formula is held as proprietary by the material manufacturer; however, published values for the tensile strength of epoxy place the value between 27.5 and 82.7 MPa (4,000–12,000 lb/in.^{2}). Therefore, the choice of 34.4 MPa (5,000 lb/in.^{2}) is reasonable. During the development of the models several values as high as 82.7 MPa (12,000 lb/in.^{2}) were utilized to determine the effect of this value. When a high value is chosen the delamination damage does not occur and all plies remain in tied contact. If a low value is taken then the plies completely delaminate early on in the simulation and the results do not agree with the experimental results. More work is planned to investigate the most efficient way to model the delamination parameters but is this work is outside the scope of the current study.

5.5.5 Summary: Flat Plates

The response of flat E-Glass/Epoxy composite plates subjected to underwater shock loading was also studied. Correlations are presented between the experiments and computational simulations in terms of strain gage data and post mortem damage levels. Additionally the fluid structure interaction at the plate water interface is described through the use of the finite element results. The Russell error measures for three out of four of the compared gages had excellent correlations. The panels showed significant damage areas including fiber/matrix breakage as well as internal delamination for which the corresponding finite element simulations were able to simulate appropriately. This work has shown the ability of the LS-DYNA material model Mat_Composite_Failure_Solid_Model to realistically model the behavior of a composite material under shock loading conditions.

5.6 Response of Curved Composite Plates

The response of doubly curved E-Glass/Vinyl Ester plates subjected to far field explosive loading is presented in the following section. The results consist of (1) a discussion of the fluid structure interaction at the water/composite interface, (2) comparisons between point-wise and full-field deformation histories obtained through CST experiments and corresponding finite element simulations, and (3) the post mortem material damage extents in the plates resulting from the testing and predicted by the simulations. Additional details involving this work are found in LeBlanc and Shukla (2011).

5.6.1 Composite Material

The material used in this phase of the study is an E-Glass/Vinyl ester composite with a 0°–90° biaxial layup. The glass fabric is a balanced construction of 0° and 90° fibers with the two layers being stitched together rather than woven. The areal weight of the dry fabric is 0.406 kg/m^2 (12 oz/yd^2). The panels utilized in the study consisted of three plys of the fabric, with each ply oriented in the same direction,

Table 5.2 E-Glass/vinyl ester biaxial laminate—mechanical properties

	MPa (lb/in.^{2})
Tensile modulus (0°)	15.8×10^3 (2.3×10^6)
Tensile modulus (90°)	15.8×10^3 (2.3×10^6)
Tensile strength (0°)	324 (47,000)
Tensile strength (90°)	324 (47,000)

Table 5.3 AOC hydropel R015 vinyl ester—mechanical properties

	MPa (lb/in.^{2})
Tensile modulus	3.44×10^3 (500,000)
Tensile strength	72.4 (10,500)

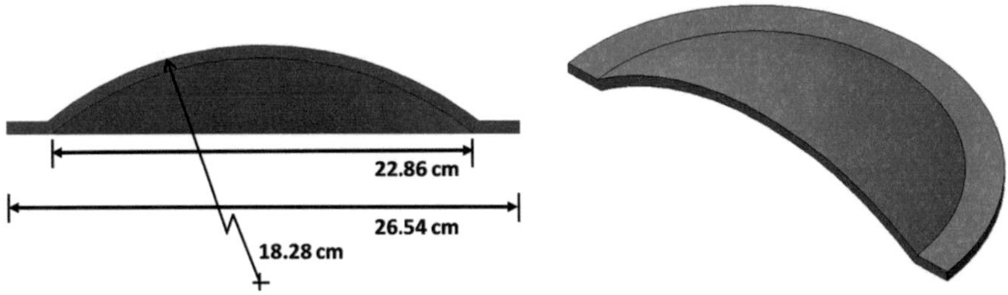

Fig. 5.17 Curved composite plate geometry

i.e., the 0° fibers in each ply were parallel. The panels were manufactured using the vacuum infusion process with a vinyl ester resin, AOC Hydropel R015-AAG-00. The finished part thickness was 1.37 mm (0.054 in.) and had a fiber content of 62 % by weight. The mechanical properties for the material are provided in Table 5.2. Furthermore, the mechanical properties of the vinyl ester resin are provided in Table 5.3. The pure vinyl ester properties are key, in that they determine the through-thickness mechanical properties (modulus, strength) of the laminate. The plates had an outer diameter of 26.54 cm (10.45 in.) with a curved mid-section of radius of curvature of 18.28 cm (7.2 in.). During experiments, the plates were oriented with the convex face into the fluid body so as to represent typical geometries of submerged vehicles (Fig. 5.17).

5.6.2 *Fluid Structure Interaction*

The finite element simulation of the shock tube testing allows for a visual full field representation of the interaction between the pressure wave and the composite plate. The pressure field in the fluid as it interacts with and loads the plate is shown in the left side of Fig. 5.18, with the associated plate response on the right side of the figure. The time as shown in these figures is the analysis time, with zero taken at the initiation of the pressure field 0.508 m (20 in.) from the test sample. The FSI for the curved plates is similar to that previously discussed for the flat plates. Specifically

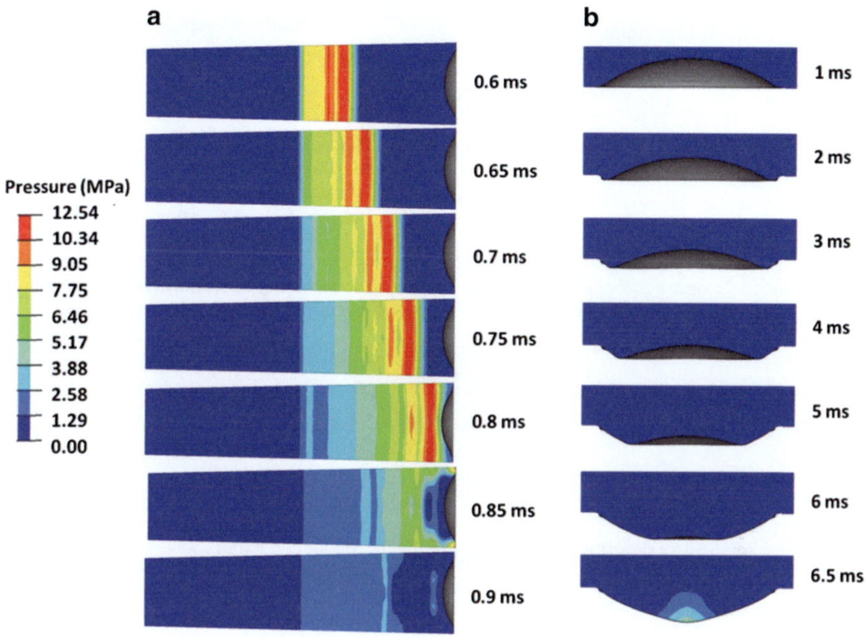

Fig. 5.18 (**a**) Fluid structure interaction, (**b**) Plate deformation progression

the loading on the plate itself is complex and not uniform, although the incident wave prior to arrival at the plate is planar in nature. Furthermore, the loading of the plate and the associated response can be separated into two distinct time regimes. These consist of the pressure wave interaction with the plate which occurs over a very short duration of time, followed by the deformation of the plate which occurs over a much longer time. Whereas the pressure wave interacts with the plate over a span of 0.2 ms, the plate does not start to deform until the wave is nearly fully reflected. The plate deformation takes approximately 5.5 ms to complete.

The plate deformation can be described as a full inversion. At 1 ms, a hinge forms at the outer edge of the plate at the clamped boundary. This hinge then continues to propagate towards the center of the plate as seen in the deformation progression images in Fig. 5.18. The deformation is completed when the plate has fully inverted itself at 6.5 ms. At this point, it is seen that there is a high pressure region that develops at the apex of the inverted shape. During the inversion process of the plate the fluid at the plate boundary is also moving along with the plate surface. Once the inversion of the plate is complete, the velocity is rapidly arrested, resulting in the development of this high pressure region. It is noted that the magnitude of this pressure is small compared to the initial shock wave pressure.

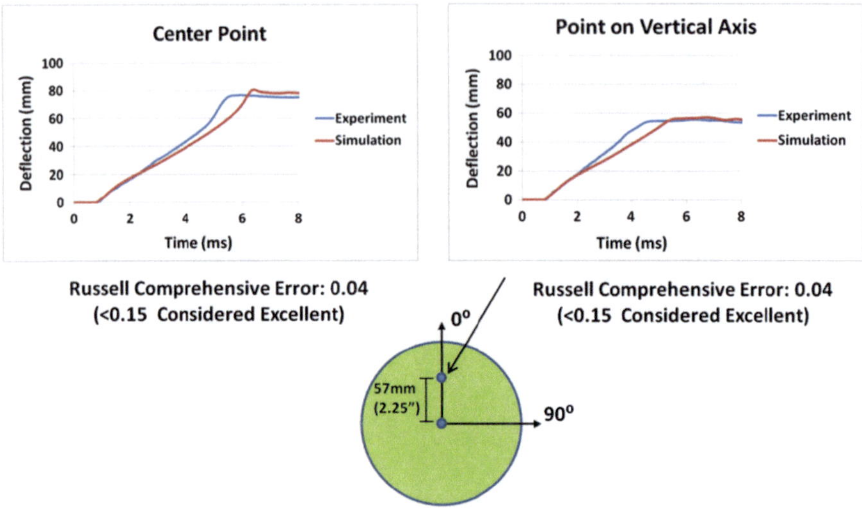

Fig. 5.19 Time history deformation comparison of experiment and simulation

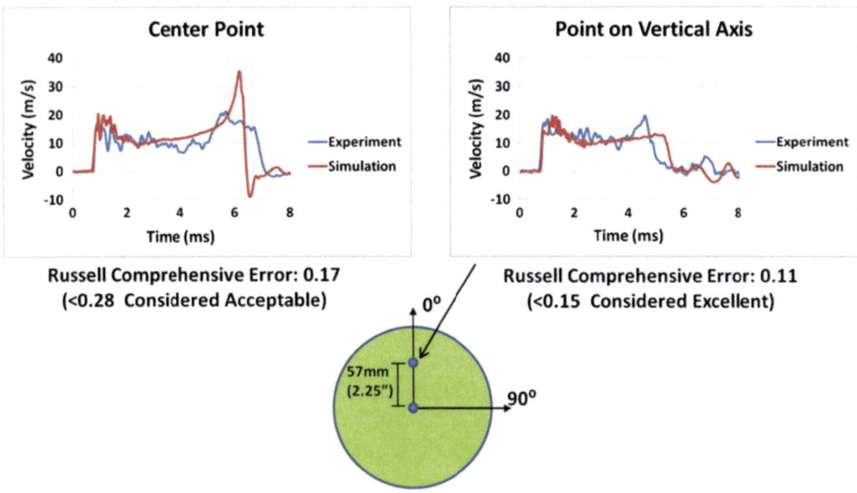

Fig. 5.20 Time history velocity comparison of experiment and simulation

5.6.3 Plate Deformation Correlation

The two variables that are used for correlation of the simulations to the experiments are the out-of-plane displacement and velocity. Time histories are extracted from the DIC data for two points on the back face of the plates: (1) the center of the plate, and (2) a point located half way between the center of the plate and the clamped boundary along the vertical axis. The displacement comparisons are shown in Fig. 5.19 and the velocity comparisons are provided in Fig. 5.20. A summary of the

Table 5.4 Russell error summary

	Magnitude Error	Phase Error	Comprehensive Error
Center Point Deflection	0.03	0.03	0.04
Center Point Velocity	0.1	0.16	0.17
Half Way Along Vertical Axis Deflection	0.03	0.03	0.04
Half Way Along Vertical Axis Velocity	0.03	0.12	0.11

RC<0.15 Excellent
.15<RC<0.28 Acceptable
RC>0.28 Poor

Russell error for each of these comparisons is provided in Table 5.4. From these graphical comparisons and error summary, it is seen that there is a high level of correlation between the experimental results and the computational simulations. It is noted here that the times of the simulation and experiments are arbitrary but are displayed using the simulation time. The two events are matched temporally by adjusting the experiment time until the first motions of the center point align. This time offset is then held constant for all other comparisons, i.e., a different time offset was not used to align the velocity comparisons.

The displacement comparison shows that the experiment and simulation results agree nearly exactly until 2.5 ms at which point the displacement in the experiment occurs slightly faster than the simulation. This is the same for both the center point and the point halfway between the center and the boundary. It is also seen that the displacement takes place in a linear fashion. The peak deflections in both cases are just under 80 mm for the center point and 55 mm for the point along the vertical axis. In the velocity comparison, it is seen that there is an initial out-of-plane velocity of just less than 20 m/s which then settles to about 10 m/s for the remainder of the event. This near constant velocity supports the linearity of the displacement profile. It is seen that the simulation is able to capture the peak velocities seen in the experiments. Overall, it is shown that the Russell error values for the displacement comparisons show excellent correlation as well as the velocity for the point halfway between the center and the boundary. The velocity comparison at the center point falls just outside the excellent agreement criteria but well within the acceptable level (<0.28). Table 5.4 summarizes the Russell error values for magnitude, phase, and comprehensive for each data comparison.

In addition to the point wise time histories, full field comparisons are made between the experiment and simulations. A comparison of the full field, out of plane displacement evolution is shown in Fig. 5.21. From this figure, it is seen that the experiment and simulation show good correlation in terms of the displacement evolution. The experiment shows some un-symmetric behavior in that the displacements are slightly higher along a line at 135°. However, the values of these displacements are only a few millimeters more than the rest of the plate. As expected, the simulation shows symmetric behavior. It is possible that this would not be the case if initial thickness variations were built into the model. Overall, the

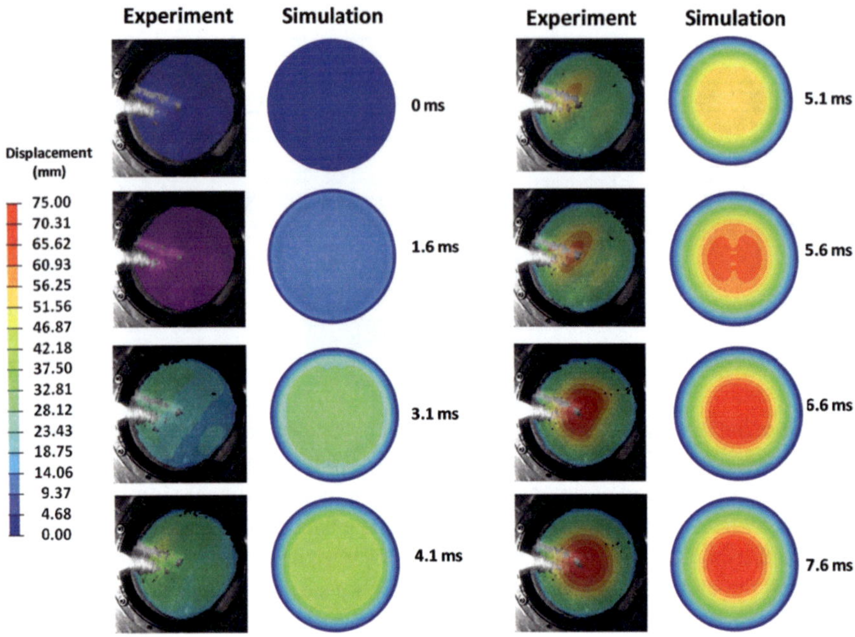

Fig. 5.21 Full field deformation comparison of experiment and simulation

displacements over the surface of the plate correlate well between the experiment and the simulation.

5.6.4 Material Damage Correlation

The relationship between the material damage observed during the experiments and the damage level predicted by the simulations is of interest as ultimately the ability to predict the evolution of material damage is the precursor to determining the residual strength of the material after a shock event. In the experiment the main damage mechanism that was observed is interlaminar delamination. There was minimal fiber rupture or matrix cracking as compared to the flat plate response. The final damage state from the shock test is shown in Fig. 5.22. Here it is seen that the extent of the delamination ranges from approximately 12.7 mm (0.5 in.) in the lower left region of the plate to 50.8 mm (2 in.) at the 3 o'clock position. In the corresponding computational model, Fig. 5.22, the delamination area is highlighted by the black area at the edge of the plate and propagates inwards from the edge 10 mm (0.4 in.). In the simulation results, there are also four larger delamination zones at the 45° directions. These areas indicate the weaker material directions (fibers run in the 0° and 90° directions) and extend 38.1 mm (1.5 in.) from the edges. Although the amount of delamination is somewhat larger in the experiment than is observed in

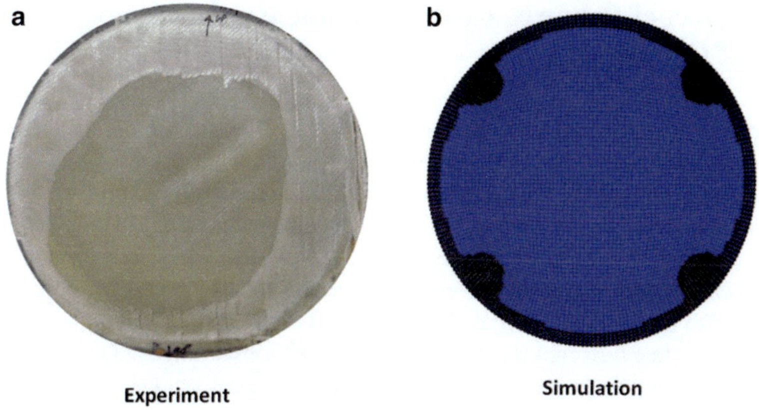

Fig. 5.22 (**a**) Material damage during test, (**b**) Material damage from simulation

the computational model, it is encouraging that the model is able to predict the onset of the delamination itself and propagate it to a comparable distance.

In the current model, the choice of a delamination criterion was taken to be 36 MPa (5,250 lb/in.^{2}) for both tensile and shear stresses. This value is one-half of the tensile strength of the pure resin, Table 5.3. The rationale for the selection of this value is the same as previously discussed in the flat plate results. The following observations are provided to aid in the development of the best delamination modeling practices but are not meant to be definitive. More work is needed in order to discover the most efficient way to model the delamination parameters but is outside of the scope of the current study.

5.6.5 *Summary: Curved Plates*

The response of doubly curved E-Glass/Vinyl Ester composite plates subjected to underwater shock loading has been studied. Correlations are presented between the experiments and computational simulations in terms of transient plate deformations (both point-wise and full field) and post mortem damage levels. Additionally, the fluid structure interaction at the plate water interface is described through the use of the finite element results. The displacement and velocity data for the center point, and a point halfway between the center and boundary, are correlated to the computational models by utilizing the Russell error. The Russell error value for the deflection correlation at these two points is excellent. The velocity correlation at the point halfway between the center and the boundary is also excellent, while the center point velocity has acceptable correlation. The full field displacement evolution is also shown to agree between the experiment and the simulations. The computational model was also able to predict the onset of delamination damage and predict its growth to a reasonable degree.

5.7 Effect of Polyurea Coatings on the Response of Composite Plates

The effect of polyurea surface coatings on the response of composite panels to underwater shock loading is presented in the following discussion. The thickness and location of the polyurea coating has been varied to determine how these parameters affect the transient response. The composite material is a 0°/90° biaxial layup and the coatings are applied to either the loaded or non-loaded faces. The determination of the effectiveness of each of the polyurea coated panels, as compared to the baseline (uncoated) panels, consists of comparison of displacement and velocity time histories, as well as full-field DIC observations. The results show that for a given polyurea thickness, better performance is obtained when the back face of the panel is coated. Similarly the performance is improved as the coating thickness is increased; however, this comes at a cost in terms of increased areal weight. The results show that there is likely an optimal coating thickness, that when located on the back face, provides a balanced tradeoff between panel performance and weight increase. Additional details involving this work are found in LeBlanc and Shukla (2013).

5.7.1 Panel Configuration

The curved composite panels presented in the previous sections serve as the baseline substrate to which the polyurea coatings are applied. The polyurea material used for coating the panels is Dragonshield-BC, which is a two-part material that is spray cast onto structures. The manufacturer provides a tensile strength of over 37.9 MPa (5,500 lb/in.^{2}) and an elongation of 344 % (Dragonshield-BC 2011). Two coating thicknesses are utilized: 2 mm (0.08 in.) and 4 mm (0.16 in.). Additionally, each coating thickness is applied to either the front (loaded) or back (non-loaded) side of the panels. These constructions are chosen to represent what would typically be found in a real-world application in which structures are retrofitted (spray coated) with this material as opposed to incorporating the material into the original design.

A summary of the panel thicknesses and areal weights is provided in Table 5.5, and a schematic of the laminate designs are shown in Fig. 5.23. The mechanical properties for the 0°/90° baseline composite are provided in Table 5.6.

Table 5.5 Thickness and areal weight of laminates

	Thickness, mm (in.)	Areal weight, kg/m^2 (oz/yd^2)
0°/90° Baseline laminate	2.54 (0.10)	4.25 (126)
0°/90° Baseline with 2 mm polyurea coating	4.57 (0.18)	6.54 (193)
0°/90° Baseline with 4 mm polyurea coating	6.6 (0.26)	8.40 (248)

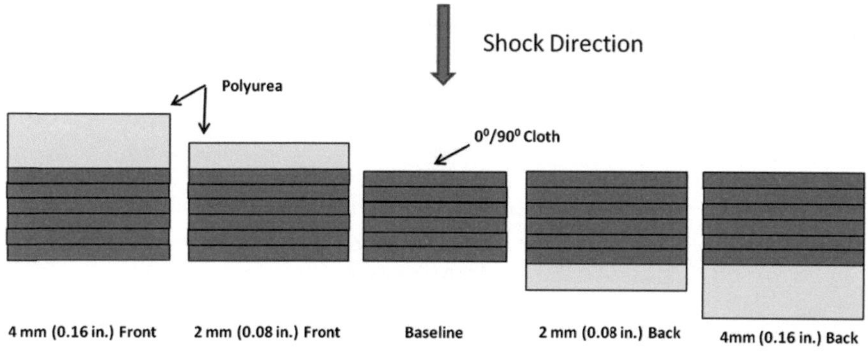

Fig. 5.23 Composite plate construction—schematic (not to scale)

Table 5.6 0°/90° Baseline laminate—mechanical properties

	MPa (lb/in.^{2})
Tensile modulus (0°)	15.8×10^3 (2.3×10^6)
Tensile modulus (90°)	15.8×10^3 (2.3×10^6)
Tensile strength (0°)	324 (47,000)
Tensile strength (90°)	324 (47,000)

5.7.2 *Effect of Coatings of Transient Panel Response*

The center point displacement–time history as obtained from the DIC data for the five panel configurations studied is shown in Fig. 5.24. In all time history plots, time zero is taken to be the time of arrival of the initial shock front at the pressure transducer in front of the panel. Using the baseline panel as a reference for comparison, it is seen that the panel with 4 mm (0.160 in.) of polyurea on the back face has the best overall performance, while the panel with 2 mm (0.08 in.) on the front face has the worst performance in terms of peak deflections. The panels coated with 4 mm on the front and 2 mm on the back have approximately the same peak deflection. It is noted that although these two panels have approximately the same peak deflections, the panel with the coating on the front has a higher overall thickness and areal weight. After the first pressure peak, the baseline laminate sustains a center point displacement of 33.5 mm (1.31 in.) while the deflection for the 4 mm back face polyurea sample is 11 mm (0.43 in.), a decrease of 67 %. In the case of the 2 mm front face panel, the center point displacement is 40 mm (1.57 in.), an increase of 20 % over the baseline. It can be seen that when the front face is coated with 4 mm of polyurea, the peak displacement is less than the baseline panel, which is in

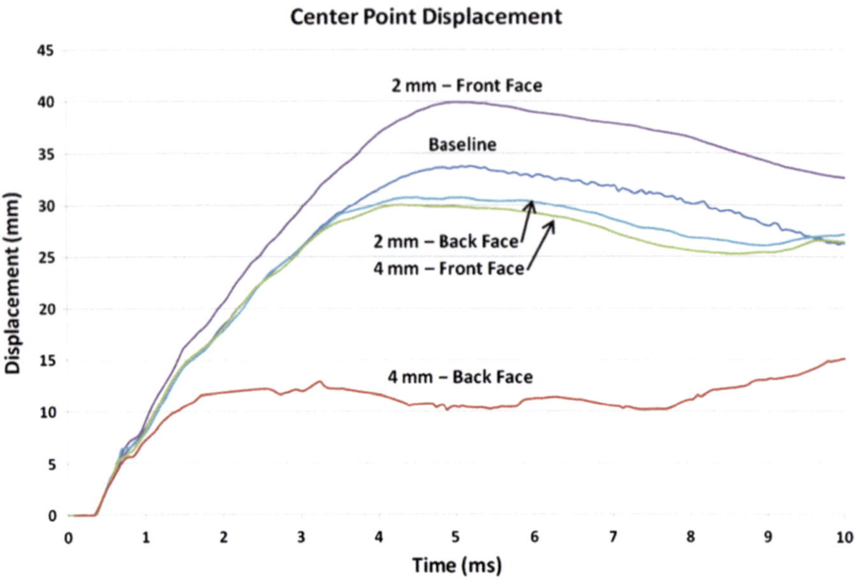

Fig. 5.24 Time history deformation comparison

contrast to the behavior displayed when just 2 mm of polyurea is applied to the front face. This is likely due to a tradeoff between the performance degrading effects seen in the 2 mm front face panel, and an increase in thickness and mass. Thin polyurea coatings on the front face appear to initially degrade the structural performance, and then as the coatings reach a critical point the effects of additional mass and thickness become the dominant contributors to the response and decrease the deflections.

From Fig. 5.24, several trends can be highlighted. The first is that when comparing the effect of polyurea location on the panel (front/back) the performance is better when the back face is coated. This is true for both the 2 mm (0.08 in.) and 4 mm (0.16 in.) coating thicknesses. The second is that when comparing the effect of thickness for a given location, the performance is enhanced as the coating thickness is increased.

In addition to the point-wise histories presented, the full-field displacement evolution for each panel configuration is shown in Fig. 5.25. In this figure, it is observed that for all of the panels the deflection profiles are not radially symmetric as would be expected for this type of panel geometry. Rather, the deflections tend to evolve from a point located along one of the 45° plate axes and then grow towards the center. For the current 0°/90° panel construction the ±45° axes are the weakest material direction and this is the likely cause of this type of deformation behavior. Additionally, because each panel is made individually there are small thickness variations as well as the possible presence of voids and resin-rich or -dry locations within a given panel. The presence of these manufacturing variables along with the

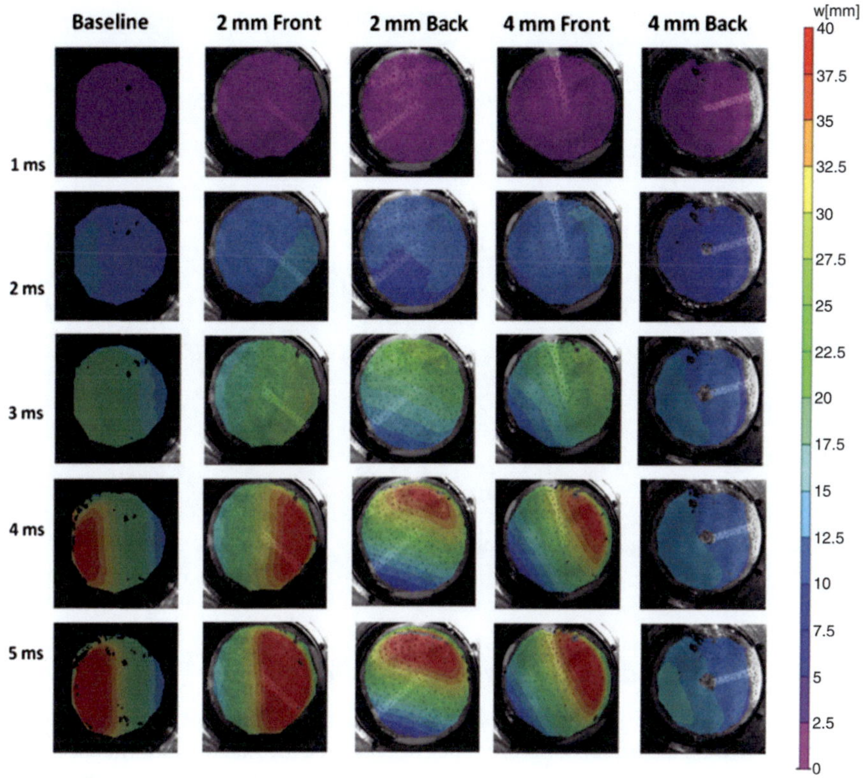

Fig. 5.25 Full-field deformation evolution

weaker material directions likely causes the deformation to initiate from one of the 45° directions and then propagate from there. On the full-field level it is seen that the panel with 4 mm (0.16 in.) of polyurea on the back shows the least overall displacements, similar to the center point displacements discussed previously.

The velocity time history comparison for the center point of each of the panel configurations is shown in Fig. 5.26. The plot focuses on the velocity profile resulting from the initial shock pressure. From the figure, it is seen that the magnitude of the kick off velocity for each of the laminates is nearly the same, approximately 16–17 m/s (52.5–55.7 ft/s). There is a difference, however, in the time that it takes for the velocity to decay back to zero for each of the panels. The velocity of the baseline laminate fully decays over 4.5 ms. The panel with 2 mm (0.08 in.) on the front face takes a slightly longer time to fully decay, ~4.7 ms to return to rest, resulting in the larger displacement previously discussed. Conversely, the back face coated panels (both 2 and 4 mm) and the 4 mm (0.16 in.) front face coated panel come to rest quicker than the baseline (~ 4 ms for the 4 mm front face and the 2 mm back face configurations, and ~2.25 ms for the 4 mm on the back face).

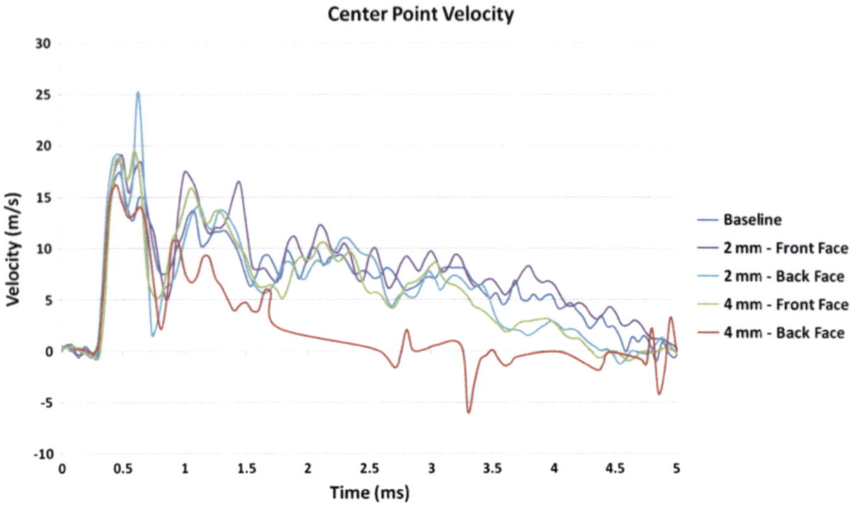

Fig. 5.26 Velocity time history comparison

5.7.3 Weight and Thickness Penalty Effects

When evaluating the effects of these types of coatings, the penalty in terms of thickness and areal weight increases should also be considered. Observations based on the raw transient displacement data do not highlight the fact that both the panel thickness and areal weights have been increased. In order to investigate this, the displacement data is adjusted based on the thickness and areal weight ratios between the coated panels and the baseline by multiplying the displacement data by the corresponding panel ratio. The thickness ratio (TR) and areal weight ratio (AWR) are defined as the coated panel value divided by the baseline value. The thickness ratios for the panels with 2 mm (0.08 in.) and 4 mm (0.16 in.) of polyurea coating are 1.8 and 2.6 respectively. Similarly the areal weight ratios for the 2 mm and 4 mm panels are 1.5 and 2 respectively. The values are not the same for thickness and areal weight due to the panels being a combination of composite and polyurea rather than a homogeneous material.

The displacement data, adjusted to account for thickness and areal mass increases, is shown in Figs. 5.27 and 5.28 respectively. In these plots, it is seen that when the displacements are adjusted to account for the cost due to increased thickness and areal weight, the baseline panel outperforms both of the panels with polyurea coatings on the front face and also the panel with 2 mm (0.08 in.) polyurea coating on the back. The 4 mm (0.16 in.) coating on the back has approximately the same thickness-adjusted performance and a better areal weight-adjusted performance than the baseline. Although the 4 mm front face coated panel had a better performance than the baseline when using the raw displacement data, the

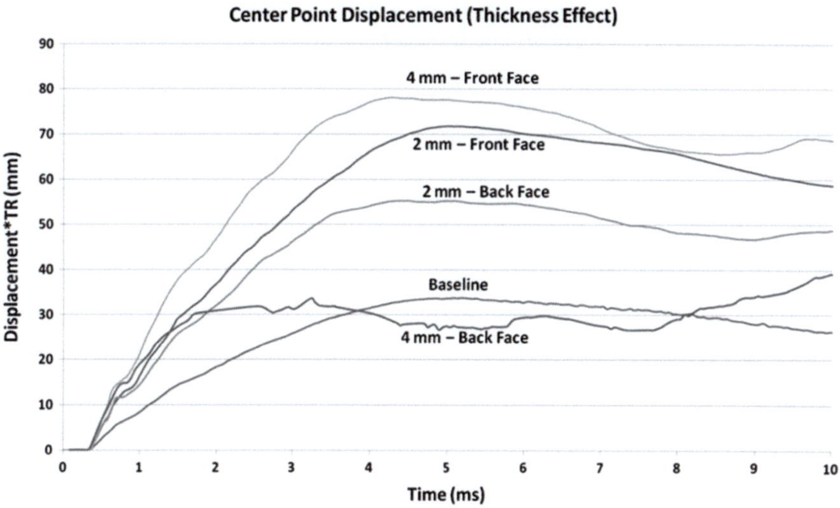

Fig. 5.27 Time history deformation comparison (thickness effect)

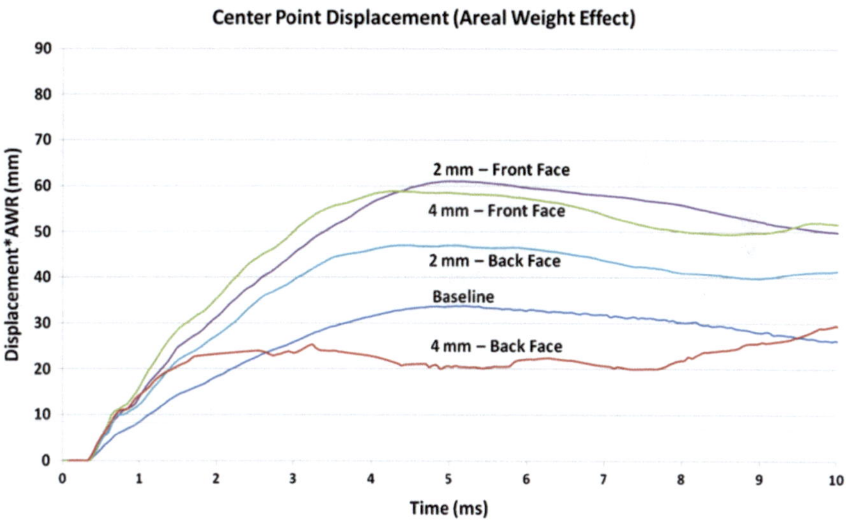

Fig. 5.28 Time history deformation comparison (areal weight effect)

performance is worse when the cost of additional thickness/areal weight is factored in. A similar effect is seen for the case of the 2 mm back face panel which, although it outperformed the baseline panel from a pure displacement perspective, it is worse than the baseline when thickness and weight considerations are made. From these figures it can also be seen that the 2 and 4 mm coatings on the front face have approximately the same relative performance when thickness and areal weight penalties are taken into account. Finally when comparing the 4 mm back face panel from an areal weight penalty perspective it still outperforms the baseline panel. This indicates that there is likely an optimal polyurea thickness, when located on the back face, that outperforms a baseline structure on both a pure displacement level as well as when increased thickness and areal weight penalties are factored in. These observations indicate that the shock performance of a baseline structure can be improved if an optimal thickness of polyurea is applied to the back face of a curved structure.

5.7.4 Summary and Conclusions: Polyurea

The effects of polyurea coatings' thickness and location on the dynamic response of E-Glass/Vinyl ester composite plates have been studied. The performance of the panel configurations is evaluated using point wise and full-field time histories of the deflection and velocity of the back face of the panels. The results are compared using baseline deflection and velocity data, as well as data which are adjusted to account for relative thickness and areal weight increases. The displacement data shows that depending on the combination of coating parameters (thickness/location), the performance of the baseline (uncoated) panel can be either improved or conversely degraded. A thin layer of polyurea located on the loaded side of the panel serves to degrade the performance of the baseline panel by increasing the transient displacement and the time to arrest the overall motion. A thicker coating on the front face was shown to decrease the overall deflection; however, with the thicker coating it is likely there is an increased contribution of thickness and mass. A similar effect is seen for the panels that were coated on the back face, in that although both coatings show improved performance with respect to the baseline, there is a much larger reduction in deflection for the 4 mm coating compared to the 2 mm coating. This is likely due to a superposition of a beneficial effect of coating the back face and the thickness effects. For a given coating thickness, it has been shown that there is a pronounced performance increase when the coating is located on the back face versus the front face. When the back face is coated the displacements are smaller as compared to a panel with an equivalent thickness coating located on the front. When the data is adjusted to account for the relative penalties due to thickness and areal weight increases associated with the coatings, it is observed that the 4 mm coating on the back face provides approximately the same response as the baseline regarding thickness effects, and an improved performance in terms of areal weight effects. Overall, it has been shown that for a given coating thickness it is preferable to coat the back face of a structure to provide optimal performance increases. Also, results indicate that there is likely an optimal coating thickness that provides a balanced tradeoff between panel performance and weight increase.

5.8 Summary

This research was performed to study the dynamic response and damage evolution of composite materials when subjected to underwater explosive loading conditions. The work consisted of both experimental work and computational simulations to aid in the understanding of the behavior of these materials. The objective of the study was to develop a better understanding of the shock response of composite materials, leading to more efficiently designed structures. The relevant findings resulting from the present study are presented below.

1. An experimental methodology has been developed which subjects composite plates of both flat and curved geometries to shock loading resulting from an underwater explosion. A conical shock tube facility is utilized which replicates the free field expansion of pressure waves resulting from the explosion. The plates are air backed and held with fully clamped boundary conditions. The dynamic response of the back face of the specimens is captured through the use of strain gages and digital image correlation. The use of digital image correlation allowed for the capture of the full field deformation and velocity histories.
2. A modeling methodology has been developed that is able to accurately simulate the response of the composite material as observed during the experiments. The modeling is performed utilizing the commercially available LS-DYNA explicit finite element code. The finite element models consist both of the water within the conical shock tube and of the composite plate. The models are able to capture the fluid structure interaction between the water and the plates as well as the transient response of the plates. Additionally the models are able to simulate the onset of material damage including both in-plane and through-thickness mechanisms.
3. The results from the shock experiments and the computational simulations have been correlated using both the dynamic response histories as well as the final damage states. The time histories obtained from the strain gages and the digital image correlation results have been correlated to the simulation results using the Russell error. This is a method that evaluates the differences in two transient data sets by quantifying the variation in magnitude and phase. The correlation between the strain, deflection, and velocities for the respective experiments were found to fall within the acceptable criteria with some comparisons being excellent. Additionally, comparisons between the final material damage state in the experiment and the simulation were agreeable both in terms of the type and extent of the individual mechanisms.
4. The effects of polyurea surface coatings were studied to determine their effects on the shock response of curved panels. The effects of coating thickness and coating location were investigated. It was found that coating the back face of the panels improved the transient response under shock loading in terms of peak deflections and velocity decay times. Furthermore, as the coating thicknesses were increased the performance was further increased but at the cost of added weight and panel thickness. Conversely, the shock response was adversely affected when the coatings were applied to the panel surface oriented towards the incoming shock fronts.

The work discussed in this chapter has provided a basis for experimental and computational techniques that can be applied to the study of the dynamic response of composite materials subjected to UNDEX loading conditions. However, there remains a significant body of work to be completed in this area before the dynamic response of these materials matures to an equivalent level of understanding as that for metallic materials. This work includes further experimental and computational studies as well as validation efforts that correlate the two. These efforts will ultimately lead to validated modeling practices that can be applied during the design phase of composite structures. Future work in this field can potentially include:

1. Investigations which include near field detonations of charges against composite panels. At smaller standoff distances the pressure waves will have pronounced spherical fronts as they interact with the plates as opposed to the previous studies in which near planar conditions were achieved, with plate loading occurring over the entire surface area. Smaller standoffs will examine the localized effects of UNDEX loading as the loading will be sufficiently removed from the boundary as to ensure that edge effects are not a dominant contributor to the structural response.
2. Studies involving complex geometries, including, oblong spheroids, cylinders, and plates with abrupt angle changes. The goal should be to incorporate real-world design shapes into the test article geometry. The current finite element modeling methodology should also be expanded to simulate these experiments to ensure it is able to accurately simulate the geometrical effects.
3. Investigate the effects of UNDEX loading on additional materials and laminate constructions. The performance of S-Glass, carbon, and Kevlar should be examined as well as the possibility of hybrid materials such as glass/carbon constructions. The performance of these materials needs to be understood as they inherently have different characteristics. For example carbon fibers have higher tensile strengths and moduli but fail much more catastrophically than glass fibers which have more elongation. Additionally the response of other laminate constructions should be examined. The current research utilized laminates that were constructed using individual layers of continuous fibers. Studies involving newer three dimensional (3D) fabrics, which include through-thickness fibers interwoven through the cloth, should be conducted. These through-thickness fibers may improve the performance of the laminates in terms of reducing the delamination.
4. The effects of large hydrostatic pressures in combination with UNDEX loading should be explored. This would provide insight as to how a composite structure would implode as compared to a corresponding metallic structure. For example the implosion process for metallic cylinders is well understood including collapse pressure and final shape. However, there is little understanding of how a composite cylinder of equivalent strength would collapse.

Acknowledgments The financial support of the Naval Undersea Warfare Center (Division Newport) In-house Laboratory Independent Research program (ILIR) directed by Dr. Anthony Ruffa is greatly acknowledged. Arun Shukla would like to acknowledge the support of Office of Naval Research under ONR Grant No. N00014-10-1-0662 (Dr. Y.D.S. Rajapakse) to the University of Rhode Island.

References

Amini MR, Isaacs JB, Nemat-Nasser S (2010a) Experimental investigation of response of monolithic and bilayer plates to impulsive loads. Int J Impact Eng 37:82–89

Amini MR, Simon J, Nemat-Nasser S (2010b) Numerical modeling of effect of polyurea on response of steel plates to impulsive loads in direct pressure-pulse experiments. Mech Mater 42:615–627

Amini MR, Isaacs JB, Nemat-Nasser S (2010c) Investigation of effect of polyurea on response of steel plates to impulsive loads in direct pressure-pulse experiments. Mech Mater 42: 628–639

Arora H, Hooper P, Dear JP (2010) Impact and blast resistance of glass fibre reinforced sandwich composite materials. Proceeding of the IMPLAST. Providence, RI, Oct 2010

Batra RC, Hassan NM (2007) Response of fiber reinforced composites to underwater explosive loads. Compos B Eng 38:448–468

Bogdanovich A, Singletary J, Mohamed M (2001) A new generation of composites reinforced with 3-D woven fabric performs. Proceeding of 22nd International SAMPE Europe Conference, Paris, France, March 27–29, 2001

Chan S, Fawaz Z, Behdinan K, Amid R (2007) Ballistic limit prediction using a numerical model with progressive damage capability. Compos Struct 77:466–474

Coombs A, Thornhill CK (1967) An underwater explosive shock gun. J Fluid Mech 29:373–383

Donadon MV, Iannucci L, Falzon BG, Hodgkinson JM, de Almeida SFM (2008) A progressive failure model for composite laminates subjected to low velocity impact damage. Comput Struct 86:1232–1252

Dragonshield-BC: Ultra high strength polyurea elastomer, Technical Data Sheet, Revised 28 July 2011, Specialty Products, Inc.

Dyka CT, Badaliance R (1998) Damage in marine composites caused by shock loading. Compos Sci Tech 58:1433–1442

Filler WS (1964) Propagation of shock waves in a hydrodynamic conical shock tube. Phys Fluid 7:664–667

Franz T, Nurick G, Perry M (2002) Experimental investigation into the response of chopped-strand mat glassfibre laminates to blast loading. Int J Impact Load 27:639–667

Gama B, Xiao J, Haque M, Yen C, Gillespie J (2004) Experimental and numerical investigations on damage and delamination in thick plain weave S-2 glass composites under quasi-static punch shear loading, A013124, Center for Composite Materials, University of Delaware

Gardner N, Wang E, Kumar P, Shukla A (2012) Blast mitigation in a sandwich composite using graded core and polyurea interlayer. Exp Mech 52:119–133

Hodge N (2004) Military experimenting with ‘spray on’ armor for humvees. Def Today 25:121

Hosseinzadeh R, Shokrieh MM, Lessard L (2006) Damage behavior of fiber reinforced composite plates subjected to drop weight impacts. Compos Sci Tech 66:61–68

Jackson M, Shukla A (2011) Performance of sandwich composites subjected to sequential impact and air blast loading. Compos B Eng 42:155–166

LeBlanc J, Shukla A (2010) Dynamic response and damage evolution in composite materials subjected to underwater explosive loading: an experimental and computational study. Compos Struct 92:2421–2430

LeBlanc J, Shukla A (2011) Dynamic response of curved composite panels to underwater explosive loading: experimental and computational comparisons. Compos Struct 93:3072–3081

LeBlanc J, Shukla A (2013) Effect of polyurea coatings on the response of curved E-glass/vinyl ester composite panels to underwater explosive loading. Compos B Eng 44:565–574. http://dx.doi.org/10.1016/j.compositesb.2012.02.038

LeBlanc J, Shukla A, Rousseau C, Bogdanovich A (2007) Shock loading of three-dimensional woven composite materials. Compos Struct 79:344–355

Matzenmiller A, Lubliner J, Taylor RL (1995) A constitutive model for anisotropic damage in fiber-composites. Mech Mater 20:125–152

McGregor CJ, Vaziri R, Poursartip A, Xiao X (2007) Simulation of progressive damage development in braided composite tubes under axial compression. Compos A Eng 38:2247–2259
Mouritz AP (1995) The effect of underwater explosion shock loading on the fatigue behaviour of GRP laminates. Composites 26:3–9
Mouritz AP (2001) Ballistic impact and explosive blast resistance of stitched composites. Compos B Eng 32:431–439
Mouritz AP, Gellert E, Burchill P, Challis K (2001) Review of advanced composite structures for naval ships and submarines. Compos Struct 53:21–41
O'Daniel JL, Koudela KL, Krauthammer T (2005) Numerical simulation and validation of distributed impact events. Int J Impact Eng 31:1013–1038
Poche L, Zalesak J (1992) Development of a water-filled conical shock tube for shock testing of small sonar transducers by simulation of the test conditions for the heavyweight MIL-S-901D (Navy). NRL Memorandum Report 7109, 10 October 1992
Russell DM (1997) Error measures for comparing transient data, part I: development of a comprehensive error measure, part II: error measures case study. Proceedings of the 68th Shock and Vibration Symposium, Hunt Valley, MD, 3–6 November 1997
Russell DM (1998) DDG53 Shock trial simulation acceptance criteria. Proceedings of the 69th Shock and Vibration Symposium, St. Paul, MN, 12–19 October 1998
Schubel PM, Luo J, Daniel I (2007) Impact and post impact behavior of composite sandwich panels. Compos A Eng 38:1051–1057
Tekalur AS, Shivakumar K, Shukla A (2008a) Mechanical behavior and damage evolution in E-glass vinyl ester and carbon composites subjected to static and blast loads. Compos B Eng 39:57–65
Tekalur A, Shukla A, Shivakumar K (2008b) Blast resistance of polyurea based layered composite materials. Compos Struct 84:271–281
Tekalur AS, Bogdanovich AE, Shukla A (2009) Shock loading response of sandwich panels with 3-D woven E-glass composite skins and stitched foam core. Compos Sci Tech 69:736–753
Tiwari V, Sutton MA, McNeill SR (2007) Assessment of high speed imaging systems for 2D and 3D deformation measurements: methodology development and validation. Exp Mech 47:561–579
Tiwari V, Sutton MA, McNeill SR, Xu S, Deng X, Fourney WL, Bretall D (2009) Application of 3D image correlation for full-field transient plate deformation measurements during blast loading. Int J Impact Eng 36:862–874
Wang L (2004) Unloading waves and unloading failures in structures under impact loading. Int J Impact Eng 30:889–900
Williams KV, Vaziri R (2001) Application of a damage mechanics model for predicting the impact response of composite materials. Comput Struct 79(2001):997–1011
Xiao J, Gama B, Gillespie J (2007) Progressive damage and delamination in plain weave S-2 glass/SC-15 composites under quasi-static punch-shear loading. Compos Struct 78:182–196
Yuan F, Tsai L, Prakash V, Rajendran AM, Dandeka D (2007) Spall strength of glass fiber reinforced polymer composites. Int J Solids and Struct 44:7731–7747
Zako M, Uetsuji Y, Kurashiki T (2003) Finite element analysis of damaged woven fabric composite materials. Compos Sci Tech 63:507–516
Zaretsky E, deBotton G, Perl M (2004) The response of a glass fibers reinforced epoxy composite to an impact loading. Int J Solid Struct 41:569–584

Chapter 6
Underwater Implosion Mechanics: Experimental and Computational Overview

James LeBlanc, J. Ambrico, and S. Turner

Abstract An implodable volume can be defined as a structural body which is acted upon by external pressure and internally contains gas at a lower pressure than the surrounding fluid. An underwater implosion occurs when the body suffers a sudden loss of structural stability and hydrostatic pressure drives the body to collapse inwardly upon itself. The result of the collapse is a rapid decrease in local pressure as the water expands to fill the void, and then a shock wave as the in-rushing water suddenly stops and is compressed. The physics of an implosion event is shown to be similar to the collapse of cavitation and underwater explosion bubbles. The pressure history resulting from an implosion event consists of several primary characteristics, namely an initial pressure drop in the surrounding fluid during the initial collapse, corresponding to the inward rush of the surrounding fluid, followed by a subsequent positive pressure spike and decay as the body collapses upon itself and the water motion is arrested. It is observed that the magnitude of both the pressure drop and subsequent peak pressures are dependent upon the hydrostatic collapse pressure. Additionally, the ductility of implodable volume material is shown to have an effect on the amount of energy released during the collapse with brittle implodables releasing larger amounts of energy into the surrounding fluid than ductile volumes. Finally, implodable volumes which are coated with energy absorbing materials display decreased energy in the implosion pulses.

J. LeBlanc (✉) • J. Ambrico • S. Turner
Naval Undersea Warfare Center (Division Newport), Building 1346/407F, 1176 Howell Street, Newport, RI 02841, USA
e-mail: james.m.leblanc@navy.mil

A. Shukla et al. (eds.), *Blast Mitigation: Experimental and Numerical Studies*,
DOI 10.1007/978-1-4614-7267-4_6, © Springer Science+Business Media New York 2014

6.1 Introduction

An *implodable volume* is defined as any structural shell or body that is acted upon by external pressure and contains internal gas at a lower pressure (or vacuum). When the volume is submerged within a fluid field, the external loading is caused by the depth pressure of the surrounding fluid. An underwater implosion is identified as the collapse of the body due to a sudden loss of structural stability. The loss of stability can be the result of many things including a bifurcation buckling of the outer shell at a critical depth, denting or deformation of the shell, or an initiation due to a nearby underwater explosion. The resulting collapse and implosion of the structure is by nature violent and results in the rapid release of energy in the form of shock pressure waves, high velocity fluid motion, and sound. The amount of energy available for release is generally taken to be the product of the internal volume of the body and the pressure differential across the structural boundary.

Structural bodies which have the potential to implode under the action of external pressures are used in a large number of applications/industries. These include, but are not limited to, naval vehicles, oil drilling, deep sea exploration craft, and high pressure research facilities. An implodable volume is of particular concern when located external to, but in close proximity to, critical equipment or submersible hull forms (Cor and Miller 2009) as the pressure pulse created during the collapse has the potential to cause damage to the surrounding structures. The risks associated with the implosion of underwater volumes/vehicles has resulted in conservative design approaches for these items, including by the United States Navy in *The Navy Unmanned Undersea Vehicle (UUV) Master Plan* (2004) which requires:

> "Ship deployed UUV systems must meet very stringent requirements to be authorized for submarine installation, including, but not limited to, shock requirements, battery (or other energy source) certification, and implodable volume requirements."

A significant example of the effect that the implosion of a volume can have on surrounding structures is the accident which occurred during the Super-Kamiokande experiment (Fukuda et al. 2003; Main report on Super-Kamiokande accident; Report of Super-Kamiokande accident study). The facility, located in Japan, consisted of a water-filled 39.3 m diameter, 41.4 m tall cylindrical tank which containing more than 11,000 photomultiplier tubes with an internal vacuum. In 2001, more than 5,000 of the photomultiplier tubes imploded due to a chain reaction set off by the implosion of a single tube. The resulting shock wave from the initial collapse damaged the surrounding tubes which subsequently imploded themselves. This event highlighted the adverse effects of an implosion occurring near surrounding structures, and the care that must be taken when locating these volumes near critical equipment.

To understand the mechanics of implosion events, some basic phenomena can be examined first. Reviewing these topics builds the foundation of physics necessary to clearly describe implosion of underwater structures. The most fundamental such phenomenon is the collapse of bubbles in water. The same physics that produce the pressure waves from collapsing bubbles also creates pressure pulse from implosion of structures. Collapsing bubbles have been studied extensively in the context of cavitation bubbles or underwater explosion gas bubble dynamics. The work in these areas is briefly reviewed here to describe the physical processes that create the pressure waves.

6.2 Bubble Collapse Dynamics

The research addressing bubble collapse events can be traced back to the early 1900s. Lord Rayleigh (1917) developed an early analytical formulation for the prediction of the pressure field resulting from the collapse of a spherical cavity in a field of incompressible fluid, in which the cavities internally contained either zero pressure or constant pressure gasses. This work highlighted the fact that pressures in the fluid surrounding a collapsed bubble could reach very high levels. The constant internal bubble pressure assumption limits the accuracy of the formulation to those collapses in which the wall speed is only a small fraction of the sound speed in the surrounding fluid. As the collapse speed increases, the compressibility of the surrounding fluid can no longer be neglected and a compressible flow assumption is required. Later work by a number of researchers has furthered the analytical formulations to include these compressible effects as well as fluid viscosity, and surface tensions. Herring (1941) introduced a first order approximation for the velocity of the collapsing bubble wall but did not consider the velocity of the fluid away from the wall. Pressure and velocity fields throughout the fluid field resulting from the collapse of a spherical bubble were later developed by Trilling (1952) but remained limited to a first order accuracy. Gilmore (1952) extended this work to include higher order compressibility effects, as well as the effects of fluid viscosity and surface tension of the bubble. There have been numerous other studies addressing the collapse of bubbles subjected to external pressure including those of Keller and Kolodner (1956) and Hickling and Plesset (1964). Keller and Kolodner presented a second order differential equation which includes the compressibility of the surrounding fluid to predict the bubble radius as a function of time which leads to the prediction of the radiated pressure wave behavior. Hickling and Plesset presented numerical solutions for the flow field surrounding a collapsing bubble with uniform internal pressure. The collapse of the bubble was considered to the point when it reached a minimum radius at which point a high pressure wave formed and is propagated into the surrounding fluid. Plesset and Prosperetti (1977) presented a comprehensive literature review of bubble dynamics and included discussions of spherical and nonspherical bubble collapses, effects of internal vapor (thermal effects), and mass diffusion effects at the bubble–liquid interface.

6.2.1 *Collapse of a Cavitiation Bubble*

In the absence of a structural shell, an underwater implosion event is comparable to the hydrostatic collapse of a bubble. There are various means by which these

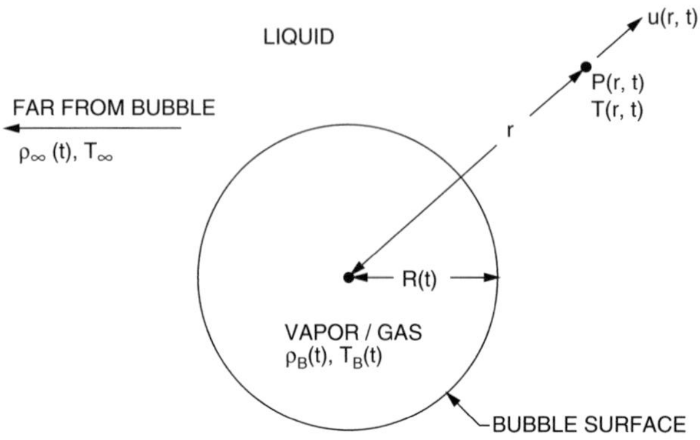

Fig. 6.1 Spherical bubble in an infinite fluid field (Brennen 1995 by permission of Oxford University Press, USA)

underwater bubbles can be caused (nucleated) including cavitations, boiling, and phase transformations and primarily result from the inability of fluids to carry tensile loads. Brennen (1995) details the dynamics of a bubble collapse the details of which are briefly discussed in the following text. Consider a bubble with radius R as shown in Fig. 6.1, which is submerged in an infinite body of fluid at pressure, P_∞, and temperature, T_∞. Internally, the bubble contains gas at uniform pressure and temperatures P_b and T_b respectively. The pressure, temperature and radial velocity in the near field fluid surrounding the bubble at a radius r measure from the bubble center are designated as $p(r, t)$, $T(r, t)$, and $u(r, t)$. Due to imbalance between the internal bubble gas pressure and the far field pressure in the surrounding fluid, the bubble will grow or collapse due to the relative pressure across the fluid boundary. The generalized Rayleigh–Plesset equation for bubble dynamics in the absence of any thermal effects is given as:

$$\frac{p_V(T_\infty) - p_\infty(t)}{\rho_L} + \frac{p_{G_0}}{\rho_L}\left(\frac{R_0}{R}\right)^{3k} = R\ddot{R} + \frac{3}{2}\left(\dot{R}\right)^2 + \frac{4\nu_L \dot{R}}{R} + \frac{2S}{\rho_L R}$$

From the above equation, the radius of the bubble at any point in time can be determined (see Brennen 1995 for full derivation). This equation can be integrated numerically provided the far field values, P_∞ and T_∞, are known along with the additional constants and the initial conditions. A typical nondimensional solution is shown in Fig. 6.2. This figure illustrates the high nonlinearity of the bubble dynamics including, an initial smooth growth of the bubble reaching maximum radius when the internal pressure is minimized, followed by a sudden collapse of the bubble. There is a subsequent oscillation of the bubble consisting of a succession of rebounds and collapses, which in the absence of the dissipative action of shock wave emission and viscosity would continue indefinitely with no decrease in maximum bubble radius.

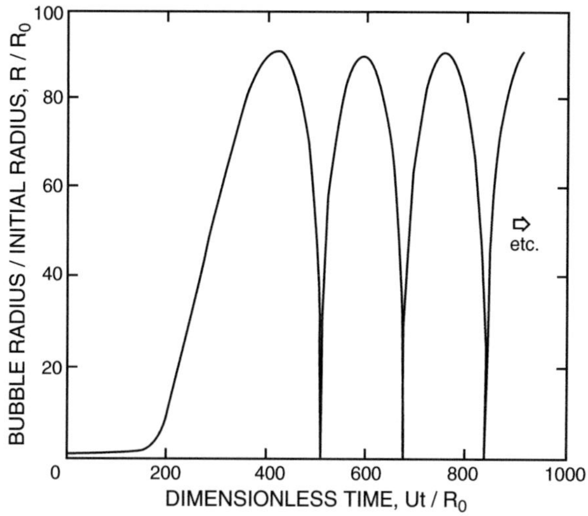

Fig. 6.2 Solution of the Rayleigh–Plesset equation for spherical bubble size (Brennen 1995 by permission of Oxford University Press, USA)

Bubble collapse is the process of most interest here because it is what creates the pressure pulse. At the instant the bubble reaches its minimum radius (full collapse), there will be a subsequent reversal of the process and the bubble will rebound. At the initiation of the rebound phase, a high-pressure wave is formed due to the arresting of the inward velocity of the fluid at the bubble–fluid interface. Keller and Kolodner (1956) modified the Rayliegh–Plesset equation to include the effects of liquid compressibility, while neglecting thermal, viscous and surface tension effects:

$$\left(1-\frac{1}{c}\frac{\mathrm{d}R}{\mathrm{d}t}\right)R\frac{\mathrm{d}^2R}{\mathrm{d}t^2}+\frac{3}{2}\left(1-\frac{1}{3c}\frac{\mathrm{d}R}{\mathrm{d}t}\right)\left(\frac{\mathrm{d}R}{\mathrm{d}t}\right)^2$$
$$=\left(1+\frac{1}{c}\frac{\mathrm{d}R}{\mathrm{d}t}\right)\frac{1}{\rho_L}\left[p_\mathrm{B}-p_\infty-p_\mathrm{c}\left(t+R/c\right)\right]+\frac{R}{\rho_L c}\frac{\mathrm{d}p_\mathrm{B}}{\mathrm{d}t}$$

where $p_\mathrm{c}(t)$ is the variable part of the pressure in the liquid at the bubble center in the absence of the bubble.

Hickling and Plesset (1964) use numerical methods to solve the compressible flow equations around a collapsing bubble in order to examine the high-pressure waves created from bubble collapse. Figure 6.3 (Hickling and Plesset 1964) illustrates the pressure in the fluid surrounding a bubble during different times of the collapse and the rebound phase of the bubble motion. After the bubble minimum (at $t=0$), a high-pressure wave can be seen to propagate outward. They showed that this pulse exhibits approximately $1/r$ geometrical attenuation as it travels away from the bubble. The figure highlights that as the minimum size of the bubble becomes smaller as compared to the initial size, the peak pressure radiated into the surrounding fluid becomes larger. The peak pressure of the radiated shock wave is approximately (Brennen 1995):

$$p_\mathrm{P}\approx 100R_\mathrm{M}p_\infty/r$$

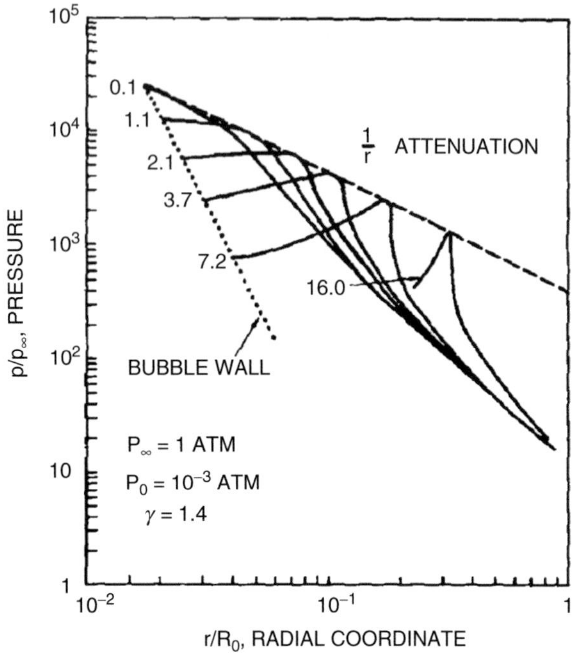

Fig. 6.3 Pressure variation with distance from the bubble wall during collapse (Hickling and Plesset 1964 by permission of the American Institute of Physics)

6.2.2 UNDEX Bubble Dynamics

A second method by which bubbles may be formed in a body of fluid is through the detonation of an explosive charge. When an explosive material is detonated underwater, the chemical reaction produces gaseous products at high pressures. As this reaction proceeds, a bubble of the detonation gasses expands extremely rapidly, at speeds greater than the wave speed of the water. This expanding bubble produces a discontinuous shock pressure wave into the water that propagates outward at the wave speed of the water. After the chemical reaction has completed (i.e., all of the explosive compound has been consumed), the gas bubble of detonation products, which is at temperatures and pressures much greater than the surrounding water, continues to expand. During the expansion phase, the internal gas pressure continually decreases as the bubble grows, moving towards an equilibrium point with the pressure in the surrounding fluid. However, the outward momentum of the bubble and the nearby surrounding water causes it to continue to grow beyond the equilibrium point. By the time the velocity of the bubble wall is brought to rest, the pressure inside the bubble is less than the surrounding water. It is interesting to note that the bubble maximum radius can be more than twice the radius where the internal bubble pressure and external hydrostatic fluid pressure are in equilibrium (Shin 2004). At this point the process reverses and the bubble begins to collapse in upon itself.

Beyond the time of the detonation bubble maximum radius, the physical processes are essentially the same as the bubble collapse dynamics described previously.

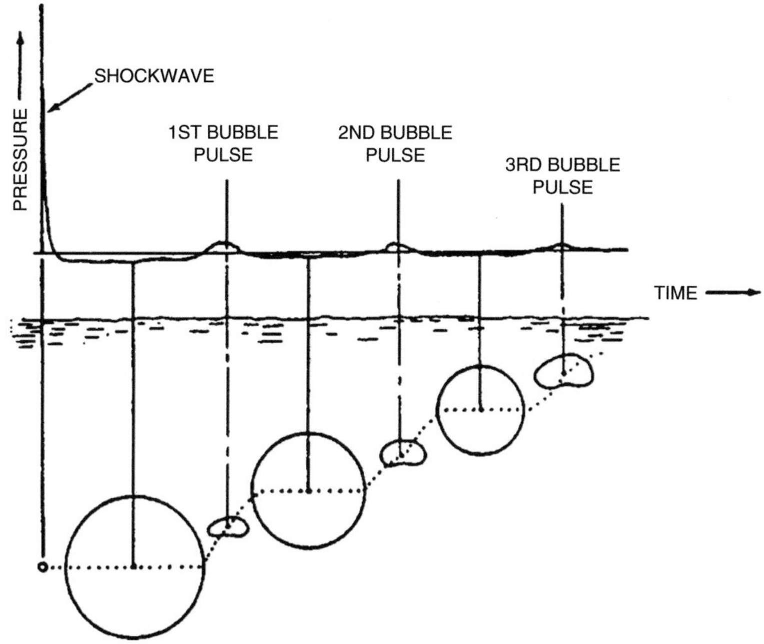

Fig. 6.4 Free field bubble motion and pressure history (Geers and Hunter 2002 by permission of the Acoustical Society of America)

During the collapse of the bubble, the inward motion of the surrounding fluid is accelerated as the volume becomes smaller, which is accompanied by an increase in the internal pressure of the bubble gasses. The momentum of the bubble and the surrounding water will cause the collapse to continue beyond the equilibrium point, over-compressing the gas until the bubble wall velocity is stopped. The process then reverses again, and the bubble begins to expand, and causes a second high pressure pulse to be propagated into the fluid field. The resulting bubble pulse is typically lower in peak pressure than the initial shock wave but is of longer duration resulting in nearly equivalent impulses. The energy associated with this secondary pulse can contain up to 50 % (Shin 2004) of the explosive energy and thus can result in damage levels on the same scale as that of the initial explosive pressure pulse. Depending upon the depth at which the explosion occurs, this cycle of bubble expansion and collapse can be repeated multiple times, Fig. 6.4. The radius of a gas bubble as a function of time for the detonation of 0.25 kg (0.55 lb) of tetryl at a depth of 91.4 m (300 ft) is shown in Fig. 6.5. In the figure, the dashed line indicates the radius at which the internal gas pressure is equivalent to the hydrostatic pressure of the surrounding fluid. From this it is evident that the bubble continues to grow and collapse beyond the point at which pressure equilibrium is achieved.

The physical behavior of the bubble resulting from an underwater explosion has been studied by numerous researchers through the use of analytical derivations,

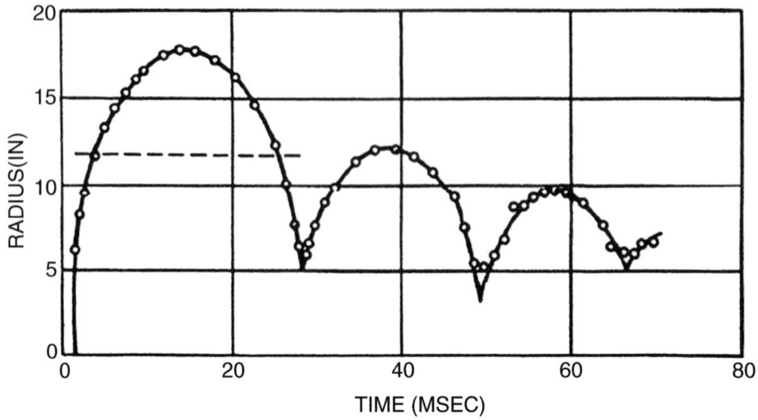

Fig. 6.5 Gas bubble radius as a function of time (Cole 1948)

experimental observations, and computational models. Cole (1948) approximated the bubble behavior with the assumption that the bubble remains spherical during collapse with no upward migration due to buoyancy effects. Application of these assumptions along with continuity at the bubble–fluid interface, Cole derives the conservation of energy for the bubble fluid system as:

$$\frac{3}{2}\left(\frac{4\pi}{3}\rho_o R^3\right)\dot{R}^2 + \frac{4\pi}{3}\rho_\infty R^3 + E(a) = Y$$

where: ρ_o—bubble equilibrium density
$E(a)$—internal bubble energy
Y—total system energy.

In the prior relationship, the first term represents the kinetic energy of the radially flowing fluid at the boundary and the second term is the work done by the bubble counteracting the hydrostatic pressure. The total system energy, Y, is a function of the type of explosive detonated and the corresponding charge weight. The internal energy of the bubble is shown to decrease as the radius becomes larger (increased volume) at a rate proportional to the cube of the radius (R^3) until the point at which the bubble reaches its maximum radius and $E(a)$ is negligible. Furthermore, it is evident that the bubble will reach its maximum radius when the outward velocity of the wall is arrested ($\mathrm{d}R/\mathrm{d}t=0$) at which point the internal energy is also nearly zero. Applying these considerations to the previous conservation of energy relationship yields the expression for maximum bubble radius, R_m:

$$Y = \frac{4\pi}{3} p_\infty R_m^{\ 3}$$

Through experimental observations of the bubble motion due to an underwater explosion, Ramsauer (1923) was able to determine relationships for the displacement-time history of the bubble up to its first maximum radius. These observations lead to the determination that the maximum radius is dependent upon the depth of detonation and the weight of the explosive charge such that:

$$\frac{4\pi}{3} p_\infty R_{\mathrm{m}}^{\;3} = \mathrm{const}\ W$$

Additionally, the period of oscillation can be expressed as:

$$T = K \frac{Y^{1/3}}{p_\infty^{\;5/6}}$$

where the constant (K) is dependent upon the exact type of explosive material utilized.

Geers and Hunter (2002) have expanded upon previous work through the inclusion of wave effects in both the surrounding fluid as well as the internal bubble gasses through the use of doubly asymptotic approximations (DAA). The inclusion of these wave effects allows for bubble wall velocities in excess of the fluid sound speed by accounting for localized compressibility of the surrounding fluid. For the case of incompressible flow (no localized wave effects), the equation of motion for the bubble surface can be described using the Rayleigh–Plesset equation. The corresponding equation of motion with the inclusion of external fluid and internal gas compressibility through the DAA method is presented as:

$$R\ddot{R}\left[1+\zeta-\left(2-\frac{\rho_g}{\rho_l}\right)\frac{\dot{R}}{c_l}\right]+\frac{3}{2}\dot{R}^2\left[1-(\gamma+1)\zeta-\left(\frac{4}{3}+\frac{\rho_g}{\rho_l}\right)\frac{\dot{R}}{c_l}+\frac{1}{3}\left(\frac{\rho_g}{\rho_l}\right)\left(1+\frac{\dot{R}}{c_l}\right)\right]$$
$$+\zeta\left(c_l+3c_g\right)\dot{R}=\rho_l^{-1}\left(P_g-p_l\right)$$

where: R = bubble radius, ρ_g = mass density of the gas, ρ_l = mass density of the liquid, c_g = speed of sound in the gas, c_l = speed of sound in the liquid, ζ = specific acoustic impedance ratio $= \rho_g c_g / \rho_l c_l$, γ = ratio of specific heats of gas, P_g = uniform internal pressure in gas from equation of state, p_l = known surface pressure of bubble = atmospheric pressure + depth pressure

6.3 Experiments with Spheres

6.3.1 Glass Spheres

Historically, attempts have been made to predict implosion using bubble collapse models. It has been assumed that, because the structure often fractures into many fragments, its effect on the implosion is small. So it is often neglected in analytical efforts, assuming that it is the water–air bubble interaction that is important. Turner (2007) conducted a series of implosion experiments on glass spheres to test this theory.

Fig. 6.6 Test stand with the glass sphere test specimen and the instrumentation installed (Turner 2007)

Glass was chosen as the material for the test articles because it will shatter and should have the smallest possible influence on the implosion.

Four underwater implosion experiments were conducted with glass spheres having an outside diameter of 7.62 cm, thickness of 0.762 mm, and an estimated buckling pressure of 7.57 MPa. The experiments were performed in a 1.52 m by 3.65 m (5-ft by 12-ft) cylindrical pressure. Three pressure sensors at 120° apart measured the pressure at a radial distance of 10.16 cm from the sphere center (refer to Fig. 6.6).

Each of the four tests was conducted with glass spheres subject to an external hydrostatic pressure of 6.996 MPa and an internal pressure of 101.3 kPa. The pressure time history for each sensor of test 1 is plotted in Fig. 6.7. The beginning of the collapse is observed as a gradual decrease in the pressure at the sensors, which occurs between 0.12 and 0.16 ms. As the surface of the glass sphere disintegrates, relieving the static pressure acting on the external surface of the glass, a rarefaction wave travels outward. The collapse phase ends when the gas reaches a minimum volume (closure). Subsequently, a large positive pressure peak occurs. This peak is created by the same physical process as described for the bubble collapse and UNDEX bubble pulse described earlier. The distance between the sensors and the wall of the pressure vessel is 66 cm. In fresh water, the time for the pressure wave to travel from the sensor to the wall and back is about 0.9 ms. Therefore, the onset of reflections in the pressure time history is expected at 1.0 ms.

During the collapse phase, the pressure time histories of sensors 1–3 are nearly identical. The curves begin to diverge as the pressure peak is reached at 0.8 ms. There is a 55 % difference in the peak pressure between all sensors, indicating some

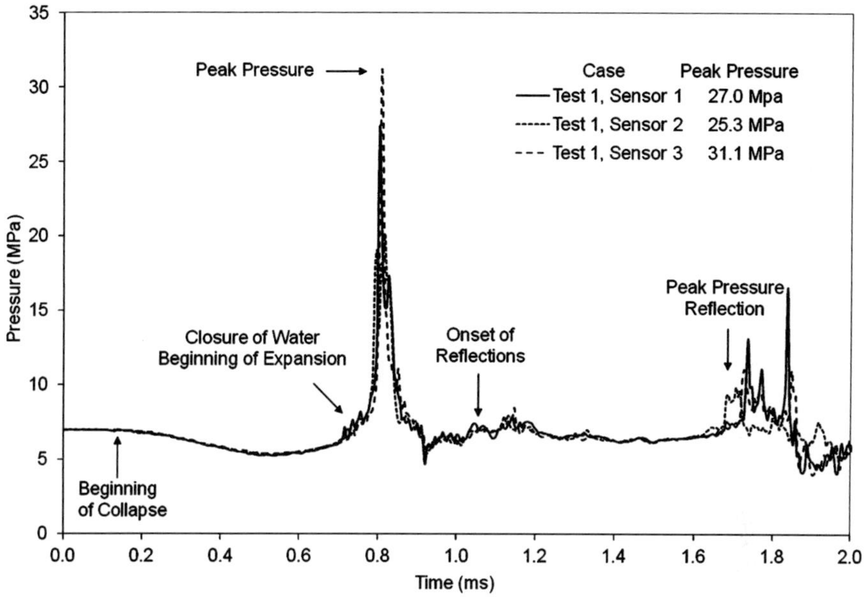

Fig. 6.7 Experimental result of the pressure pulse from the implosion of a glass sphere (Turner 2007)

asymmetry exists in the implosion. Sensor 3 recorded the highest peak pressure, but the duration of the pressure pulse was shorter than the other two sensors. It is often useful to integrate the dynamic pressure with respect to time to calculate the impulse, *I*, which is a measure of the momentum imparted to the water as the pressure wave passes through it. In Fig. 6.8, the impulse is plotted for the three pressure time histories measured in test 1. For this test, the sensor-to-sensor difference between the minimum and maximum impulse ($I_{max} - I_{min}$) is only 6 %. The impulse shows that the total momentum imparted to the water tends to be more spatially uniform, even though there is spatial variation of the pressure time histories.

Tests 2–4 were conducted with glass spheres identical to test 1 at an external hydrostatic pressure of 6.996 MPa. Implosion at the same tank pressure was achieved by initiating failure using a mechanical device that contacted the bottom of the sphere, 90° away from each sensor. The pressure time histories measured at sensor 1 for each test are shown in Fig. 6.9. During the collapse phase, there is little deviation between the pressure time histories. However, the curves show some differences in the peak pressure. Still, the magnitudes of each peak pressure are within 5 % of each other (25.8–27.2 MPa).

The implosion pressure pulse from the glass sphere test is compared with a bubble collapse calculation in Fig. 6.10. The character of the pulse is generally the same between the two. The main differences are that the pressure-drop phase begins gradually in the test and suddenly for the bubble calculation, and the peak pressure for the bubble calculation is much larger than that in the test. Turner (2007) showed that the differences are caused by the influence of the glass structure using a

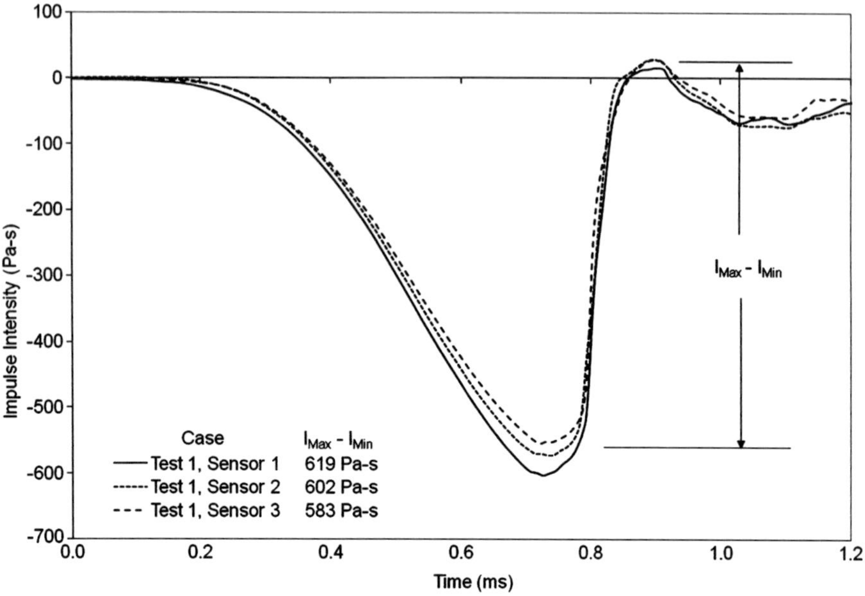

Fig. 6.8 Experimental result of the impulse per unit area of the implosion of a glass sphere (Turner 2007)

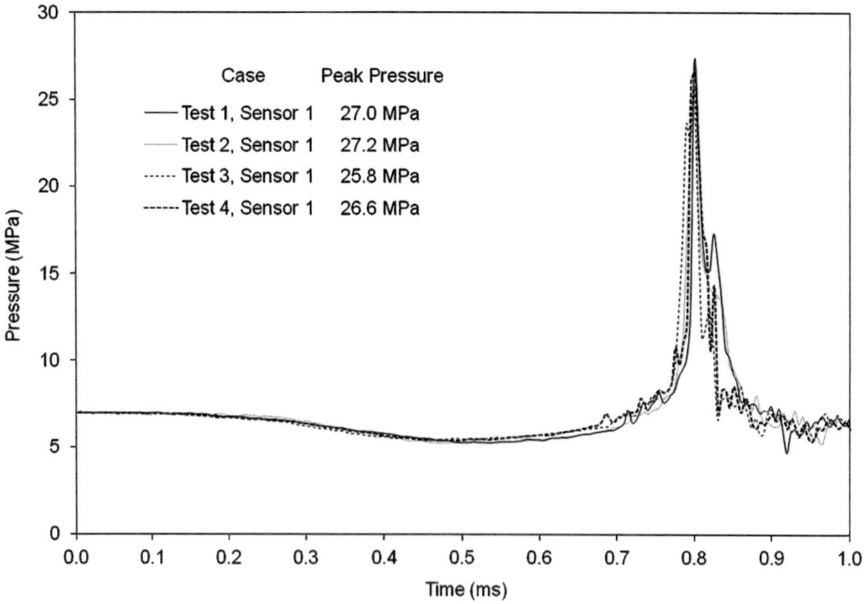

Fig. 6.9 Experimental results of the pressure pulse from the implosion of glass spheres (Turner 2007)

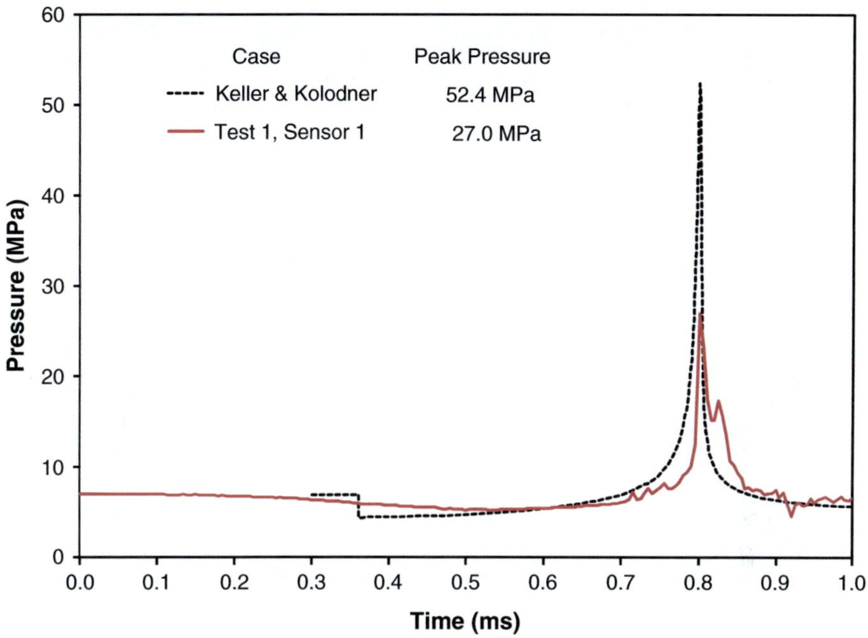

Fig. 6.10 Comparison of implosion pressure pulse from a glass sphere and from bubble collapse analysis

computational model. The conclusion is that in an implosion event, the structure always influences the pressure pulse, and neglecting it in an analysis will most likely produce an over-estimate of the pressure pulse.

6.3.2 Aluminum Spheres

Vanzant et al. (1970) conducted one of the first sets of experiments to measure the near-field pressures in the fluid surrounding an implodable volume. Their experiments studied 15.87 cm (6.25 in.) aluminum 7075-T6 spheres in three unique configurations: (1) spheres containing internal vacuum, (2) spheres containing air at atmospheric pressure and (3) spheres filled with styrofoam material. The study was conducted in a 228 cm (90 in.) diameter tank which replicates deep ocean depths in order to facilitate the measurement of the near field pressure signal characteristics. The goal of the project was to record the near field pressure profiles which resulted from the collapse of the various sphere configurations.

A photograph of the remains of one of the test spheres appears in Fig. 6.11. Clearly there has been extensive deformation and fracture during the implosion. The pressure time histories recorded during the experiments consisted of three primary phases as shown in Fig. 6.12. There is an initial drop in pressure (negative phase)

Fig. 6.11 Remains of the aluminum sphere after implosion testing (Vanzant et al. 1970)

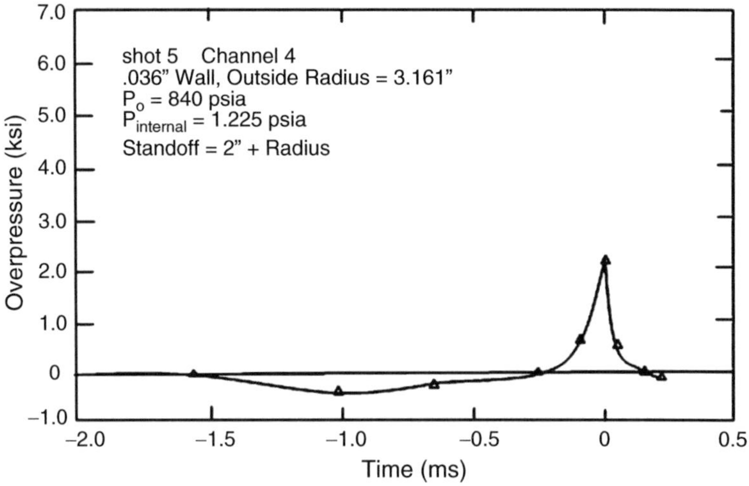

Fig. 6.12 Pressure time history for a sphere imploding at 256 m (840 ft) of depth (Vanzant et al. 1970)

corresponding to the collapse of the structure as previously discussed. This is followed by a positive pressure characterized by a sharp pressure peak caused by the rebound after the collapse reaches a minimum volume. This process then repeats with a second negative and positive phase until reflections from the tank wall appear. It should be noted that the repetition of phases occurs because the extensive

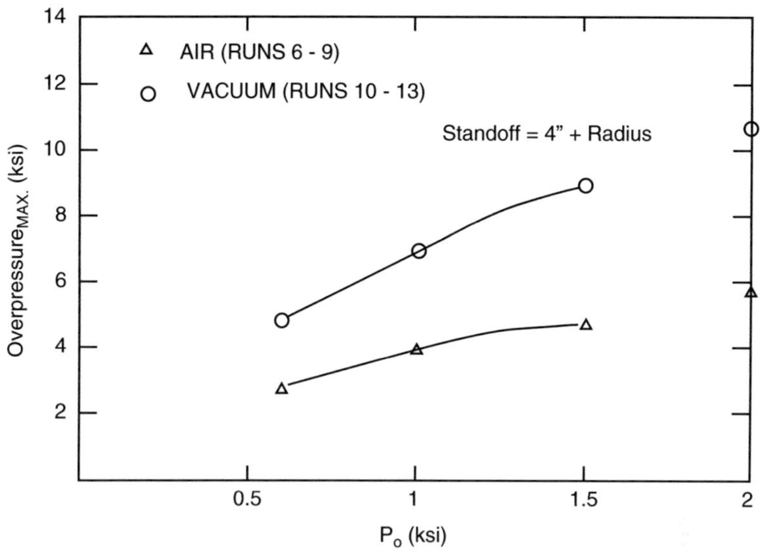

Fig. 6.13 Theoretical peak overpressure as a function of implosion pressure for a spherical body (Vanzant et al. 1970)

fragmenting of the aluminum sphere allows the air inside the sphere to interact with the water and oscillate like the collapsing bubbles discussed previously. Comparison of the time histories were made to theoretical bubble collapse equations of Gilmore (1952), and indicate that the peak overpressures pressures predicted by bubble collapse theory are much larger with a shorter duration than those observed during testing. The authors attribute the increased time duration of the positive pressure to fragmenting of the spheres during collapse and the associated acceleration of the fragments. Computation of the pressure impulses found that these two effects offset each other and the positive impulses are comparable between the analytical bubble collapse models and experimental results.

The experiments demonstrated several trends in implosion behavior of spheres. In terms of absolute peak overpressures, it was observed that there is an increase in magnitude with increasing collapse pressure, Fig. 6.13, with a corresponding decrease in the time duration of the positive pressure phase, Fig. 6.14. Furthermore, the pressure drop during the initial under pressure phase increases as a function of increasing depth, Fig. 6.15. In the experiments, there is a decay in the magnitude of the under and over pressures, as well as the unit impulse of the positive pressure phase, which is proportional to the reciprocal of the distance from the implosion center. The time duration of these phases only displayed a slight radial distance dependency. Further, the experiments show that the impulses per unit area for both the initial under pressure and the subsequent over pressure phases are not highly dependent upon the collapse pressure of the structure. For the case of the spheres pulled to internal vacuum, there is a slight increase in impulse with increase in depth pressure, Fig. 6.16, but for the case of a Styrofoam filled sphere, the impulse is nearly constant.

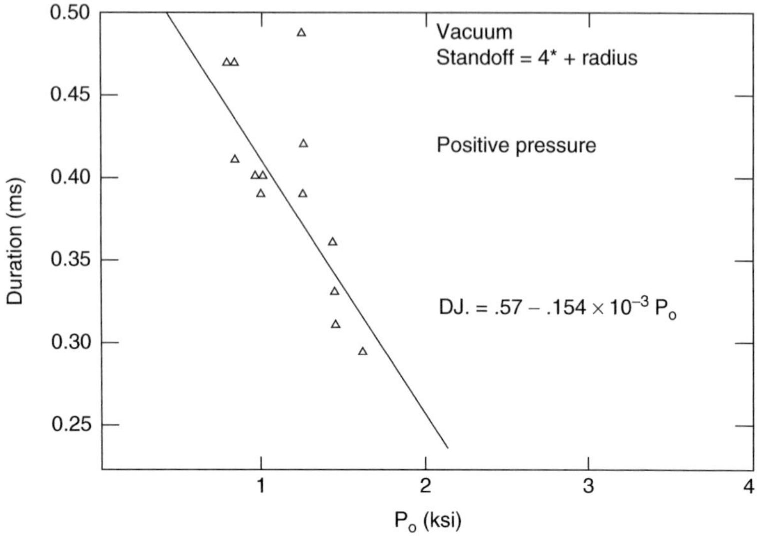

Fig. 6.14 Time duration of positive pressure phase as a function of implosion pressure (Vanzant et al. 1970)

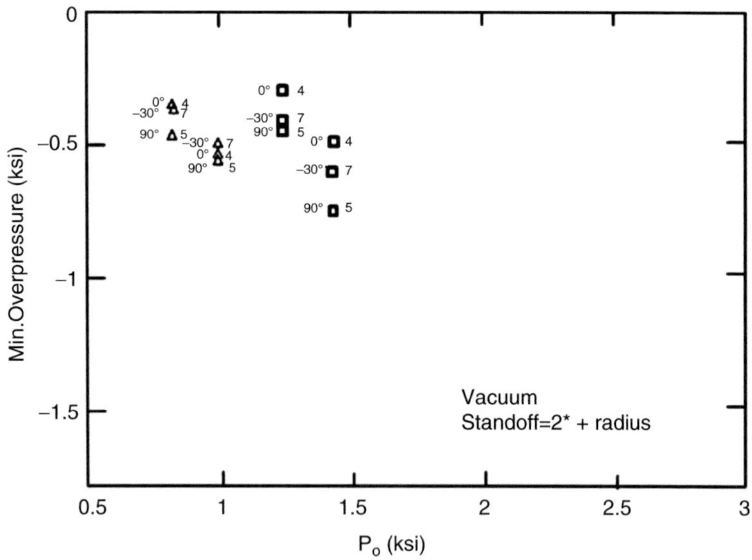

Fig. 6.15 Negative pressure magnitude as a function of implosion pressure for a spherical body (Vanzant et al. 1970)

In several of the experiments, a flat surface was machined into the surface of the spheres which caused non symmetrical collapse of the volumes. From these experiments, it was shown that for the cases of nonsymmetrical collapses there is a directionality of the pressure pulse in the near field surrounding the structure. These tests

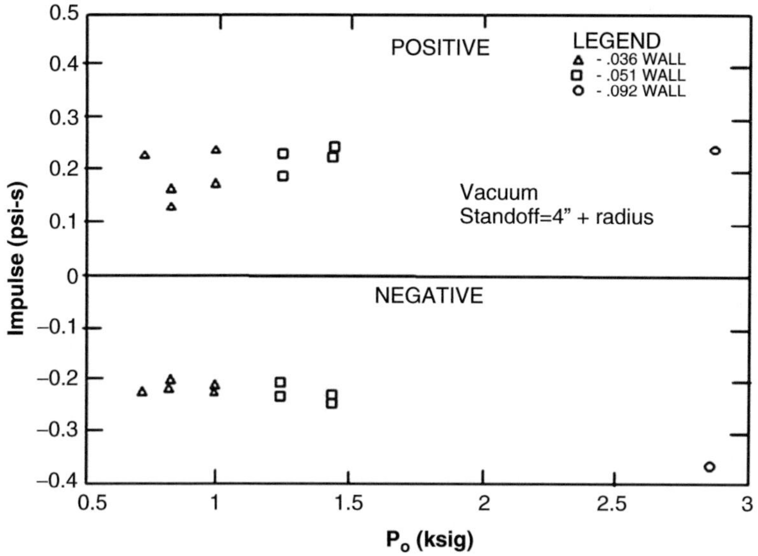

Fig. 6.16 Positive and negative impulse as a function of implosion pressure for a spherical body (Vanzant et al. 1970)

illustrated pulse shapes that would be expected in an operating environment where actual structures typically include small defects or variations that would cause un-symmetric collapses.

6.4 Experiments with Long Cylinders

6.4.1 Aluminum Cylinders

Several series of experiments were conducted at the Naval Undersea Warfare Center with un-stiffened aluminum alloy 6061-T6 cylinders (Turner 2004; Turner and Ambrico 2013) containing air at 101.3 kPa. One test series compared the response of cylinders imploded at their natural buckling pressure of 7.00 MPa (1,000 psig) with cylinders caused to implode at 6.31 MPa (900 psig) by mechanical initiation. The length to diameter ratio of these cylinders was approximately 10, resulting in a buckling mode of 2. A schematic of cylinders for tests 5 and 6 is shown in Fig. 6.17. The cylinders used in tests 7 and 8, were a little shorter and had shorter endplugs, resulting in an internal length 0.60 cm (0.24 in.) shorter than tests 5 and 6. In the experiment, the cylinder collapses into a mode-2 flattened shape, as seen in Fig. 6.18. A dynamic pressure sensor located in the water 10.16 cm (4 in.) from the centerline at the middle of the cylinder length was used to measure the pressure during the implosion event. The pressure time histories measured in tests 5 through 8 is plotted in Fig. 6.19.

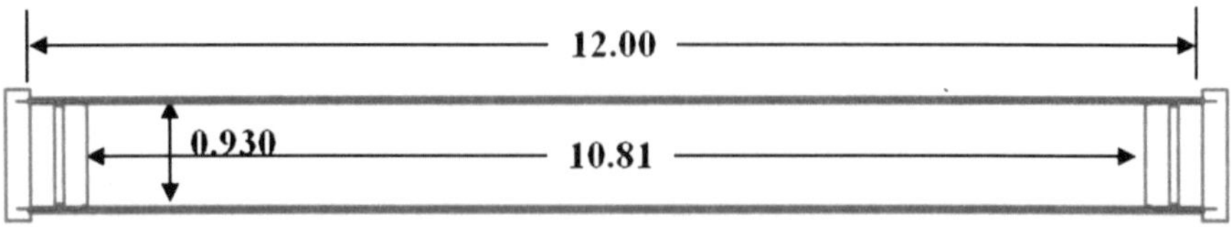

Fig. 6.17 Schematic of the aluminum cylinder test sample used in implosion experiments (Turner 2004)

Fig. 6.18 Collapsed shape of the aluminum cylinder after the implosion experiment

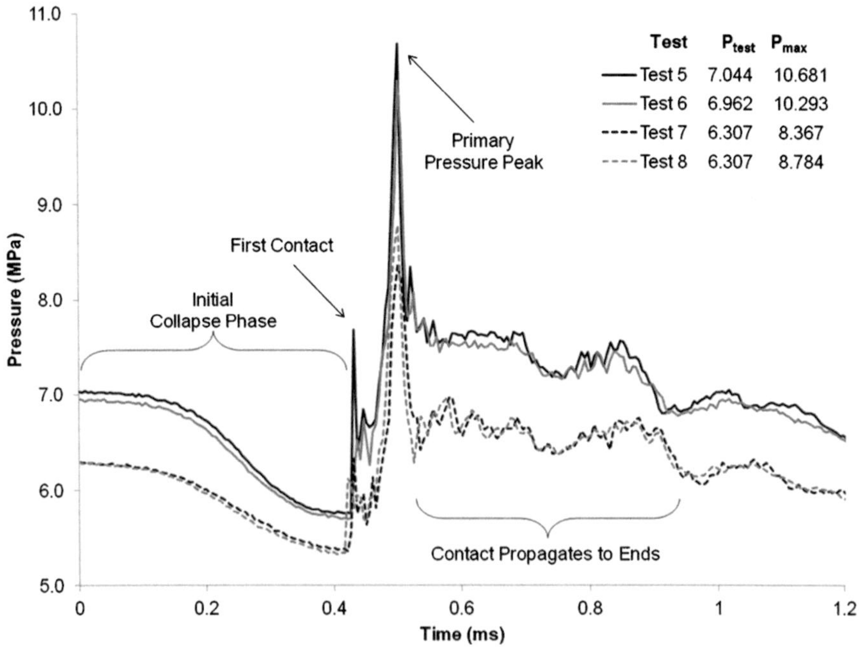

Fig. 6.19 Experimental pressure pulse for the implosion of long aluminum cylinders (Turner and Ambrico 2013)

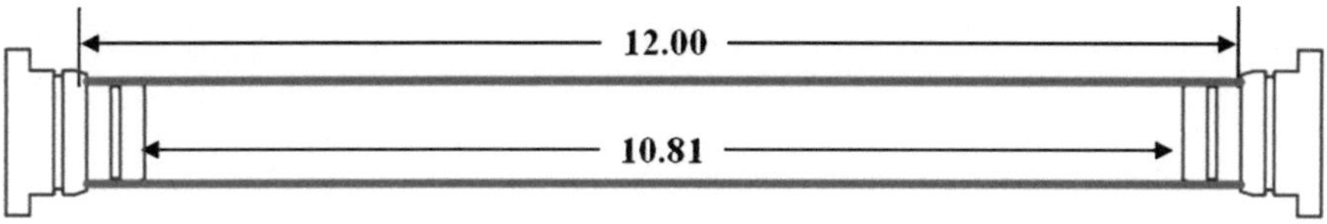

Fig. 6.20 Schematic of the glass cylinder test sample used in implosion experiments (Turner 2004)

The general behavior of the implosion pressure pulse can be summarized by four significant features. (1) When the external pressure applied to the cylinder reaches a critical point and the cylinder becomes unstable, the collapse phase begins. During the collapse phase, and prior to any metal-to-metal contact, the pressure decreases smoothly from the hydrostatic pressure to about 80–85 % of the hydrostatic pressure (when measured at 4 diameters away from the centerline). (2) At the moment the opposing sides make contact a short duration pressure spike is produced; followed by (3) a primary pressure peak as the metal-to-metal contact spreads from initial contact at a single point across the diameter of the cylinder at the middle of the length. (4) The primary pressure peak is followed by a "tail" of positive pressure, which is caused as the buckling event propagates from the center to the ends of the cylinder. This lasts about 0.4 ms in these tests before dropping down to the hydrostatic pressure.

The data plotted in Fig. 6.19 show that the magnitude of the peak pressure is dependent on the hydrostatic pressure at the time of collapse. When the test pressure was 7.00 MPa (1,000 psig), the peak pressure is approximately 3.44 MPa (500 psi) over test pressure; whereas for the 6.3 MPa (900 psig) tests the peak pressure is 2.17–2.58 MPa (315–375 psi) over the test pressure. The figure also shows the similarity in the pressure response between implosion at the natural buckling pressure and implosion initiated at 90 % of the natural buckling pressure. Details of the experimental procedure and a complete description of the implosion mechanics have been documented by Turner and Ambrico (2013).

6.4.2 *Glass Cylinders*

Two additional tests were conducted with borosilicate glass cylinders having the same length, diameter, and wall thickness as the aluminum cylinders of tests 5 and 6 (see Fig. 6.20). The natural buckling pressure of these two glass cylinders is 7.444 and 7.299 MPa (1,065 and 1,044 psig), with internal air pressure of 101.3 kPa. A photograph of the glass test specimen in the test stand is shown in Fig. 6.21. The photograph shows the position of the three pressure sensors at 10.16 cm (4 in.) from the centerline of the cylinder (one on each side, and one above). After the implosion, only a fine glass powder remained; there were no large pieces of glass. The pressure time histories for each of the sensors are shown in Fig. 6.22. First, it can be seen that the pressure profiles are very similar at each of the sensor locations.

Fig. 6.21 Test stand after glass cylinder implosion experiment (Turner 2004)

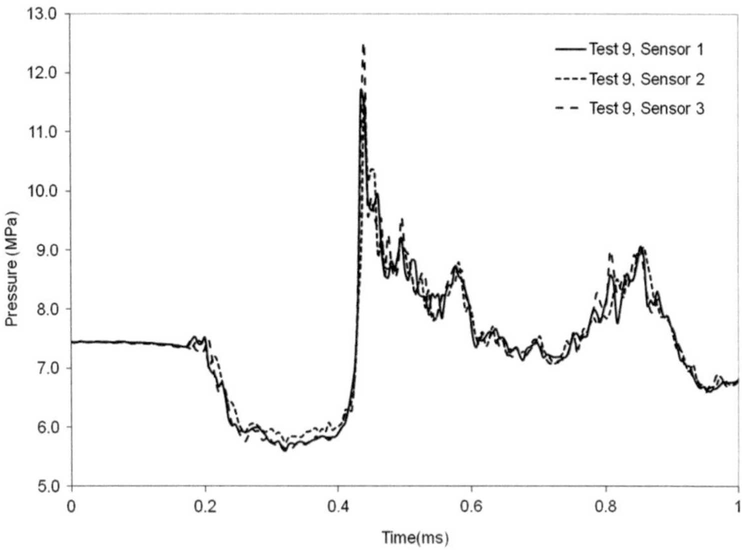

Fig. 6.22 Experimental pressure pulse for glass cylinder implosion experiments (Turner 2004)

Directional effects around the circumference of the cylinder appear to be small, although there is some variation in the magnitude of the primary peak pressure. Second, the pressure-drop phase begins more abruptly than for the spheres or the aluminum cylinders and is a bit shorter in duration.

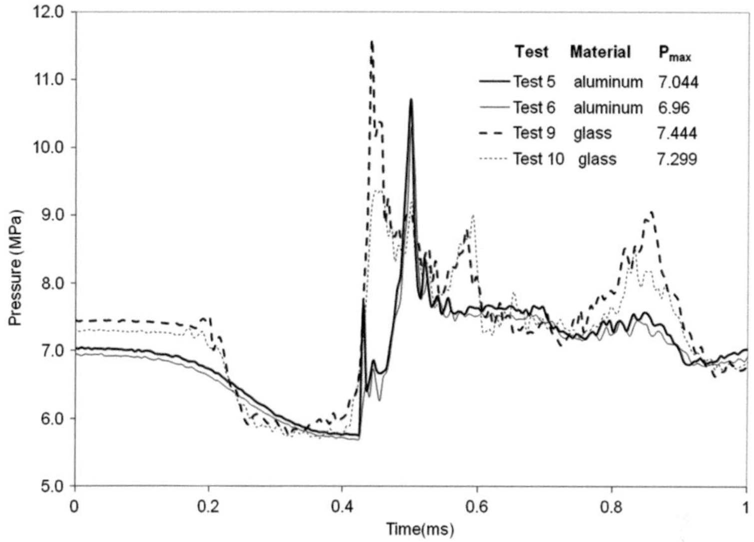

Fig. 6.23 Experimental pressure pulses for the glass and aluminum implosion tests at sensor 1 (Turner 2004)

6.4.3 Comparison of Aluminum and Glass Cylinder Implosions

In each of the four tests [aluminum cylinders tested at 7.044 and 6.96 MPa (1,007 and 995 psig) and glass cylinders tested at 7.444 and 7.299 MPa (1,065 and 1,044 psig)], the position of sensors 1 and 2 remained the same. The pressure time history at sensor 1 is plotted for each of the four tests in Fig. 6.23. First, there is an obvious difference between the glass and aluminum samples during the collapse phase. For the glass, the pressure drops quickly, and remains level for a short time, until the gas bubble reaches a minimum volume and a pressure pulse is produced. For the aluminum cylinder, the pressure gradually decreases until the two opposing sides touch, and then the pressure pulse is produced. The authors believe that upon initiation of failure, the glass loses its structural stiffness, enabling a more rapid collapse of the volume. In contrast, the aluminum retains some stiffness during collapse, which causes a longer, more gradual collapse, compared to the glass cylinders.

The aluminum and glass cylinders produce peak pressures of about the same magnitude, even though the collapsed shapes look very different. However, after the peak pressure, the pressure time histories look a little different. The aluminum collapse starts at the center of the length, and then propagates toward the endplugs. So, for the aluminum, after the peak pressure, the pressure stays above hydrostatic pressure until the collapse reaches the endplugs, then the pressure settles out to the hydrostatic pressure. When the glass cylinder fails, a mixture of water and glass particles rush to the center of the cylinder, compressing the gas bubble. The pressure

Table 6.1 Experimental impulses per unit area for the glass and aluminum implosion tests

Test	Sensor	I_{min} (Pa s)	$I_{min}-I_{max}$ (Pa s)
5, aluminum	1	265	165
5, aluminum	2	245	155
6, aluminum	1	258	156
6, aluminum	2	253	157
9, glass	1	320	232
9, glass	2	294	221
10, glass	1	306	216
10, glass	2	287	200
Avg aluminum		255	158
Avg glass		302	217

peak is seen at about 0.5 ms and the bubble pulse at about 0.9 ms (since the gas is no longer restrained by the glass, it can oscillate).

To help assess the influence of material ductility on the measured dynamic pressure, each of the curves is integrated in time to determine the impulse per unit area of the dynamic pulse. Specifically, the collapse impulse, I_{min}, is determined by integrating pressure from the time at which the dynamic pressure decreases below zero until it increases from negative to positive for the primary pressure peak (see Fig. 6.8). The impulse of the implosion pulse is determined by subtracting the maximum impulse from the minimum impulse. For tests 9 and 10, the end of the response time is 0.18 ms. For direct comparison with the glass tests, the response impulse of the aluminum tests (5 and 6) is integrated over a time of 0.18 ms. The impulses are listed in Table 6.1 for sensors 1 and 2 of the four tests. A trend is clearly apparent: the ductile aluminum absorbs more energy than the glass during the collapse and creates a less energetic implosion pressure pulse in the water. Again, the structure's behavior is a significant contributor to the implosion event.

6.4.4 Experiments with Steel and Composite Cylinders

A series of large-scale implosion experiments was conducted in 2006 by Texas Research Institute (Dingus 2007). These experiments tested 25.4–66 cm (10–26 in.) cylinders made from aluminum, steel or composite in an open-ocean environment. Each test article was attached to a test cage with an array of pressure sensors located around it to measure the near-field pressures during the implosion. The rig was the lowered into the ocean until the cylinder collapsed from depth pressure alone. Although all test articles had been designed to collapse at the same depth, there is quite a spread in the actual collapse depths. Fortunately, one steel and one composite 45.7 cm (18 in.) cylinder collapsed at nearly the same depth, which allows for direct comparison of the implosion. The steel cylinder is made from API 5 L Grade B commercial steel pipe. The composite cylinder has a filament wound carbon fiber epoxy resin construction. Figures 6.24 and 6.25 show the collapsed shape for these steel and composite cylinders for the at-sea tests. The ductile steel collapses into a flat, mode 2 buckling shape, whereas the composite fractures into many pieces, small and large.

Fig. 6.24 Collapsed shape of the 45.7 cm (18 in.) steel cylinder after the at-sea implosion test

Fig. 6.25 Collapsed shape of the 45.7 cm (18 in.) GRP composite cylinder after at-sea implosion testing

The pressure pulse measured one radius from the surface of the cylinder is plotted for each cylinder in Fig. 6.26. The curves have the same basic shape, with a few differences. The pressure drop, the peak pressure and the tail all have larger magnitude for the composite implosion as compared to the steel. The collapse impulse and the peak impulse (calculated as described earlier) are nearly the same for the respective test, with the steel cylinder producing a 3.51 kPa s (0.51 psi s) impulse while the

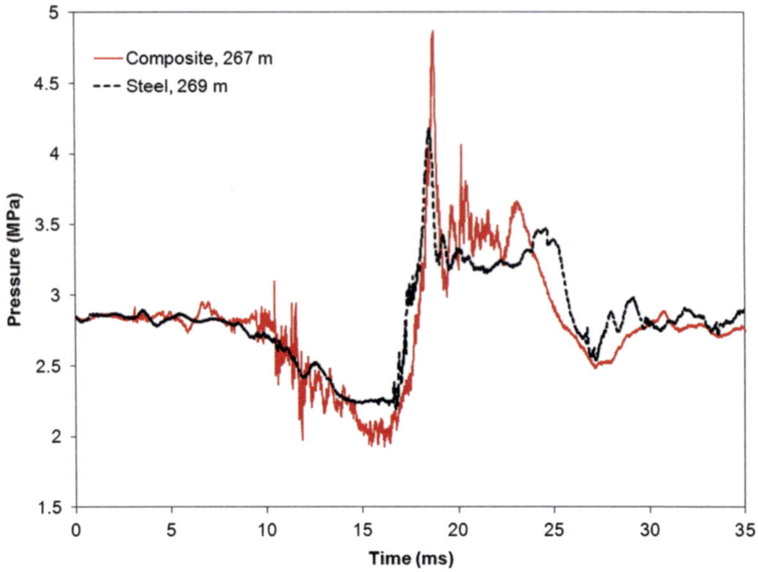

Fig. 6.26 Pressure pulse from the implosion of an 18-in. steel and composite cylinder tested at-sea

composite produces a 4.41 kPa s (0.64 psi s) impulse. Again, the ductility of the implodable volume structure affects the implosion pulse produced. In this case, the difference between steel and composite implosion pulses is not quite as large as between glass and aluminum cylinders as shown previously.

6.4.5 Experiments with Coated Aluminum Cylinders

Because the structure has been shown to affect the impulse, a strategy to mitigate the implosion pulse is investigated next. Since ductile materials seem to absorb more energy, another test series was conducted to determine if the blast mitigating polyurea material coated on the outside of a cylinder can significantly reduce the pressure response of an implosion. Tests 5 and 6 from Fig. 6.23 are a baseline pressure response from implosion of uncoated cylinders. Then a set of identical cylinders was sent to the Carboline Company for application of the polyurea coating. First, the outer surface was abrasive blasted, leaving a 0.076–0.1 mm (3–4 mil) anchor profile; and then spray-coated with Polyurea (Polybrid 705). The coating thickness is systematically varied between 0.38 and 1.14 mm (15 and 45 mils) to experimentally determine the effect coating thickness on the implosion pressure pulse. Each cylinder contained air at 101.3 kPa, and was loaded to 6.301 MPa (900 psig) hydrostatic pressure; then the implosion was slowly initiated by mechanical means.

The pressure, measured 10.16 cm (4 in.) from the center of the cylinder is plotted in Fig. 6.27 for the two uncoated cylinders and two cylinders with 0.38–0.51 mm (15–20 mil) thickness of polyurea. The primary observation is that the coated cylinders

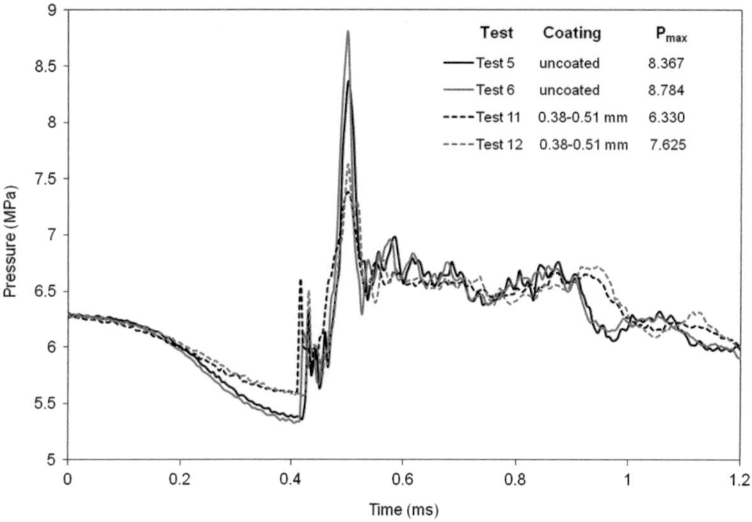

Fig. 6.27 Experimental pressure pulse from aluminum cylinders with and without a polyurea coating

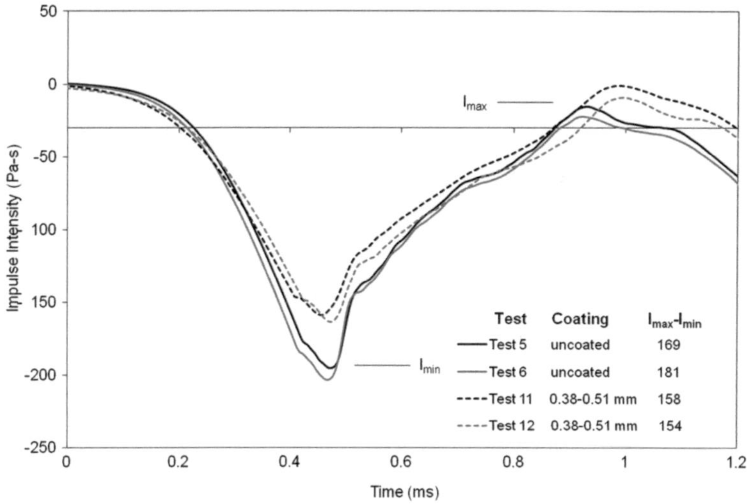

Fig. 6.28 Experimental impulse per unit area for aluminum cylinders with and without a polyurea coating

exhibit a smaller pressure drop during the collapse phase and a smaller peak pressure than the uncoated cylinders. Yet, it does not appear that the coated cylinders collapsed at a slower rate. Figure 6.28 shows the impulse plotted for the two uncoated cylinders and two cylinders with a 0.38–0.51 mm (15–20 mil) coating of polyurea. Prior to the collapse, the dynamic pressure and the impulse are zero. The difference between I_{max} and I_{min} is considered a measure of the strength of the implosion pressure pulse at the sensor. It can be seen in Fig. 6.29 that the impulse is lower for the coated samples.

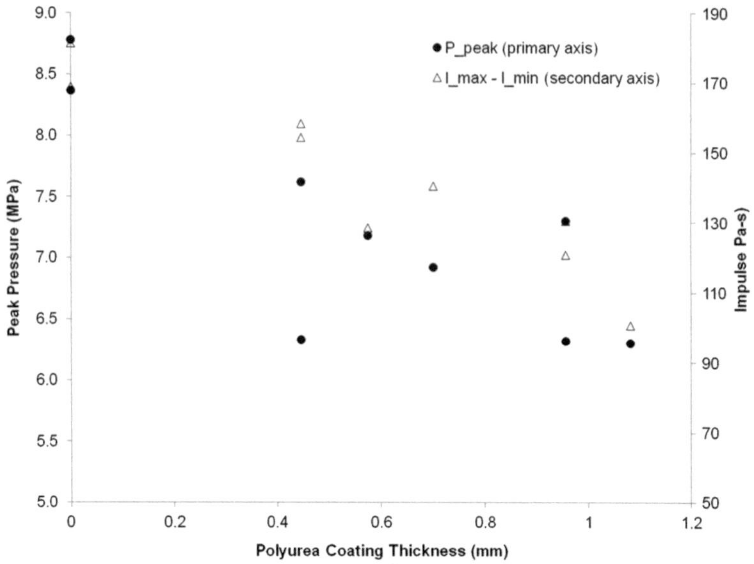

Fig. 6.29 Summary of the peak pressure and the impulse dependence on the polyurea coating thickness

Instead of comparing the pressure time histories of each test, the peak pressure and the impulse, $I_{max} - I_{min}$ has been plotted against the polyurea coating thickness in Fig. 6.29. While the coating thickness has been reported as a range, the average thickness has been used for Fig. 6.29. It is clear that as the coating thickness increases, both the peak pressure and the impulse of the implosion pressure pulse decreases. Indeed, using a thin polyurea coating on the outside of an implodable volume is an effective method to reduce the energy radiated into the water by an implosion.

6.5 Conclusions

An underwater implosion is the rapid inward collapse of a structure subjected to depth pressure. This rapid collapse produces a pressure wave that is radiated out into the water. The basic physical process that produces the wave is the same as that from cavitation bubble collapse or the bubble pulse from an underwater explosion bubble pulse. The rapid inward collapse causes a low pressure wave to propagate outward. Once the implodable volume collapses to a minimum volume, the momentum of the in-rushing water causes the water to over-compress and a high-pressure wave propagates outward. Although these physics are essentially the same as for bubble models, using such models to predict implosion over-predicts the peak and the impulse of the pressure wave. The reason for this is that the structure's response has a significant effect on the implosion that cannot be neglected. Comparisons of the implosion pressure pulse from glass and aluminum and from composite and steel cylinders show that increasing the ductility of the implodable volume's material effectively absorbs energy from the implosion and produces a smaller implosion. It is important to note also that the structure does not need to fracture and that air–water interaction is not necessary to create an implosion pressure pulse. Finally, coating the implodable volume with an energy absorbing material such as a polyurea is an effective means to decrease the energy in the implosion pulse.

References

Brennen CE (1995) Cavitation and bubble dynamics. Oxford University Press, New York, NY

Cole RH (1948) Underwater explosions. Princeton University Press, Princeton, NJ

Cor JJ, Miller TF (2009) Theoretical analysis of hydrostatic implodable volumes with solid inner structures. J Fluid Struct 25:284–303

Dingus M (2007) Pressure vessel implosion sea test - final report, SBIR phase II topic N01-106 final report, contract no. N00024-04-C-4145

Fukuda Y et al (2003) Nucl Instrum Meth A 501:418

Geers TL, Hunter KS (2002) An integrated wave-effects model for an underwater explosion bubble. J Acoust Soc Am 111:1584–1601

Gilmore FR (1952) The growth or collapse of a spherical bubble in a viscous compressible liquid, report no. 26-4, Hydrodynamics Laboratory, California Institute of Technology, Pasadena, CA

Herring C (1941) Theory of the pulsations of the gas bubble produced by an underwater explosion, OSRD report no. 236

Hickling R, Plesset MS (1964) Collapse and rebound of a spherical bubble in water. Phys Fluid 7:7–14

Keller JB, Kolodner II (1956) Damping of underwater explosion bubble oscillations. J Appl Phys 27:1152–1161

Main report on Super-Kamiokande accident (2001) Available at: http://www.icrr.u-tokyo.ac.jp/ICRR_news/ICRRnews48.pdf (in Japanese) and at http://www-sk.icrr.u-tokyo.ac.jp/cause-committee/1st/report-nov22e.pdf (English translation)

Plesset MS, Prosperetti A (1977) Bubble dynamics and cavitation. Annu Rev Fluid Mech 9:145–185

Ramsauer C (1923) Ann d Phys 4(72):265

Rayleigh L (1917) On the pressure developed in a liquid during the collapse of a spherical cavity. Phil Mag 34:94

Report of Super-Kamiokande accident study (2002) Available at: http://www-sk.icrr.u-tokyo.ac.jp/~totsuka/nov-12/report-2/report_2.ppt (Japanese) and at http://superk.physics.sunysb.edu/nngroup/publication_theses/SKaccident02.ps (English translation)

Shin YS (2004) Ship shock modeling and simulation for far-field underwater explosion. Comput Struct 82:2211–2219

Trilling L (1952) The collapse and rebound of a gas bubble. J Appl Phys 23:14–17

Turner SE (2004) Small-scale implosion testing of glass and aluminum cylinders, NUWC-NPT technical memorandum 04-061, Naval Undersea Warfare Center Division, Newport, RI

Turner SE (2007) Underwater implosion of glass spheres. J Acoust Soc Am 121(2):844–852

Turner SE, Ambrico JM (2013) Underwater implosion of cylindrical metal tubes. J Appl Mech 80:011013

United States Department of the Navy (2004) The navy unmanned undersea vehicle (UUV) master plan

Vanzant BW, Schraeder AL, Edlund CE, DeHart RC (1970) Pressure pulses generated by the sudden collapse of structures subjected to external hydrostatic pressure, final report project no. 03-1938, U.S. Naval Underwater Weapons Research and Engineering Station, Newport, RI

Chapter 7
Dynamic Buckling and Fluid–Structure Interaction of Submerged Tubular Structures

Neal P. Bitter and Joseph E. Shepherd

Abstract Dynamic buckling of submerged structures is a challenging problem for which experimental data is scarce and generalized theoretical models are difficult to employ. In addition to the complexities of dynamic buckling, this problem features additional difficulties due to the strong fluid–solid interaction that is characteristic of structures submerged in a dense fluid. This chapter reviews some recent experiments in which time-resolved measurements of pressure and strain were made during the buckling of submerged tubes. This data clarifies the buckling behavior over a useful range of conditions and provides a means to validate theoretical models with a rigor not possible using post-collapse measurements alone. Observations from the experiments are then used to develop simple models of buckling and fluid–structure interaction; comparisons with the experimental data demonstrate good agreement in spite of the many simplifications used in the modeling.

7.1 Introduction

Designing submerged structures to withstand blast loads is a challenge encountered in numerous military and industrial applications. The fundamental difference between blast loads in a gaseous medium and blast loads in a dense fluid, like water, is the strength of the coupling between the fluid and solid motion. This strong coupling complicates dynamic failure predictions since the fluid and solid motion cannot in general be solved independently. The presence of a dense fluid can play an

N.P. Bitter (✉) • J.E. Shepherd
California Institute of Technology, Pasadena, CA 91125, USA
e-mail: nbitter@caltech.edu; joseph.e.shepherd@caltech.edu

A. Shukla et al. (eds.), *Blast Mitigation: Experimental and Numerical Studies*,
DOI 10.1007/978-1-4614-7267-4_7,

important role in the failure of the structure. For instance, structural deformation can produce local volume changes in the fluid domain which reduce the intensity of the pressure load. In addition, the added mass of the fluid can alter the timescales of structural motion, and radiation of energy into the surrounding fluid provides an additional damping mechanism not present for freely vibrating structures. To address these complications, the effects of fluid–structure interaction (FSI) are carefully considered as we study dynamic buckling in this chapter.

One of the most common and detrimental failure modes of submerged structures is buckling of hollow components due to external pressure. Components of cylindrical shape are frequently used in marine applications, and this shape is also a useful idealization of more complicated real geometries. Accordingly, this chapter considers dynamic buckling in the specific context of cylindrical tubes, focusing in particular on tubes loaded by axisymmetric shock waves which travel along the tube axis. More general load cases, such as a blast wave impinging on the side of the tube or at an angle relative to the tube axis, are beyond the scope of this review, but some experimental data (Lindberg and Sliter 1969) are available for the case of blast waves in air. For experimental convenience, we also do not consider the effects of uniform hydrostatic pressure, assuming instead that the magnitude of the blast load is much greater than the hydrostatic component of pressure. However, the theoretical models presented here are easily extended to include hydrostatic effects that are encountered in practice.

Static buckling of cylindrical tubes due to external pressure has been studied for several decades, and classical theories such as the linear theory of Timoshenko and Gere (1961) provide reasonable estimates of the buckling threshold under many conditions; the accuracy of these predictions is especially good for long tubes (Batdorf 1947). Much effort has also been committed toward modeling the effects of boundary conditions (Galletly and Bart 1956, Sobel 1964), shape imperfections (Hutchinson and Koiter 1970, Kempner et al. 1957, Koiter 1945), and nonlinear pre-buckling deformation (Stein 1964, Yamaki 1969), and considerable progress has been made. Review papers on static buckling have been written by Simitses (1986) and Teng (1996), and a detailed description of the most common theoretical techniques for elastic static buckling is given by Yamaki (1984).

For transient pressure loads, the analysis is in some ways more complicated. Due to inertial effects, structures can usually withstand loads significantly exceeding the static buckling limit, provided that the duration is sufficiently short. As a result, the buckling threshold depends on both the amplitude of the applied load and its duration, or equivalently, its impulse. For time-varying loads, there is also potential for large deformations due to resonance, which may be caused either by periodic forcing or by autoparametric resonance (Goodier and McIvor 1964, Lindberg 1974). In this review, such resonance is not considered because our experiments show that the effects of FSI and internal material damping limit the amplitude of axisymmetric vibrations and inhibit autoparametric excitation.

In the 1960s, an extensive research effort was initiated by the Air Force Weapons Laboratory to study dynamic buckling of tubes subjected to blast waves in air (Abrahamson et al. 1966, Lindberg and Firth 1967, Lindberg and Florence 1987, Lindberg et al. 1965, Lindberg and Sliter 1969). These experiments used several different configurations of explosives to measure the dynamic buckling threshold over a range of pressures and impulses, and fair agreement with two-dimensional, plane-strain models was found.

As an outcome of these experiments, Lindberg (1964) reported on the buckling of very thin shells ($a/h = 480$) subjected to "impulsive loads," that is, loads which have a duration much less than the response time of the tube. He found that the tube's buckling behavior was controlled purely by elastic effects. Abrahamson et al. (1966) studied the corresponding problem of thick shells ($a/h = 10-30$) subjected to impulsive loads and showed that the response is governed primarily by effects of plasticity. Anderson and Lindberg (1968) investigated the response of tubes under "quasi-impulsive" loads, i.e., loads having a duration comparable to the response time of the tube. They found that both elastic and plastic effects can be important and developed a model which takes this factor into account.

A common challenge in all of these studies was making a direct comparison between experiments and theoretical models. In many of the experiments, the response was evaluated only through post-collapse analysis, so the predictions of models during the intermediate stages of buckling could not be evaluated. In a few instances (Lindberg et al. 1965, Lindberg 1974) dynamic strain measurements were recorded; however, only limited data was reported and the strain traces exhibited many features which could not be easily explained by the theoretical models.

Recently we have developed an experimental facility for studying buckling of submerged tubes in which the loads are applied using shock waves in water rather than blast waves in air. This approach enables dynamic strain measurements to be made while the tube buckles, which provides significantly more information about the mechanics of failure than can be obtained from post-collapse inspection alone. In addition, our experimental facility can operate in regions in the parameter space that are not easily reachable using explosives, and can also explore the effects a dense fluid surrounding the tube. As a result, our new experiments nicely complement previous results and allow for a more thorough evaluation of theoretical dynamic buckling models.

The first part of this chapter presents some recent experimental measurements of the buckling behavior of submerged tubes. The specimen tubes are relatively thick, with ratios of radius to wall thickness in the range of 13–25. The duration of the applied pressure load is 50–100 times the period of axisymmetric vibration, so the response is considered "quasi-static." However, the peak pressures required to buckle the tubes are still five to ten times greater than the static buckling pressure, demonstrating the importance of inertial effects. For the load conditions considered in this chapter, the buckling behavior is controlled by elastic effects alone.

The second part of this chapter describes an approximate theoretical model of the tube's buckling behavior. Observations from experiments are used to justify several simplifications, including the treatment of the tube as a 2D ring in a state of plane strain along its axis. This model is then used to predict the dynamic response of tubes and generate buckling threshold curves, and good agreement with experimental data is found in both cases.

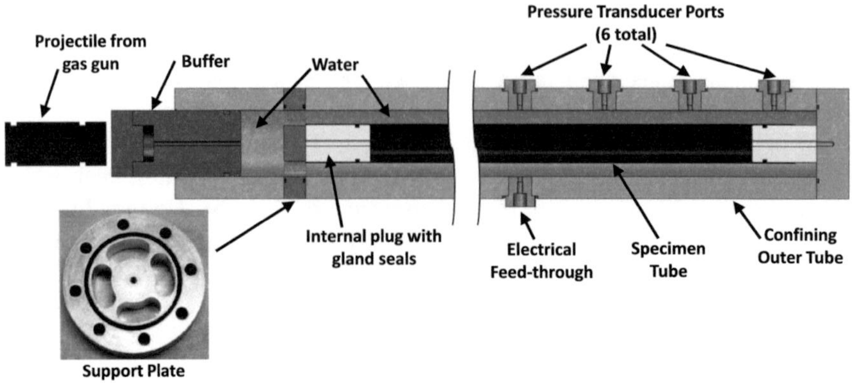

Fig. 7.1 Diagram of experimental apparatus. Confining outer tube is 1.1 m long overall and 127 mm in outer diameter. Inset: photograph of support plate (127 mm outer diameter) used to prevent axial loading of the specimen tube

7.2 Experimental Results

Figure 7.1 shows a diagram of our experimental facility, which is an adaptation of the apparatus developed by Damazo et al. (2010). A thin-walled specimen tube is mounted concentrically inside of a larger cylindrical vessel made of 4140 high strength steel and having an inner radius of 38.1 mm and wall thickness of 25.4 mm. The ends of the specimen tube are supported and sealed using internal plugs with gland seals. These plugs restrict inward radial motion of the specimen tube, but do not constrain axial displacement except through the friction between the tube and plug that is generated as the tube compresses. The upper plug (left side of Fig. 7.1) is fixed to a support plate so that pressure waves in the water do not produce an axial load on the specimen tube. As shown in the inset photograph, this support plate features four holes which allow pressure waves to pass through.

The annular space between the specimen tube and the thick-walled outer tube is filled with water, and a small volume of water is also present above the support plate. Shock waves in the water are generated using a vertical gas gun, which is described in more detail by Inaba and Shepherd (2010). The gas gun fires a steel projectile into a buffer, which generates a stress wave that is then transmitted into the water and travels along the outside of the specimen tube. The resulting pressure pulse in the water is approximately exponential in shape, and the rate of decay is dictated by the wave mechanics in the projectile and the buffer. As a result, the duration of the pressure pulse can be controlled by changing the lengths and materials of these two components. The results reported in this chapter were obtained using a 1.5 kg steel projectile that is 120 mm long and 50 mm in diameter. The buffer is made from either aluminum or steel and is 130 mm long. The buffer is capped with a 25.4 mm thick steel striker plate to prevent damage during the projectile impact.

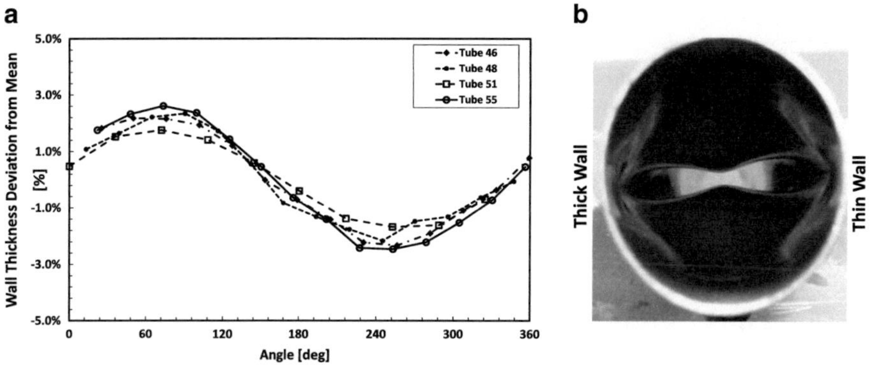

Fig. 7.2 (**a**) Variations in wall thickness around the circumference of several specimen tubes. Variations are reported as a percentage of the mean wall thickness. (**b**) Photograph of a tube buckled in mode 2 orientation. Tube is 6061-T6 aluminum with $a = 15$ mm and $a/h = 17.4$

The characteristics of the pressure wave, such as speed, peak pressure, and impulse, are measured using a row of six piezo-electric pressure transducers (PCB model 113A23) along the side of the thick-walled outer cylinder. The transducers have a response time of less than 1 μs and a resonant frequency greater than 500 kHz, and are sampled at 1 MHz. Since the radii of the specimen tube and the outer cylinder differ by only 15–20 mm, the radial transit time of pressure waves through the annulus of water is about 10–15 μs, which is quite short compared to the other timescales involved in the problem. As a result, the measured pressure is expected to be very close to that at the surface of the specimen tube.

Deformation of the specimen tube is measured using bonded strain gauges, which are coated with a compliant sealant (Vishay PG, M-Coat D) to eliminate electrical interference by the water. Signals are amplified using a Vishay 2310B signal conditioners operated in wide-band mode (−3 dB at 250 kHz) and digitized at 1 MHz. The strain gauges are usually installed in the hoop direction, and by installing multiple gauges around the circumference of the tube at a given axial location, the number of lobes (usually 2–3) in which the tube vibrates or buckles can be determined.

The specimen tubes are all 0.91 m long and have radii a of 15, 20, or 22 mm and wall thicknesses h of 0.9, 1.25, or 1.5 mm. The combinations of radius and wall thickness that were tested give a/h in the range of 13–25 and L/a in the range of 40–60. Tubes were made from 6061-T6 aluminum, 3003-H14 aluminum, or 304 stainless steel. The steel tubes were of welded construction, while the aluminum tubes were extruded (seamless).

We have found that systematic variations in wall thickness of the specimen tubes significantly affect the orientation of the buckles. For extruded tubes, the wall thickness typically varies sinusoidally around the circumference, as depicted in Fig. 7.2a where measurements were made using a round-tipped micrometer. These variations in wall thickness appear to be uniform or slowly varying along the axis of the tube.

It is important to point out that other types of imperfections, such as noncircular shape or residual stresses, may also be present. However, as shown in Fig. 7.2b, our experiments indicate that variations in wall thickness are the dominant imperfection since the major axis of mode 2 buckles always aligns with the point of minimum wall thickness. This consistent behavior also facilitates dynamic strain measurements: the strain gauges can be placed along the thinnest side of the tube to measure the maximum strains that occur.

7.2.1 Linear Elastic Deformation

For pressure waves of small enough amplitude or short enough duration, the tube's response is elastic and primarily axisymmetric. Examples of pressure and strain traces, plotted on an x-t diagram, are shown in Fig. 7.3. An incident wave characterized by a jump in pressure followed by an exponential decay travels along the tube with an approximately constant speed and wave profile. This wave then reflects off of the bottom end of the test fixture and traverses the specimen tube once more. Note that the maximum pressure is greater than the static buckling pressure by a factor of 1.4, yet the motion remains axisymmetric. This occurs because the duration of the pressure pulse is short and buckles do not grow rapidly enough at this pressure to be observable.

It is interesting to note that the tube's response for low pressures is analogous to that of a waterhammer event occurring in an internally pressurized tube. Such a situation was examined using a very similar projectile-impact facility by Inaba and Shepherd (2010), who recorded pressure and strain traces that look very much like those in Fig. 7.3. This suggests that the axisymmetric deformation that occurs prior to buckling can be predicted using models that have been developed and thoroughly evaluated by the waterhammer community (Shepherd and Inaba 2010, Tijsseling 1996, Wiggert and Tijsseling 2001, Wylie and Streeter 1993).

7.2.1.1 Simple Model of Pressure Wave

One of the simplest of these waterhammer models, first devised by Korteweg (1878) and experimentally validated by Joukowsky (1900), is based on the assumption that radial inertia of the tube and fluid, as well as the effects of bending of the tube along its axis, are small. These assumptions are strictly valid only for low frequencies and long wavelengths, but experiments by Inaba and Shepherd (2010) have demonstrated that the performance of this model is quite good for many practical conditions. As discussed by Shepherd and Inaba (2010), the speed c_k of the coupled fluid–solid wave predicted by the Korteweg–Joukowsky theory is given by:

$$c_k = \frac{c}{\sqrt{1+\beta_k}} \tag{7.1}$$

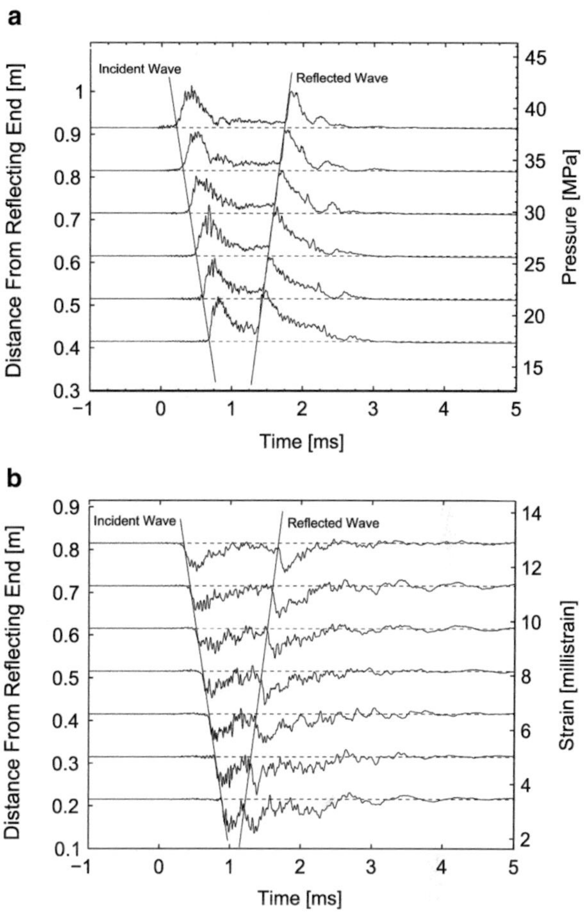

Fig. 7.3 Pressure (**a**) and strain (**b**) histories for a tube with $p_{max} / p_{static} = 1.4$. Tube is 6061-T6 aluminum with radius $a = 15$ mm and $a/h = 17.4$

where

$$\beta_k = c^2 \rho_w \frac{2a}{Eh}.$$

Here β_k is a dimensionless parameter that describes the extent of fluid–solid coupling, which increases as β_k increases. Given the success of this simple model for analyzing pressure waves inside of tubes, this theory will be extended for use in the present annular geometry.

If the width of the fluid annulus is small enough that the radial reverberation time of pressure waves is small compared to other timescales, then it is reasonable to model the fluid pressure as averaged across the annular gap. This is equivalent to neglecting the radial inertia of the fluid. Using this assumption, the linearized equations of conservation of mass and momentum (Lighthill 1978) of the fluid can be combined and written in the following form:

$$\frac{\partial^2 p}{\partial x^2} = \frac{1}{c^2}\frac{\partial^2 p}{\partial t^2} + \frac{\rho_w}{A_o}\frac{\partial^2 A'}{\partial t^2} \tag{7.2}$$

where ρ_w is the fluid density, A_o is the original cross-sectional area of the annulus, and A' is the change in area due to deformation. As this equation shows, the system obeys the usual linear wave equation for the pressure with an additional term involving the change in cross-sectional area A of the tube. This additional term models the compliance of the tube wall and, as will be shown, has the effect of reducing the speed of pressure waves.

Deformation of both the inner and outer tubes of the annular channel are modeled. Deflection of the outer tube (radius b) is denoted w_o and is taken positive outward, while deformation of the inner tube (radius a) is denoted w_i and assumed positive inward. The area A of the channel can then be written in terms of these deformations.

$$\begin{aligned} A(x,t) &= A_o + A'(x,t) \\ &= \pi(b^2 - a^2) + 2\pi(aw_i + bw_o). \end{aligned} \tag{7.3}$$

The Korteweg–Joukowsky hypothesis neglects bending and inertia of the tube, treating it as a massless membrane. Using this approximation, the radial deformations of the inner and outer tubes are proportional to the pressure:

$$w_i = \frac{pa^2}{E_i h_i} \qquad w_o = \frac{pb^2}{E_o h_o} \tag{7.4}$$

where E and h are the elastic modulus and wall thickness. Substituting (7.3) and (7.4) into (7.2) produces a wave equation with a modified wave speed:

$$\frac{\partial^2 p}{\partial x^2} = \left[\frac{1}{c^2} + \frac{2\rho_w}{b^2 - a^2}\left(\frac{a^3}{E_i h_i} + \frac{b^3}{E_o h_o}\right)\right]\frac{\partial^2 p}{\partial t^2}. \tag{7.5}$$

The effective wave speed can be expressed in the form of (7.1) if a modified fluid–solid coupling parameter β_{mod} is used:

$$\beta_{\text{mod}} = \frac{2\rho_w c^2}{b^2 - a^2}\left(\frac{a^3}{E_i h_i} + \frac{b^3}{E_o h_o}\right). \tag{7.6}$$

The predictions of this theory are compared with our present experimental measurements in Fig. 7.4. The experimental wave speeds were obtained by recording the time at which the pressure wave reached each of the six pressure transducers and fitting a line to these data points. In all cases, the experimentally measured wave speeds are within 1–4 % of the theoretical predictions; the fact that the experimental

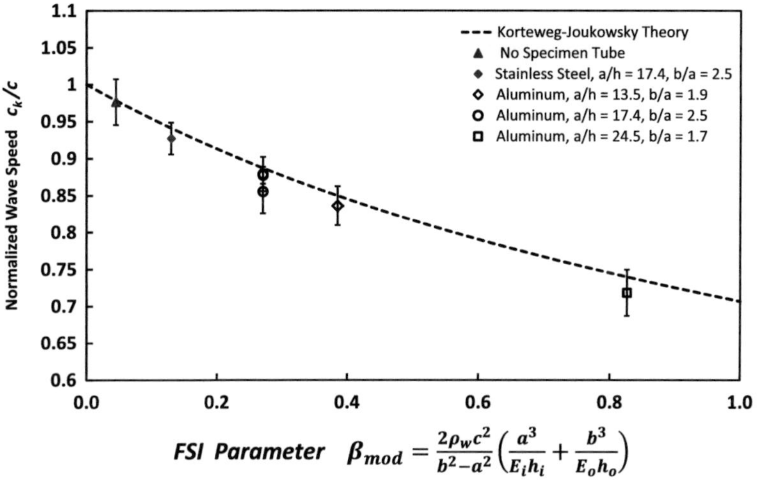

Fig. 7.4 Comparison of predicted and measured wave speeds for several tube materials and sizes

speeds are consistently low is likely due to air bubbles in the water which accumulate when the vessel is filled.

7.2.1.2 Coupling Between Fluid and Solid Motion

The high frequency oscillations observed in the pressure and strain traces of Fig. 7.3 are not unidentifiable "noise," but rather are repeatable axisymmetric vibrations that are excited by the sharp pressure wavefront. This fact is shown more clearly for a different tube in Fig. 7.5, where pressure and strain traces at a single axial location are directly compared with one another. At this axial location, three hoop strain gauges have been installed: one is mounted on the thinnest side of the tube, while the other two are offset by 90° and 180° around the circumference. The phase difference between these gauges then indicates whether or not the motion is axisymmetric. Upon arrival of the incident pressure wave at $t=0$, the tube begins to vibrate axisymmetrically, as indicated by the close agreement of all the strain gauges. At about $t=0.5$ ms, the reflected pressure wave from the bottom of the tube arrives and further increases the pressure and strain. At this point a slight departure from axisymmetric motion is observed. As the pressure wave dies away, the tube continues to vibrate in a mode 2 shape, as indicated by the fact that strain gauges positioned 180° apart are in phase while gauges 90° apart are out of phase.

Figure 7.5 leads to several important conclusions about FSI in this geometry. The first observation is that as long as the deformation remains axisymmetric, the pressure is very nearly proportional to the strain. In fact, on the time interval of 0–0.5 ms the instantaneously measured strain $\varepsilon(t)$ is within about 10 % of the predicted static value based on the instantaneously measured pressure $p(t)$:

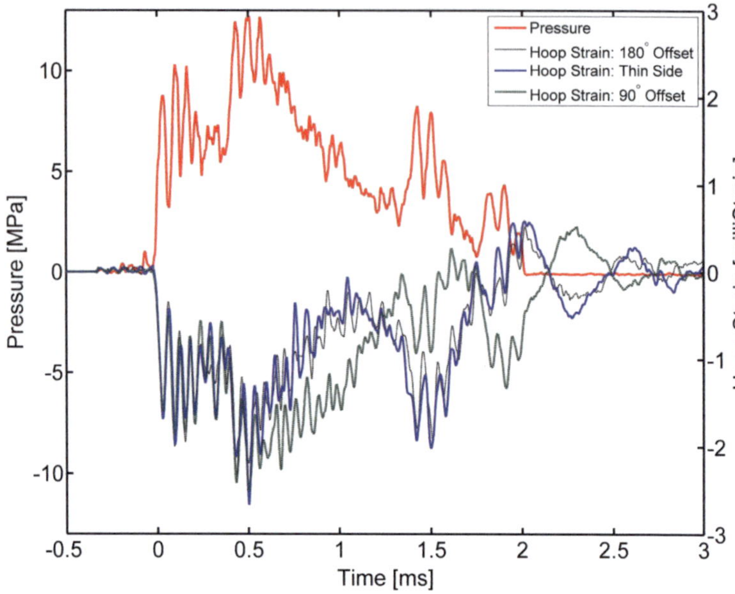

Fig. 7.5 Comparison of pressure and strain traces 250 mm from the reflecting end wall. Tube is 6061-T6 aluminum with $a = 20$ mm, $a/h = 13.5$, and $p_{max}/p_{static} = 2.5$

$$\epsilon(t) = \frac{p(t)a(1-\nu^2)}{Eh}. \tag{7.7}$$

This suggests that the effects of bending along the tube axis are small and the use of a membrane-type model for the pre-buckling motion of the tube is reasonable.

A second important observation is that the amplitude of axisymmetric vibrations is fairly small compared to the quasi-static component of strain, and the vibrations die away on the order of tens of cycles. Furthermore, as the specimen tube is made thinner (such as in Fig. 7.3), the vibratory component of strain becomes even smaller. This differs from situations in which the effects of FSI are negligible, such as a tube subjected to an internal shock or detonation wave. In those cases, the amplitude of vibration is about twice the quasi-static strain and the vibrations can persist for many cycles without significant change in amplitude (Shepherd 2009). Based on these observations, it is reasonable to expect that autoparametric excitation due to the interaction between the axisymmetric and non-axisymmetric modes will not be significant in our present experiments.

A third observation from Fig. 7.5 is that unlike the axisymmetric vibrations, non-axisymmetric vibrations are not strongly coupled to the fluid pressure. This is clearly demonstrated by examining the pressure and strain traces after $t = 2$ ms, where the tube continues to vibrate in a mode 2 shape but the pressure remains constant. This behavior suggests that the non-axisymmetric vibrations are nearly volume-preserving, at least provided that these vibrations are small. The smallness criterion can be estimated by considering an annulus of water of outer radius b and inner radius a, with the inner tube subjected to a radial deformation w in the form of a Fourier series:

$$\mathrm{w}(\theta,t) = \mathrm{w}_o(t) + \sum_{n=1}^{\infty} \mathrm{w}_n(t)\cos(n\theta). \tag{7.8}$$

If it is assumed that one particular mode $n=N$ is dominant, then the relative change in volume of the annulus due to the deformation (7.8) is given by:

$$\frac{\Delta V}{V} = \left(\frac{2w_o}{a} + \frac{w_o^2}{a^2} + \frac{w_N^2}{2a^2}\right) / \left(\frac{b^2}{a^2} - 1\right). \tag{7.9}$$

Typically $w/a << 1$, so the linear term is dominant and the change in volume depends primarily on the axisymmetric motion. It will be shown in Sect. 7.3.2.1 that the non-axisymmetric vibration frequencies are low enough that the radiated pressure field is negligible, which means that changes in pressure scale with the volume change:

$$\Delta p = K_f \frac{\Delta V}{V} \tag{7.10}$$

where K_f is the bulk modulus of the liquid. As a result, the change in pressure due to non-axisymmetric vibrations alone (i.e., for $w_o=0$) can be expressed as:

$$\Delta p = \frac{K_f}{2(b^2/a^2 - 1)} \left(\frac{w_N}{a}\right)^2. \tag{7.11}$$

For a practical situation, such as a steel tube submerged in water and deformation restricted to less than the yield strain, this pressure is less than about 50 kPa unless b/a is very small (less than about 1.05). This result shows that non-axisymmetric deformation has only a small effect on the pressure in the fluid surrounding the tube, and hence the speed and amplitude of pressure waves depend primarily on the axisymmetric component of deformation. This fact will also be demonstrated experimentally in the following sections where experimental results at higher pressures are considered. The implication is that in developing buckling models, the motion of the pressure wave can initially be decoupled from the buckling deformation.

7.2.2 Nonlinear Elastic Deformation

As we have already observed in Fig. 7.5, higher pressures begin to excite non-axisymmetric modes of deformation. As the pressure or impulse of the load increases

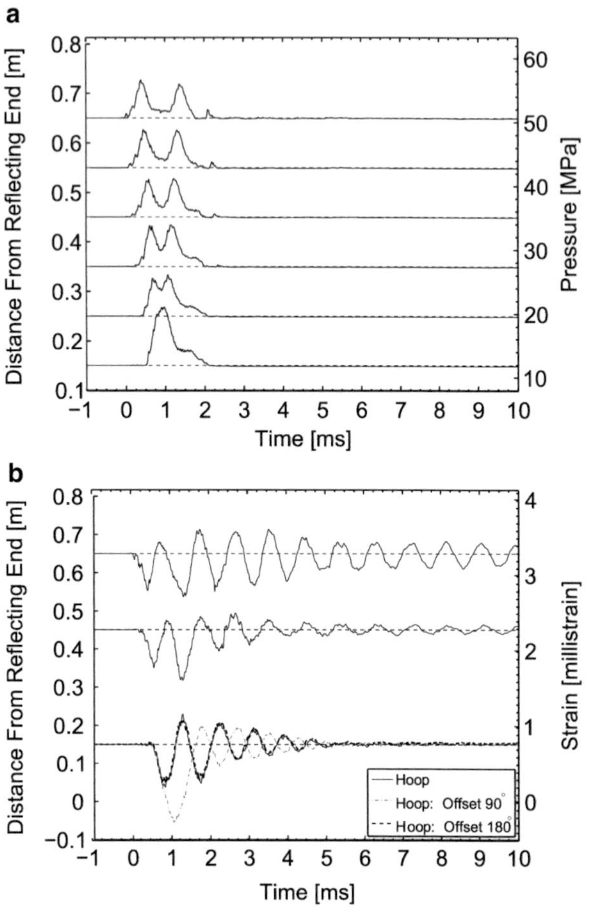

Fig. 7.6 Pressure (**a**) and strain (**b**) histories with multiple strain gauges around the circumference at a single axial location. Tube is 304 stainless steel with radius $a = 22$ mm and $a/h = 18.5$. $p_{\max} / p_{\text{static}} = 1.2$

further, these non-axisymmetric motions grow rapidly and become much larger than the axisymmetric component of deformation. An example of this is shown in Fig. 7.6.

The speed and shape of the pressure wave resemble those shown for lower pressures (e.g., Fig. 7.3), but the response of the tube is quite different. The strain traces feature a series of large amplitude, mode 2 vibrations which persist for much longer than the duration of the pressure load. Despite the excitation of non-axisymmetric displacement, the motion remains elastic and no permanent deformation is observed.

The three strain gauges located at the bottom of the tube (150 mm from the reflecting end) are spaced 90° apart around the circumference. The two strain gauges on opposing sides of the tube are exactly in phase with one another, while the strain gauge between them is exactly out of phase. This confirms that the deformation consists of two lobes. Note that the two gauges on opposite sides of the tube have

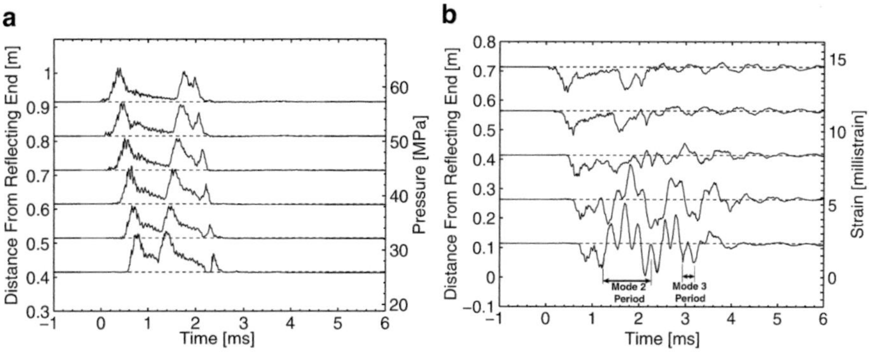

Fig. 7.7 Pressure (**a**) and strain (**b**) traces showing superposed deformation of modes 2 and 3. Tube is 6061 aluminum with radius $a = 15$ mm and $a/h = 17.4$. $p_{max} / p_{static} = 2.0$

been attached at the thickest and thinnest points on the tube wall. Thus the major axis of vibration is exactly aligned with the points of maximum and minimum wall thickness, indicating that this imperfection plays a key role in determining the orientation of the vibration.

It is interesting to note that in Fig. 7.6 the maximum pressure was 1.2 times the static buckling threshold and non-axisymmetric deformation was observed, while in Fig. 7.3 no non-axisymmetric deformation occurred despite having a larger value of $p_{max} / p_{static} = 1.4$. The reason for this behavior is that the load duration is the same in both cases, but due to differences in material properties (steel vs. aluminum) and tube diameter (22 mm vs 15 mm), the natural frequency of the tube in Fig. 7.3 is about 1.5 times that of the tube in Fig. 7.6. Thus when normalized by the tube's response time, the load duration for the tube in Fig. 7.6 is about 1.5 times longer and hence non-axisymmetric deformation is more strongly excited.

For thinner tubes and higher pressures, vibration in mode numbers greater than 2 can be observed. An example of mode 3 vibrations is shown in Fig. 7.7. The strain traces exhibit a superposition of vibrations in mode 2 (longer period of about 1 ms) and mode 3 (shorter period of about 0.3 ms). These vibration periods agree closely with theoretical predictions (described in more detail in Sect. 7.3.2.1), which helps confirm that the behavior has been correctly interpreted. This superposition of multiple excited modes is a well-known feature of dynamic buckling and is more pronounced for shorter impulses and thinner tubes (Lindberg and Florence 1987, Lindberg 1974). Note that the mode 3 vibrations are excited only near the reflecting end of the tube where the pressure is greatest; elsewhere the pressure is insufficient to destabilize these modes and only mode 2 vibrations are (slightly) excited.

In many of these experiments, axial strains have also been measured in order to determine whether or not bending along the tube axis plays an important role in the buckle development. The data has revealed that axial bending is negligible, since the measured axial strains are almost exclusively caused by Poisson coupling with the hoop strain. This is demonstrated in Fig. 7.8 where the measured axial

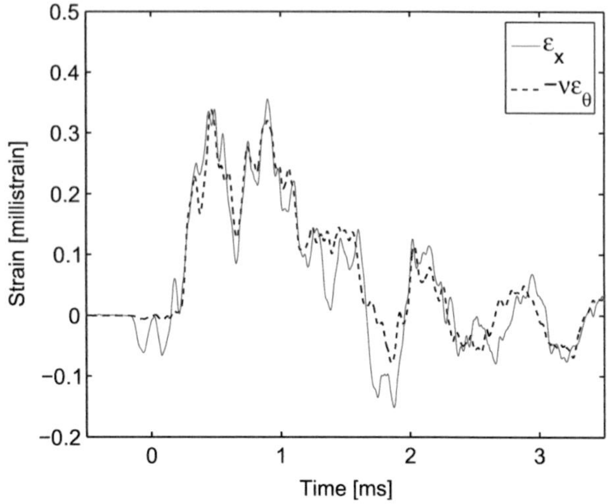

Fig. 7.8 Comparison of hoop (ε_θ) and axial (ε_x) strains 250 mm from the reflecting end of the tube. Gauges are spaced 90° apart circumferentially. Tube is 3003-H14 aluminum with $a = 15$ mm, $a/h = 17.4$, and $p_{max}/p_{static} = 1.8$

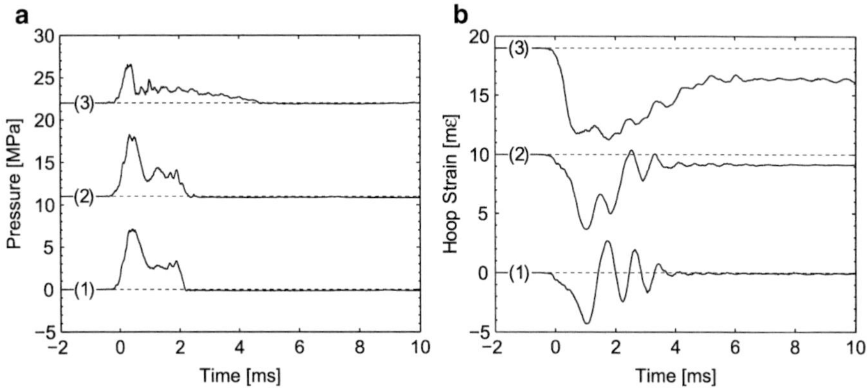

Fig. 7.9 Comparison of pressure (**a**) and strain (**b**) traces for three consecutive shots. For clarity, the traces have been offset vertically from one another. Traces are recorded 0.25 m from the reflecting end. Tube is 6061-T6 aluminum with $a = 15$ mm and $a/h = 17.4$

strain ε_x is compared with the axial strain due to Poisson coupling: $-\nu\varepsilon_\theta$. The only place where these two quantities differ substantially is in the precursor wave, which is to be expected since the precursor wave is a longitudinally dominant wave (Shepherd and Inaba 2010). Over the rest of the trace, the agreement is quite good, which indicates that the effects of bending along the tube axis (which would otherwise show up prominently in the axial strains) are quite small. This result adds further support for the use of a membrane-type model to predict the tube motion prior to buckling.

7.2.3 Onset of Plastic Deformation

Further increases in the load's pressure or duration lead to plastic deformation and the formation of plastic hinges at the extremities of the buckle where the strain is the greatest. This behavior is demonstrated in Fig. 7.9, where pressure and strain traces are plotted for three consecutive shots using the same tube from shot to shot.

The first shot, marked (1), produced purely elastic deformation with a peak strain of 4.2 millistrain (which is very close to the yield strain of about 4.5 millistrain for 6061 aluminum); the pressure and strain traces in this shot look very similar to those shown in previous elastic cases (e.g., Fig. 7.6).

The second shot, marked (2), has a peak strain of 6.3 millistrain, which exceeds the yield strain. As a result, slight permanent deformation is observed as indicated by the final strain offset of about 0.9 millistrain due to the residual stresses. Note that the pressure trace in this shot looks very similar to that of the preceding shot, which again supports the conclusion that the pressure wave is unaffected by the non-symmetric elastic vibration, even for relatively large strains.

In the third shot, marked (3), the peak strain is 7.7 millistrain and additional permanent deformation is observed. The strain is no longer oscillatory since much of the kinetic energy is dissipated via plastic work. In this case, the pressure trace looks quite different from that of the preceding shot. Despite a 5 % increase in projectile velocity from the gas gun, the peak pressure dropped by about 40 % relative to the preceding shot. Furthermore, the pressure wave spreads out and decays more slowly over time. This dispersion of the pressure wave and reduction in peak pressure are likely caused by the residual plastic deformation from the preceding shot. Evidently plastic deformation is not volume preserving, so pressure waves disperse as they encounter the volume change in regions of plastic deformation. This result suggests that our prior conclusion that the pressure wave can initially be decoupled from non-axisymmetric motion is only valid until plastic deformation occurs.

7.2.4 Single-Shot Collapse

The final regime of motion considered in these experiments is single-shot collapse, which occurs when the pressure and impulse are high enough to catastrophically fail the tube during a single pressure pulse. Examples of pressure and strain traces for a collapsing tube are shown in Fig. 7.10. Unlike the results shown previously, in which pressure and strain were measured near the bottom end of the tube where the pressure is greatest, in this shot the pressure and strain were measured 150 mm from the top end of the tube. For high pressure shots, this is the location at which collapse first occurs.

The strain trace exhibits a series of vibrations superimposed over an offset that steadily increases with time. The frequency of these vibrations is the same as those shown in Fig. 7.7, which indicates that these are mode 3 vibrations. The permanent

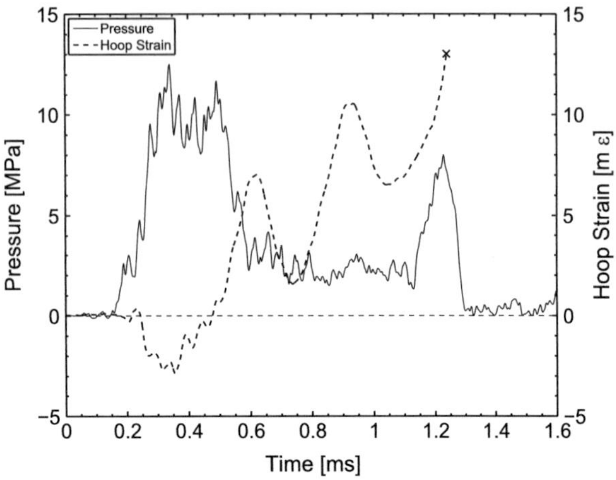

Fig. 7.10 Pressure and strain traces for a collapsing aluminum tube of radius $a = 15$ mm and $a/h = 17.4$. The static buckling pressure for this tube is 3.7 MPa. Pressure and strain were measured 150 mm from the top end of the tube. The strain gauge de-bonded from the surface of the tube at the point marked by an **x**

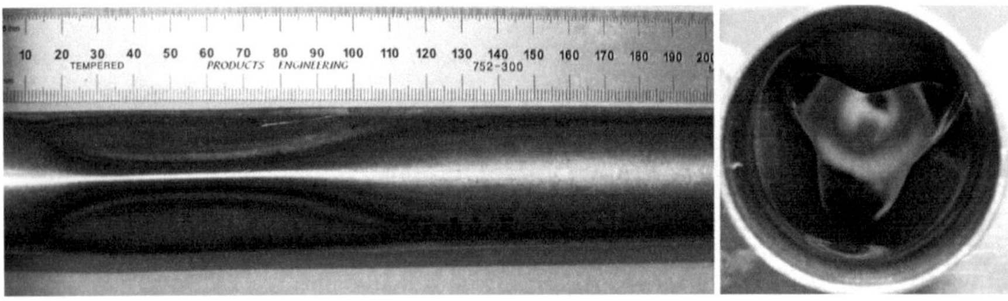

Fig. 7.11 Post-collapse photographs of tube following the test which generated the data shown in Fig. 7.10 and illustrating buckling in mode 3. Tube is 6061-T6 aluminum with mean radius $a = 15$ mm and $a/h = 17.4$

deformation observed after the test, shown in Fig. 7.11, was also mode 3, which supports this interpretation. The increasing offset of this strain trace appears to be the plastic component of strain. The fact that the peak-to-peak amplitude of the vibrations is equal to twice the yield strain lends further support for this conclusion. An interesting observation is that the buckle grows on a timescale that is significantly slower than that of the pressure wavefront. As a result, the leading edge of the pressure wave is not affected by the buckle. However, the pressure drops off sharply behind the incident wave, and the reflected pressure wave at about 1.2 ms is substantially attenuated.

It is interesting to consider whether the permanent deformation observed in these results is associated with flexure due to buckling alone, or consists also of hoop yielding. For loads with duration longer than the period of axisymmetric vibration, the pressure p_y beyond which the tube is brought to hoop yield is given by:

$$p_y < \frac{\sigma_y h}{a} \tag{7.12}$$

where σ_y is the yield stress. For the conditions used in Fig. 7.10, the pressure limit for hoop yielding is $p_y \sim 17$ MPa. The measured pressure is well below this value, and the overshoot of axisymmetric vibrations due to dynamic loading appears to be small; hence the plastic deformation observed in the strain traces is due primarily to flexure due to buckling.

The data in Fig. 7.10 are just one example of the possible types of behavior that can occur at higher pressures. As shown in the strain traces, the effects of plasticity become important almost immediately, so the development of the buckle is dictated by both elastic and plastic effects rather than elastic effects alone. An additional difficulty is that as the applied pressure increases, the mode number of the buckles increases and their length decreases. As a result, the buckles become more spatially localized and it is more difficult to measure their development using bonded strain gauges. Interpretation of the resulting strain measurements is also more challenging. Characterizing the buckling behavior under high pressures is a subject of ongoing investigation in our laboratory, but in this chapter we focus on the behavior at lower pressures and higher impulses where elastic effects are dominant.

7.2.5 Measurements of the Buckling Threshold

A convenient experimental criterion for the buckling threshold is the onset of plastic deformation. This criterion is practically meaningful since the tube's strength is greatly reduced after plastic deformation occurs, and this metric can also be applied quite repeatably in experiment using bonded strain gauges. Even if a strain gauge is not located at the point of peak strain, any plastic deformation that occurs produces residual stresses and strains which are readily detected.

For dynamic loads, the buckling threshold is a function of both peak pressure and impulse (Lindberg and Florence 1987). In principle, then, one could measure the buckling threshold by fixing one of these variables and gradually increasing the other until plastic deformation occurs, which would reveal a single point along the buckling threshold. By repeating this process, one could then map out the buckling threshold over some range of pressures and impulses.

In our experiments, we cannot vary the pressure and impulse independently; rather, the load duration is approximately fixed and as the speed of the projectile from the gas gun increases, the pressure and impulse increase in proportion to one another with the load duration as the constant of proportionality. The load duration is governed by the wave dynamics between the buffer and projectile used to generate the pressure pulse (see Fig. 7.1), so adjustment of the pulse duration is accomplished by changing the materials and dimensions of these components. Our measurements of the buckling threshold are then made by fixing the load duration and increasing the pressure and impulse until plastic deformation occurs.

An example of the measured buckling threshold for three identical tubes is shown in Fig. 7.12. Each line of data points represents a sequence of consecutive shots for which the peak pressure and impulse were gradually increased, and the slopes of these lines were adjusted by changing the duration of the load. The dark x's mark

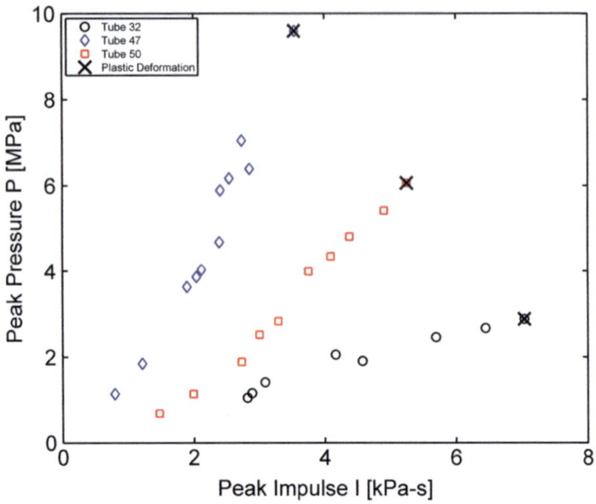

Fig. 7.12 Pressure vs. impulse curves for three identical tubes under loads of various ratios P/I. Tubes are 6061-T6 aluminum with mean radius $a = 22$ mm and $a/h = 24.5$. The classical static buckling pressure for these tubes is 1.3 MPa. The bottom curve was generated using an alternate apparatus which features windows for visualization of the buckling process; the compliance of these windows facilitated the small slope *P/I* achieved in that data set

the points at which plastic deformation first occurred, and hence they map out a segment of the buckling threshold.

This data exhibits the expected trend that the tube is able to withstand greater pressures when the impulse is shorter. The classical static buckling threshold for this tube is 1.3 MPa, so pressures more than seven times this value were needed to fail the tube at the lower impulses. This demonstrates that despite the classification of the load as "quasi-static," effects of inertia are still quite important.

As described by Lindberg and Florence (1987), for very large impulses the buckling threshold curve tends asymptotically toward a horizontal limit that is close to the static buckling pressure. For very short impulses, the threshold curve approaches a vertical limit at a particular critical impulse. However, as the impulse of the load is decreased, the effects of plasticity eventually become important in determining the initial buckling behavior (Abrahamson et al. 1966). We have not yet made measurements in this regime, but some data are available from Lindberg and Sliter (1969) which support these expectations.

7.3 Theoretical Models

7.3.1 Background

The general theory of dynamic buckling due to pressure loads that vary along the tube axis remains rather undeveloped. A few researchers have considered this and related problems and made some progress, but usually the results are restricted to very specific situations or are not readily applied to engineering problems. Instead, progress has been made by applying simplifications that are valid for restricted sets of load conditions or tube geometries. Here we briefly review several noteworthy investigations in order of increasing level of approximation.

To the authors' knowledge, only one researcher has reported on the buckling of cylindrical shells due to transient external pressure loads that are not uniform along the tube axis. Hegemier (1966, 1967) analyzed a class of steady-state traveling loads and demonstrated a reduction in the buckling pressure as the speed of the load increased. He also postulated the existence of a critical load speed at which, in the absence of damping, the tube buckles under arbitrarily low pressures. At this critical speed, the group, phase, and load velocities are all equal, so energy introduced by the load cannot disperse away from the load front. Such behavior is quite similar to the resonance that occurs in tubes loaded internally by shock or detonation waves, which results in large displacements when the speed of the shock or detonation reaches the same critical velocity (Beltman and Shepherd 2002, Schiffner and Steele 1971, Simkins et al. 1993, Tang 1965).

Several researchers have employed a considerable simplification by assuming that the pressure load is spatially uniform, but still modeling axially varying pre-buckling displacements. Thus the boundary conditions are strictly enforced at the ends of the tube and the effects of bending along the tube are explicitly modeled. Lockhart and Amigazo (1975) assumed that uniform pressure was instantaneously applied to the tube and held constant for all subsequent time. They obtained the interesting result that in the limit of small imperfections, the dynamic buckling threshold could be obtained directly from the classical buckling load (i.e., for a perfect tube) and the static buckling load (i.e., for an imperfect tube). However, their result is only applicable for loads that are a step function in time and does not apply to finite-duration loads. Their approach is similar to that of Budiansky and Hutchinson, who developed approximate relationships between the dynamic and static buckling loads for general structures rather than for a specific geometry (Budiansky and Hutchinson 1965, Hutchinson and Budiansky 1966).

The same problem of a tube subjected to uniform pressure, incorporating axially varying pre-buckling displacements, was studied by McIvor and Lovell (1968) using Fourier series in both the axial and circumferential directions. This produces a set of ordinary differential equations for the Fourier modes which, due to the nonlinearity of the shell equations, are highly coupled. However, for small deformations, many of the nonlinear terms were found to be negligible and the initial buckling behavior was shown to be governed by a Mathieu-type equation for each Fourier mode. The stability characteristics of this type of equation are quite well known (McLachlan 1964).

Due to the difficulty of the general theory of dynamic buckling, most researchers have simplified their analysis to the case of a two-dimensional ring. Goodier and McIvor (1964) wrote a landmark paper on the elastic stability of thin shells using this assumption, and Abrahamson and Goodier (1962) developed the corresponding theory for plasticity-dominated motions which occur for thick tubes or very low impulses. The intermediate range in which both elastic and plastic effects are important was considered by Anderson and Lindberg (1968).

The objective of this chapter is not to discuss the general theory of dynamic buckling, but rather to describe some simple theories which still provide useful predictions that agree well with experimental results. Such simple theories also give insight regarding the parameters that affect buckling. In accordance with this goal, a theory that is very similar to that of Goodier and McIvor (1964) is adopted in which the tube is treated as a plane-strain ring. This model is then modified to allow the tube to be loaded by an external pressure rather than an initial radial velocity, as well as to account for the effects of FSI.

7.3.2 *Elastic Two-Dimensional Model*

The tube is modeled as a ring in a state of plane strain along the tube axis. This 2D approximation and the model that follows are only valid if several constraints on the geometry and load conditions are satisfied:

Load Speed: In order to apply the 2D approximation, axial motion of the tube and bending along the tube axis must be negligible. This places restrictions on the speed of the applied load. Mann-Nachbar (1962) performed a parameter study which suggests that if a shock wave traveling along a tube moves sufficiently rapidly, then bending stresses are very small and the deformation can be predicted quite accurately using a membrane-type model, which would be equivalent to the present 2D ring. For the membrane model to be accurate, the load must travel faster than the "critical velocity" studied by Beltman and Shepherd (2002). For loads which travel at speeds very close to this critical speed, the displacements and stresses are highly amplified, but as the load speed is raised above the critical velocity, the contribution of axial bending to the total response of the tube decreases rapidly. To a good approximation, this critical speed υ_{cr} can be estimated by Jones and Bhuta (1964):

$$\upsilon_{\mathrm{cr}} = \left[\frac{E^2 h^2}{3\rho^2 (1-\nu^2)^2 a^2} \right]^{1/4}. \tag{7.13}$$

A second constraint is that the load speed must remain below the shear wave velocity $\upsilon_s = \sqrt{\kappa G / \rho}$ and the dilatational wave velocity $\upsilon_d = \sqrt{E / \rho(1-\nu^2)}$; if these conditions are not met then the effects of transverse shear deformation and rotary inertia become important and the membrane-type model is no longer valid (Chonan 1977, Lin and Morgan 1956, Naghdi and Cooper 1956). Fortunately, for metal tubes of practical sizes, the lower limit on the load speed given by (7.13) is on the order of 500–1,000 m/s and the shear wave speed is on the order of 3,000 m/s. Many realistic load cases, such as shock waves in water or detonations in gaseous

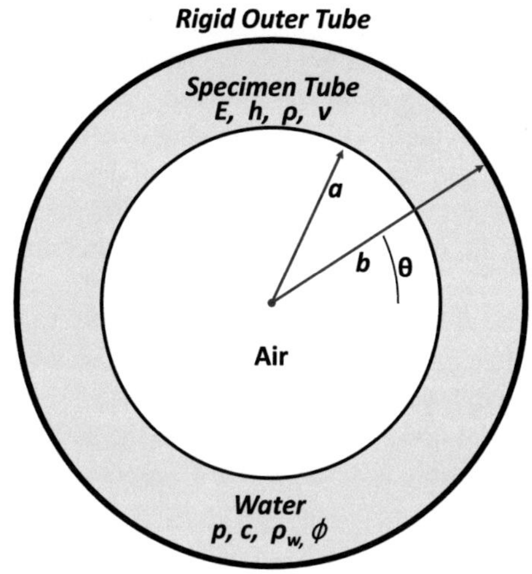

Fig. 7.13 Diagram of two-dimensional ring with surrounding annulus of fluid

media, travel at velocities which fall between these bounds. For instance, in the present experiment the critical velocity is about 900 m/s, the shear wave speed is about 2,800 m/s, and the measured pressure wave speeds are 1,200–1,500 m/s.

The restriction that the load speed fall between υ_{co} and υ_s provides a general guideline regarding the applicability of membrane-type models. To ensure that the membrane model is sufficiently accurate under a specific set of conditions, one can always solve the axisymmetric problem using a shell theory which includes effects of bending, shear deformation, and rotary inertia (such as the theory of Tang 1965) and compare the results with the predictions of a membrane model.

Tube Length: The 2D approximation also requires that the length of the specimen tube be much greater than its diameter, so that the local reinforcement due to boundary conditions at the ends of the tube is negligible. In the present experiments, the length of the tube is 0.91 m and the tube radius is 15–20 mm, so this condition is satisfied.

Wavelength of Pressure Wave: In the sections that follow, FSI is analyzed using a 2D fluid model, which is only valid if the axial wavelengths of the pressure load is much greater than the width of the annular gap between the specimen tube and the rigid outer tube. In the present experiments, the spatial extent of the load (duration times speed) is on the order of 1–2 m and the width of the fluid annulus is at most 23 mm, so this criterion is met.

Magnitude of Pressure Pulse: A final restriction on the pressure load is that the maximum pressure must not bring the tube to hoop yield. The maximum pressure which satisfies this condition can be estimated using (7.12). Abrahamson et al. (1966) have developed a shell theory which is appropriate if this criterion is violated, but since the focus of this chapter is on elastic effects and the corresponding experiments do not involve hoop yield, such a model is not pursued here.

A diagram of the 2D tube model is shown in Fig. 7.13. The deformable tube under consideration has a radius a, density ρ, elastic modulus E, and Poisson's ratio ν. It is surrounded by an annulus of dense fluid of sound speed c, pressure p, and density ρ_w, which is contained in a rigid outer cylinder of radius b. Radial deformation of the specimen tube is denoted w and is taken positive inward.

The equations of motion that follow are based on a model that was first introduced by Goodier and McIvor (1964) and later described in more detail in the monograph by Lindberg and Florence (1987). In those references, the buckling model was developed for impulsive loads which were represented by applying an initial inward radial velocity to the surface of the tube. In what follows, this model is suitably adapted to allow pressure loads to be simulated and the effects of FSI included. For conciseness, some of the details are omitted when they have been discussed elsewhere; a more comprehensive discussion of these details is available in the previously mentioned references.

For a ring in a state of plane strain, the total hoop strain ε_θ consists of a component due to stretching of the middle surface and a component due to bending:

$$\varepsilon_0 = \varepsilon_0 + \kappa z. \tag{7.14}$$

Here $\varepsilon_o(\theta)$ is the hoop strain at the middle surface of the tube, $\kappa(\theta)$ is the curvature, and z is the normal distance from the middle surface. The membrane strain ε_o and curvature κ are modeled by choosing an appropriate set of strain–displacement relations. Many different strain–displacement models have been proposed (Leissa 1973, Mushtari and Galimov 1961), but in the present case the following relations, based on the work of Lindberg and Florence (1987), are selected:

$$\varepsilon_0 = \frac{1}{a}\frac{\partial v}{\partial \theta} - \frac{w}{a} - \frac{w}{a^2}\frac{\partial \mathrm{v}}{\partial \theta} + \frac{1}{2a^2}\left(\frac{\partial w}{\partial \theta}\right)^2 + \frac{1}{a^2}\left(\frac{\partial w}{\partial \theta}\frac{\partial w_i}{\partial \theta} - w_i\frac{\partial \mathrm{v}}{\partial \theta} - w\frac{\partial \mathrm{v}_i}{\partial \theta}\right) \tag{7.15a}$$

$$\kappa = \frac{1}{a^2}\left(w + \frac{\partial^2 w}{\partial \theta^2}\right). \tag{7.15b}$$

In these equations, w and v are the radial and circumferential displacements. Displacements with subscript i describe the shape imperfection, that is, the initial deviation of the tube from a circular shape when no external loads are present. Thus (7.15) models the changes in membrane strain and curvature that are produced when the tube deforms relative to an initially noncircular shape.

Using the above models, the strain energy U and the kinetic energy T_K per unit length of the tube are expressed as:

$$U = \frac{1}{2}\frac{Eha}{(1-\nu^2)}\int_0^{2\pi}\left(\varepsilon_o^2 + \frac{h^2}{12}\kappa^2\right)d\theta \tag{7.16a}$$

$$T_K = \frac{1}{2}\rho h a \int_0^{2\pi}\left(\frac{\partial w}{\partial t}\right) + \left(\frac{\partial v}{\partial t}\right)^2 d\theta \tag{7.16b}$$

To account for loading by the external fluid pressure, the work W_p per unit length done by the pressure as the tube deforms must also be included.

$$W_p = \int_0^{2\pi}(pw)\,a\,d\theta. \tag{7.17}$$

This model is much more conveniently described in terms of several dimensionless parameters:

$$[W, V, W_i, V_i]^T = \frac{1}{a}[w, v, w_i, v_i]^T \qquad \beta^2 = \frac{h^2}{12a^2}$$
$$T = t\sqrt{\frac{E}{\rho(1-\nu^2)a^2}} \qquad P = \frac{pa(1-\nu^2)}{Eh}. \tag{7.18}$$

In terms of these dimensionless parameters, the strain energy, kinetic energy, and pressure work are given by:

$$U = \frac{1}{2}\frac{Eha}{(1-\nu^2)}\int_0^{2\pi}\left(\varepsilon_o^2 + \beta^2(\kappa a)^2\right)d\theta \tag{7.19a}$$

$$T_k = \frac{1}{2}\frac{Eha}{(1-\nu^2)}\int_0^{2\pi}\left(\dot{W}^2 + \dot{V}^2\right)d\theta \tag{7.19b}$$

$$W_p = \frac{Eha}{(1-\nu^2)}\int_0^{2\pi} PW\,d\theta \tag{7.19c}$$

where dots above the variables denote differentiation with respect to the dimensionless time T. The deformations, shape imperfections, and pressure can now be expanded using Fourier series:

$$W = W_o(T) + \sum_{n=2}^{\infty} W_n(T)\cos(n\theta) \tag{7.20a}$$

$$P = P_o(T) + \sum_{n=2}^{\infty} P_n(T)\cos(n\theta) \tag{7.20b}$$

$$V = \sum_{n=2}^{\infty} V_n(T)\sin(n\theta) \tag{7.20c}$$

$$\{W_i, V_i\} = \sum_{n=2}^{\infty} \{\delta_n, \alpha_n\}\cos(n\theta) + \{\gamma_n, \beta_n\}\sin(n\theta). \tag{7.20d}$$

The sine components of the Fourier series for the displacements W and V have been neglected since the frame of reference can always be rotated such that these components are zero. However, both terms in the expansions for the shape imperfections must be retained since there is no reason to expect that they have the same phase angle as the deformations. In addition, the $n=1$ mode of deformation has been excluded from the summations since it corresponds to a rigid translation of the tube.

Goodier and McIvor (1964) proposed a significant simplification in which the only extensional component of deformation is W_o, the axisymmetric part. All non-axisymmetric components of deformation are assumed to be inextensible. This assumption greatly simplifies the analysis by allowing the circumferential displacements V to be described in terms of the radial displacements W. This condition of inextensionality can be expressed as:

$$W - \frac{\partial V}{\partial \theta} \approx W_o. \tag{7.21}$$

Physically, this approximation assumes that, to first order in displacements, flexure of the tube produces no net change in length of the tube's middle surface. This aspect of the model is similar to the inextensional static buckling model described by Timoshenko and Gere (1961), with the exception that they assumed the motion to be completely inextensional, i.e., $W_o=0$. The inextensional approximation typically improves as the tube is made thinner or the wavelength of flexural motion grows longer (i.e., for low mode numbers n). In terms of the Fourier coefficients of (7.20), the inextensional condition implies that $nV_n = W_n$, $n\beta_n = \delta_n$, and $n\alpha_n = -\gamma_n$.

Using the inextensional model, Lindberg and Florence (1987) showed that the integrals for the strain energy and kinetic energy given in (7.19) evaluate to:

$$T_K = \frac{Eha\pi}{(1-v^2)} \left\{ \dot{W}_o^2 + \frac{1}{2}\sum_{n=2}^{\infty}\left(\frac{n^2+1}{n^2}\right)\dot{W}_n^2 \right\} \tag{7.22}$$

$$U = \frac{Eha\pi}{(1-v^2)} \left\{ W_o^2 + \frac{1}{2}\sum_{n=2}^{\infty}\left[\beta^2(n^2-1)^2 - W_o(n^2-2)\right]W_n^2 \right.$$
$$\left. - \sum_{n=2}^{\infty} W_o(n^2-2)\delta_n W_n \right\} + O(W_n^4). \tag{7.23}$$

In calculating these integrals, many of the terms involved in the strain–displacement equations integrate identically to zero due to the orthogonality of the trigonometric functions. Because of this property, only one of the Fourier coefficients for the shape imperfections, namely δ_n, plays a role in the response. A number of 4th order terms arising in (7.23) have been listed as $O(W_n^4)$ and will be excluded from the subsequent analysis since they are small during the initial stages of buckling.

The integral (7.19c) for the pressure work can be expressed as:

$$W_p = \frac{Eha}{(1-v^2)} \int_0^{2\pi} \left[P_o W_o + \sum_{n=2}^{\infty}\left[P_o W_n \cos(n\theta) + W_o P_n \cos(n\theta)\right] \right.$$
$$\left. + \sum_{n=2}^{\infty}\sum_{m=2}^{\infty} P_n W_m \cos(n\theta)\cos(m\theta) \right] d\theta \tag{7.24}$$

The first-order terms integrate to zero and the second-order terms are nonzero only if $m=n$, due to the orthogonality of the cosines. The resulting expression for the pressure work is:

$$W_p = \frac{Eha\pi}{(1-v^2)}\left[2P_o W_o + \sum_{n=2}^{\infty} P_n W_n \right]. \tag{7.25}$$

Having obtained expressions for the kinetic energy, strain energy, and work due to external loads, ordinary differential equations for the Fourier modes W_o and W_n can now be obtained from Lagrange's equation of motion:

$$\frac{d}{dT}\left(\frac{\partial T_K}{\partial \dot{q}_i}\right) - \frac{\partial T_K}{\partial q_i} + \frac{\partial U}{\partial q_i} = \frac{\partial W_p}{\partial q_i} \tag{7.26}$$

where q_i is a generalized displacement, which may be either W_o or W_n. The respective equations of motion for $q_i = W_o$ and $q_i = W_n$ are the following:

$$\ddot{W}_o + W_o \quad = P_o + O(W_n^2) \tag{7.27a}$$

$$\left(\frac{n^2+1}{n^2}\right)\ddot{W}_n + [\beta^2(n^2-1)^2 - W_o(n^2-2)]W_n = P_n + W_o(n^2-2)\delta_n + O(W_n^3) \qquad (7.27b)$$

The first equation (7.27a) describes the axisymmetric component of the deformation. As shown, this equation contains a term which is quadratic in W_n so that (7.27a) and (7.27b) are coupled. Goodier and McIvor (1964) have considered retaining this nonlinear coupling and predicted a cyclic transfer of energy between the axisymmetric vibrations W_o and non-axisymmetric vibrations W_n. However, in our experiments we have observed that the small amplitude and rapid damping of axisymmetric vibrations preclude autoparametric excitation of this sort, and hence the nonlinear terms are neglected.

The second equation (7.27b) describes the growth of non-axisymmetric Fourier modes W_n, the motion of which is forced by W_o. This equation also contains nonlinear terms of cubic order which couple the motion amongst the non-axisymmetric modes (Lindberg and Florence 1987). However, this coupling can be neglected on the grounds that cubic terms are small during the initial stages of buckling, an assumption which is supported by our experiments which show no evidence of coupling between modes.

To solve (7.27), one requires a model for the shape imperfection δ_n. The shape imperfection can sometimes be measured, although this is difficult for small imperfections or large mode numbers n. A common alternative is to assume shape imperfections of unit amplitude and analyze the amplification of buckles, that is, their growth relative to the size of the shape imperfection (Lindberg and Florence 1987). Another technique is to choose random imperfections that satisfy an appropriate statistical distribution (Lindberg 1988).

In our experiments, we have found that wall thickness variations are the dominant imperfection rather than shape imperfections. While shape imperfections describe an initial noncircularity of the tube in the absence of applied loads, wall thickness variations are characterized by a circumferentially varying wall thickness that is present in a perfectly circular tube. In practice, both types of imperfections are always present.

7.3.2.1 Self-Induced Pressure Loads

In (7.27b), the parameter P_n is the acoustic pressure load due to vibration of the tube and resonance in the surrounding annulus of water. To model P_n, the fluid motion is described using the linear wave equation for the velocity potential ϕ:

$$\frac{1}{c^2}\frac{\partial^2\phi}{\partial t^2} = \frac{\partial^2\phi}{\partial r^2} + \frac{1}{r}\frac{\partial\phi}{\partial r} + \frac{1}{r^2}\frac{\partial^2\phi}{\partial\theta^2}. \qquad (7.28)$$

If the velocity potential is represented using a Fourier series expansion in θ and harmonic vibrations are assumed with frequency ω, then (7.28) reduces to a Helmholtz equation for each Fourier mode ϕ_n:

$$r^2\frac{\partial^2\phi_n}{\partial r^2}+r\frac{\partial\phi_n}{\partial r}+\left(\frac{r^2\omega^2}{c^2}-n^2\right)\phi_n=0. \tag{7.29}$$

In general the solutions to this equation are either Bessel functions (for a bounded domain, such as an annulus) or Hankel functions (for an unbounded domain). However, in many situations the frequency of vibration is low enough that an incompressible approximation is applicable. Referring to (7.29), the criterion for which this approximation can be made is:

$$\frac{r^2\omega^2}{c^2n^2}<<1. \tag{7.30}$$

In our current experiments, this quantity is less than about 0.01, so the frequency term in (7.29) can be neglected. A physical interpretation of this approximation is that the reverberation time of pressure waves is much shorter than the period of structural vibration, so the radiated pressure field can be treated as uniform. This is also equivalent to taking the limit of $c\to\infty$ in (7.28), such that the fluid is incompressible. After applying this low-frequency approximation, (7.29) reduces to a Cauchy–Euler equation:

$$r^2\frac{\partial^2\phi_n}{\partial r^2}+r\frac{\partial\phi_n}{\partial r}-n^2\phi_n=0 \tag{7.31}$$

which has solutions $r^{\pm n}$. The relevant boundary conditions at the rigid outer wall ($r=b$) and the deforming inner wall ($r=a$) are that the radial fluid velocity must match the radial speed of the wall:

$$\frac{\partial\phi}{\partial r}\bigg|_{r=b}=0\quad\frac{\partial\phi}{\partial r}\bigg|_{r=a}=-\frac{\partial w}{\partial t}. \tag{7.32}$$

The minus sign in the second boundary condition is needed because the tube displacement is defined positive inward. Applying these boundary conditions, the solution for the velocity potential is:

$$\phi_n(r,t)=\frac{\partial w_n}{\partial t}\frac{a}{n}a^nb^n\left[\frac{(r/b)^n+(b/r)^n}{b^{2n}-a^{2n}}\right]. \tag{7.33}$$

Within the linear theory used to model the fluid (Lighthill 1978), pressure is related to the velocity potential by:

$$p = -\rho_w \frac{\partial \phi}{\partial t}. \tag{7.34}$$

As a result, the pressure at the tube wall ($r=a$) is found to be:

$$-p_n = \left[\frac{\rho_w a}{n} \left(\frac{(b/a)^{2n}+1}{(b/a)^{2n}-1} \right) \right] \frac{\partial^2 w_n}{\partial t^2} \equiv m \frac{\partial^2 w_n}{\partial t^2}. \tag{7.35}$$

The constants which pre-multiply the tube acceleration $\ddot{w}_n$ have been labeled m since they describe the effective added mass of the fluid. Since the self-induced pressure p_n contains only terms proportional to $\ddot{w}_n$, the effect of the liquid is purely mass-like and contributes only to a change in vibration frequency with no damping. Such behavior is characteristic of incompressible fluid models; however, the same purely mass-like behavior is observed even when the exact solution to (7.29) is used instead of the incompressible approximation. To obtain this exact solution, one would express solutions to (7.29) in terms of Bessel functions and solve the eigenvalue problem which arises when boundary conditions (7.32) are applied. To each of the resulting eigenfrequencies there corresponds a purely mass-like contribution to the pressure. This property is a consequence of the confined, annular geometry for which no energy-radiation mechanism is present. In contrast, when a tube is submerged in an unbounded medium, the FSI produces terms that are both mass-like (proportional to $\ddot{w}$) and resistive (proportional to $\dot{w}$) which results in damping of vibrations (Junger and Feit 1972).

The pressure from (7.35) can now be made dimensionless using the variables listed in (7.18). It is convenient also to define a dimensionless added mass coefficient M as follows:

$$M \equiv \frac{m}{\rho h} \frac{n^2}{n^2+1} = \left[\frac{\rho_w a}{\rho h n} \left(\frac{(b/a)^{2n}+1}{(b/a)^{2n}-1} \right) \right] \left(\frac{n^2}{n^2+1} \right). \tag{7.36}$$

Using this definition of the added mass coefficient, (7.27) can be rewritten as follows:

$$\ddot{W}_o + W_o = P_o \tag{7.37a}$$

$$\ddot{W}_n + \left[\Omega_n^2 - F_n^2 W_o(T) \right] W_n = F_n^2 W_o(T) \delta_n \tag{7.37b}$$

where

$$\Omega_n^2 = \frac{n^2 \beta^2 (n^2-1)^2}{(1+M)(n^2+1)} \tag{7.38}$$

Table 7.1 Comparison of measured and predicted natural frequencies. "No FSI" indicates frequencies calculated using (7.38) with $M=0$, while values in the column labeled "FSI" were computed with the values of M listed in the table. The percent error is computed between the measured frequencies and the frequencies calculated with the inclusion of FSI

	a/h	h	n	M	— Frequency (kHz) —			Error
Material		(mm)			No FSI	FSI	Measured	(%)
6061 Alum.	13.5	1.47	2	2.3	2.45	1.35	1.36	1.0
6061 Alum.	17.4	0.89	2	2.7	2.46	1.28	1.21	5.6
6061 Alum.	17.4	0.89	2	2.7	2.46	1.28	1.20	6.5
6061 Alum.	17.4	0.89	2	2.7	2.46	1.28	1.24	3.1
6061 Alum.	17.4	1.24	2	3.2	1.76	0.86	0.85	1.4
304 SS	17.4	0.89	2	0.9	2.36	1.70	1.82	6.3
6061 Alum.	24.5	0.89	2	4.5	1.24	0.53	0.54	2.4
6061 Alum.	17.4	1.24	3	2.1	4.98	2.84	2.82	0.9
6061 Alum.	17.4	0.89	3	2.0	6.97	4.06	3.97	2.3
6061 Alum.	24.5	0.89	3	2.9	3.50	1.77	1.85	4.5

$$F_n^2 = \frac{n^2(n^2-2)}{(1+M)(n^2+1)} \tag{7.39}$$

The parameter Ω_n is the frequency of free vibration for mode number n, while the parameter F_n models the interaction between the axisymmetric mode W_o and the nonsymmetric modes W_n.

Verification of this model was obtained by comparing the natural frequency predictions of (7.38) with some experimental measurements, as shown in Table 7.1. Experimental values were found using the same setup that was used to examine buckling; in this setup we can also measure the long-time frequencies of elastic vibration which occur when tubes are subjected to loads near, but not above, the buckling threshold (see, for example, Fig. 7.6). Frequencies were extracted from the strain traces either by using fast fourier transforms or by counting periods, depending on how many cycles were present in the data. Only mode numbers 2 and 3 are included in Table 7.1 since only these modes were observed in experiment. In some cases (such as Fig. 7.7), vibration frequencies from more than one mode could be extracted from a single strain trace.

The results in Table 7.1 demonstrate fair agreement between the predicted and measured frequencies. This validates a number of assumptions involved in the model, including the 2D plane strain approximation, the neglect of axial bending, and the incompressible FSI model. Good agreement with the experiments was still found even for vibrations of large amplitude approaching yielding of the tube, which suggests that the shell model described above is reasonable over the whole range of elastic deformations that we seek to model. This result also confirms that our neglect of geometrical nonlinearities (quadratic and cubic terms in the shell equations (7.27)) is reasonable.

7.3.2.2 Static Buckling

Before considering dynamical solutions to (7.37), it is interesting to explore the static buckling limit of this equation. Taking the time derivatives equal to zero, we find that $W_o = P_o$ and displacements W_n become unbounded when $\Omega_n^{\,2} = F_n^{\,2} P_o$. This criterion yields the static buckling pressure for the nth mode:

$$P_{\mathrm{crn}} = \beta^2 \frac{(n^2 - 1)^2}{n^2 - 2}. \tag{7.40}$$

When written in terms of dimensional variables, the static buckling limit becomes

$$p_{\mathrm{crn}} = \frac{Eh^3}{12(1 - \nu^2)a^3} \frac{(n^2 - 1)^2}{n^2 - 2}. \tag{7.41}$$

This result differs slightly from the classical static buckling threshold for an infinitely long tube derived by Timoshenko and Gere (1961):

$$p_{\mathrm{crn}} = \frac{Eh^3}{12(1 - \nu^2)a^3}(n^2 - 1). \tag{7.42}$$

The discrepancy arises primarily because in the development of (7.27), the pressure force was taken in the direction of the undeformed normal rather than that of the deformed normal. This problem has been encountered in other shell formulations, as evaluated by Simitses and Aswani (1974). This difference is smaller for larger values of n; for $n \geq 5$, the error is less than 5 %.

7.3.2.3 Direct Integration of Fourier Modes

For a general pressure load $P_o(T)$, (7.37) cannot be solved analytically. However, numerical solutions to this pair of uncoupled ordinary differential equations can be easily integrated with modern computational tools. To demonstrate the predictions of this model, experimentally measured pressure traces were fed into the model as P_o and the numerical solution was computed for the first ten Fourier modes. To make comparisons with experimental measurements, it was necessary to estimate the numerical value of the shape imperfections δ_n in (7.37). This was done by multiplying the measured variations in wall thickness (described in Sect. 7.2) by a factor of a/h, an approach which was motivated by comparing the relative contributions of shape imperfections and wall thickness variations to the total strain energy of the tube.

After solving for the dimensionless displacements W_o and W_n, hoop strains at the outer surface of the tube were calculated using (7.14) and (7.15). Strains were evaluated at angles of 0°, 90°, and 180° around the circumference, which is where the corresponding strain gauges from experiment were located. The results are compared with the experimental strain measurements in Fig. 7.14.

Fig. 7.14 Comparison of simulations and experiments for a steel tube very near the buckling threshold. Traces have been offset for clarity. Tube is 304 stainless steel with radius $a = 22$ mm and $a/h = 17.4$

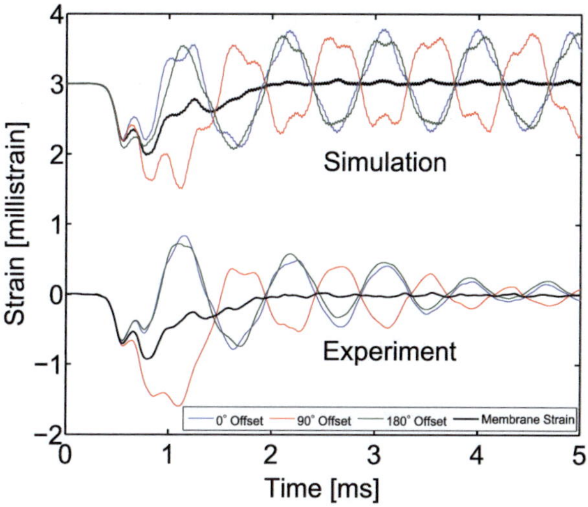

Both the simulation and the experiment feature a superposition of vibrations of modes 2 and 3. The amplitude and frequency of both modes is well predicted, with the exception that the numerical solution does not predict any damping. This behavior is a consequence of the incompressible fluid model and the omission of mechanical damping within the tube material.

The membrane strain from the experiments was estimated by averaging the measurements from two strain gauges located 90° apart. Since these guages were located on the major and minor axes of the vibration, averaging their measurements cancels out any bending strain due to mode 2 vibration (though bending strains from modes 3 and higher are not necessarily canceled). Good agreement between the measured and simulated membrane strains is also observed. These results demonstrate that in spite of the numerous simplifications used to develop this buckling model, the predictions are in remarkably good agreement with experimental measurements.

7.3.2.4 Generalized Loads

Although integrating the dynamic response of a tube for a given pressure load is often of interest, in many cases it is preferable to study the response to a class of generalized dynamic loads rather than to a specific load. Several types of loads that are of practical interest are a square wave of finite duration, a triangular wave, and an exponentially decaying wave. These pulse shapes can be used as idealized representations of many practical loads, such as shock or blast waves. Each of these pulse shapes can be characterized by a peak pressure P_i and a total impulse I, which is the integral of the pressure distribution over time. It is also convenient to define a characteristic load duration, $\tau \equiv I/P_i$.

In Sect. 7.2.1 it was observed experimentally that prior to buckling, the strain consists of small vibrations around the quasi-static value. Since these vibrations are small and damp out quickly, it is reasonable to neglect the axisymmetric breathing motion altogether and assume that the hoop strain consists only of its quasi-static component. In terms of the present nondimensional variables, this means that $W_o \approx P_o$.

It should be recognized that neglecting the dynamics of axisymmetric vibration eliminates the possibility of autoparametric excitation. Such excitation has been the subject of much research (Goodier and McIvor 1964, Lindberg 1964, 1974), but is usually not observed in experiments due to the internal damping that is present in real materials. Lindberg (1974) considered this problem in detail and found that since autoparametric buckling often occurs only after a large number of vibration cycles, it is mitigated by even small amounts of damping. As a result, he was able to observe experimental evidence of autoparametric buckling only under very carefully chosen test conditions. In our present experiments, the small amplitude and fast damping of axisymmetric vibrations due to FSI add further evidence that for submerged structures, autoparametric excitation is likely to be insignificant and axisymmetric vibrations can be neglected. Disregarding the axisymmetric vibration is also appropriate only if the load duration is longer than the period of these vibration, which is the case here.

To study the buckling of tubes under square, triangular, and exponential loads, it is necessary to choose a criterion for the buckling threshold. A convenient choice is to consider the tube to be buckled when the maximum displacement divided by the initial shape imperfection exceeds a particular amplification, A. The buckling threshold can then be determined by choosing a peak pressure P_i and integrating (7.37) numerically, gradually increasing the load duration τ until one of the Fourier modes exceeds the chosen amplification A. This value of τ then marks a point on the buckling threshold. By repeating this process for many values of P_i, the buckling threshold can be mapped out. This same approach was taken by Lindberg and Florence (1987), who investigated the effects of pulse shape and amplification A on the computed buckling threshold and concluded that both effects are small.

Examples of buckling thresholds for $a/h=20$ obtained using this method are plotted in Fig. 7.15. An amplification of $A=100$ was chosen, so these curves represent contours along which the maximum deformation is 100 times the size of the initial imperfection. The numbers along the curves indicate the most unstable mode number. Although the amplification value of $A=100$ has been arbitrarily chosen, it is representative of realistic values and the qualitative features of the results are not sensitive to the precise value of A.

In Fig. 7.15a, the buckling threshold is plotted in terms of the dimensionless pressure and impulse. This plot reveals that the buckling threshold tends toward two asymptotes: there are a minimum pressure (the static buckling pressure) and a minimum impulse below which the tube is always stable. The differences between the three load shapes are quite small along these asymptotes, but near the knee in the curve larger differences are observed. As will be seen, this is the region in which our experimental measurements were recorded.

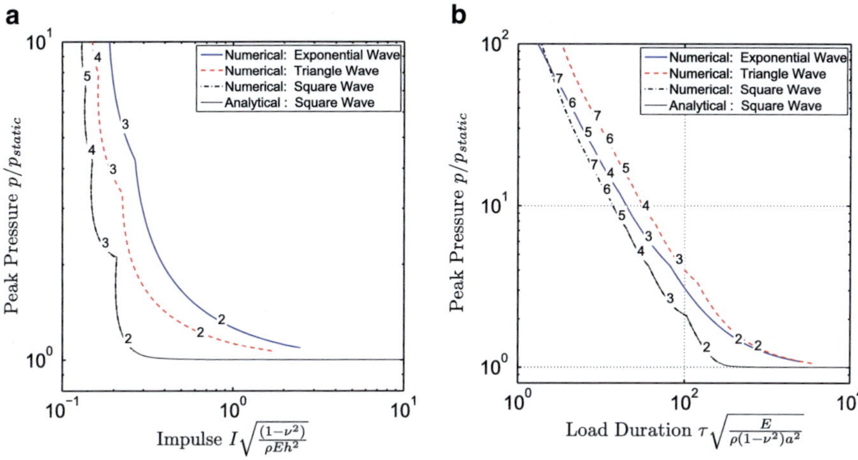

Fig. 7.15 (**a**) Buckling threshold in terms of pressure and impulse. (**b**) Buckling threshold in terms of pressure and load duration. Numbers indicate the most unstable mode number n_{cr}. Curves were computed with $a/h = 20$ and an amplification of 100

In Fig. 7.15b, the same buckling threshold data is represented on a plot of peak pressure vs. dimensionless load duration τ. In this scaling, the period of axisymmetric vibration falls at 2π on this plot, which is far below the knee in the curve. This reveals that the dynamic buckling pressure can be more than an order of magnitude greater than the static buckling pressure even for loads that are classified as "quasi-static."

7.3.3 Effects of FSI

Recall from the FSI model developed in Sect. 7.3.2.1 that the dimensionless added mass of the fluid is given by:

$$M = \frac{1}{n}\frac{\rho_w}{\rho}\frac{a}{h}\left[\frac{(b/a)^{2n}+1}{(b/a)^{2n}-1}\right]\left(\frac{n^2}{n^2+1}\right). \tag{7.43}$$

For a tube submerged in an infinite medium, the term in brackets tends to 1 and the added mass depends only on the ratio of densities, the thickness of the tube, and the mode number. Since the added mass is always greater for lower mode numbers, this indicates that FSI effects are most pronounced near the knee in the buckling threshold curve, where the mode numbers are smallest.

Examples of buckling threshold curves are plotted in Fig. 7.16 for tubes made from several practical materials and submerged in air and water. The tubes are

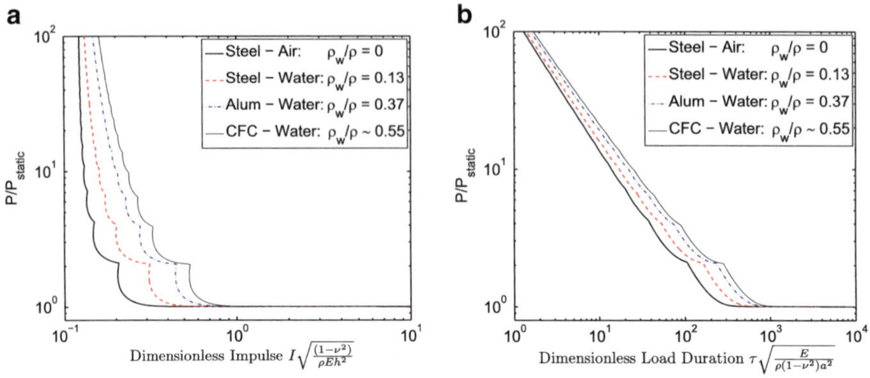

Fig. 7.16 Effect of added mass on buckling threshold for a square-wave pressure pulse with $a/h = 20$ and several tube materials. CFC is carbon fiber composite. (**a**) Bucking threshold in terms of pressure and impulse. (**b**) Buckling threshold in terms of pressure and load duration

subjected to a square-wave pressure pulse, and the amplification chosen to define the buckling threshold is $A = 100$. As expected, the effect of the fluid is greatest near the knee in the curve and is largest for tubes of low density.[1] For comparison with our experiments, the buckling thresholds in Fig. 7.16 were computed for relatively thick tubes, with $a/h = 20$. Under these conditions, the added mass coefficient for an aluminum tube with $n=2$ is about $M=3$ and the buckling threshold is shifted to the right by about a factor of $\sqrt{1+M} = 2$. However, for very thin shells ($a/h \sim 100$) and lightweight materials, or for tubes in a confined space, the added mass coefficient can be on the order of 30–40 and can shift the buckling threshold by a factor as high as 5–7.

Because of the primarily mass-like influence of FSI, the only effect of the fluid is to decrease the frequency of vibrations or the growth rate of buckles. In fact, by rescaling the time variable T in (7.37b) by a factor of $\sqrt{1+M}$, one can eliminate the added mass from the equation entirely. This would suggest that if one computes the buckling threshold for a tube in vacuo, then the corresponding threshold curve for a submerged tube can be obtained by rescaling the load duration τ by a factor of $\sqrt{1+M}$. Since M is a function of mode number, this rescaling would have to be done separately for each segment of the buckling threshold.

A nuance to this approach is that the presence of FSI shifts the transition points at which the most unstable mode number changes from one to the next. However, as shown in Fig. 7.16, the transition points do not shift very much. This suggests that the naïve approach of simply shifting the in vacuo buckling threshold to the right by a factor of $\sqrt{1+M}$ will still provide reasonable predictions. In fact, this has been

[1]For composite materials, the non-isotropic stiffness can affect the buckling behavior in other ways, but the added mass effect is expected to follow the same scaling as for metal tubes.

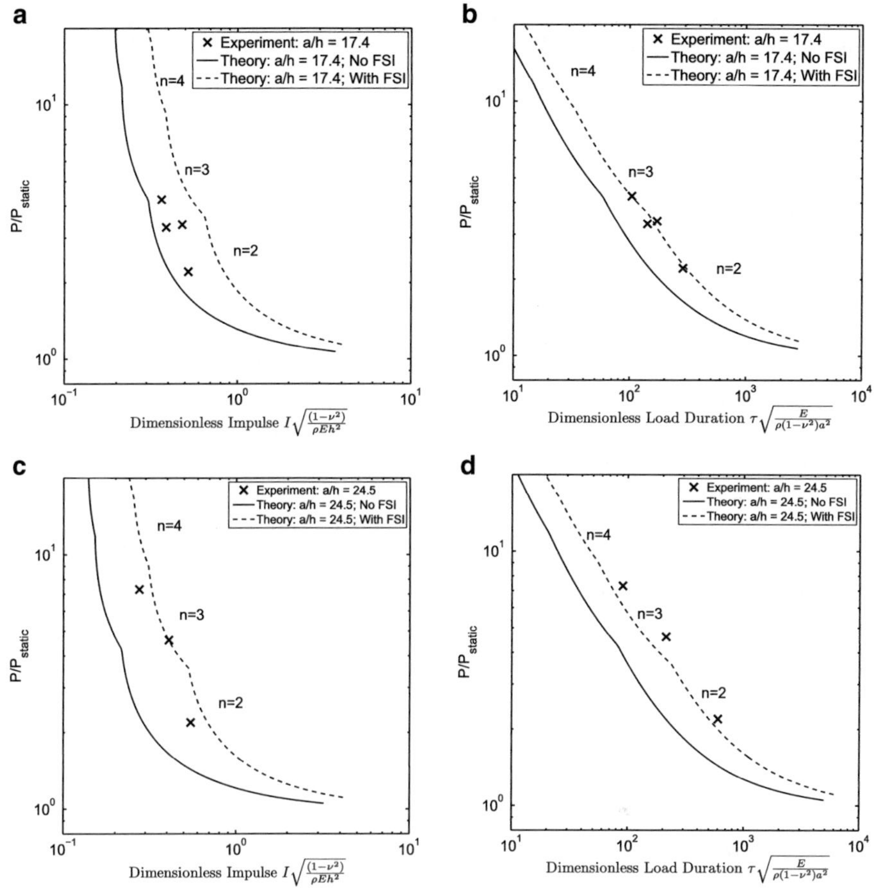

Fig. 7.17 Comparison of predicted and experimental buckling thresholds 6061-T6 aluminum tubes. Theoretical predictions are computed assuming an exponential pulse shape and an amplification of $A = 100$. (**a**) Buckling threshold for $a/h = 17.4$ in terms of load impulse. (**b**) Same as (**a**), but in terms of load duration. (**c**) Buckling threshold for $a/h = 24.5$ in terms of load impulse. (**d**) Same as (**c**), but in terms of load duration

done for the tube shown in Fig. 7.16, and the results (not shown) were nearly indistinguishable from the curves plotted in the figure.

7.3.4 Comparison with Experiments

Having developed a theoretical model for the buckling of cylindrical tubes, we now consider how well its predictions compare with experimental measurements. This comparison is made in Fig. 7.17 for aluminum tubes with $a/h = 17.4$ and 24.5.

For each tube size, the data is represented using both the impulse and the load duration as independent variables. The theoretical curves were computed by assuming an exponential load shape and an amplification of $A = 100$, while experimental data points correspond to the conditions at which plastic deformation was first observed. Good agreement between the experiments and predictions is observed.

For each value of a/h, Fig. 7.17 includes buckling thresholds both with and without the effects of FSI. In most cases, the experimental data points lie closer to the curve for which the effects of FSI are included. However, since the tubes are fairly thick, the effects of FSI are rather small and demonstration of the added mass effect based on these results is by no means conclusive.

A final observation from Fig. 7.17 is that some data points lie near the boundary at which the most unstable mode transitions from 2 to 3. Such behavior was confirmed in the experiments: tubes subjected to loads near this boundary exhibited excitation of both modes 2 and 3. However, the theoretical curves were computed by assuming that each mode is excited independently, whereas in the experiments tube failure occurred via a superposition of multiple excited modes (see, for instance, Fig. 7.7). This demonstrates that the most excited mode does not transition abruptly, but rather a gradual transition is made over a range of pressures and impulses in which two or more modes are strongly excited.

7.4 Summary

This chapter has considered the dynamic buckling of tubes with an emphasis on the complexities that arise when the tubes are submerged in a dense fluid. Experimental measurements of dynamic strains have clarified the behavior of the tube during the intermediate stages of buckling, a subject which has not been studied in detail in the literature. These results have confirmed many of the theoretical expectations that have been developed in the past, including the shape of the buckling threshold on a pressure-impulse diagram and the increase in the most unstable mode number as the applied pressure is increased. The experiments have also demonstrated that variations in wall thickness around the circumference of the tube have a significant and repeatable effect on the orientation of the tube's vibration and buckling which is in good agreement with theoretical predictions.

The experimental measurements have also clarified several of the effects of FSI on the dynamic response of the tubes. Measured vibration frequencies were in good agreement with theoretical added-mass models, and an incompressible fluid model was found to capture the added-mass effects quite adequately due to the relatively low frequency of non-axisymmetric vibrations. Due to FSI, axisymmetric vibrations were found to play only a small role in the response of the tube and measured hoop strains were quite close to those predicted by static models. The coupling between non-axisymmetric motion and the fluid pressure was found to be small due to the negligible volume changes involved in these vibrations; however, at the onset of plastic deformation, these volume changes were no longer negligible.

A two-dimensional buckling model was developed by making approximations based on the observations from experiments, and good agreement with both dynamic strain measurements and buckling threshold measurements was obtained. This model showed that the main effect of FSI is a reduction in the growth rate of buckles; depending on the extent of the FSI, this reduced growth rate can increase the impulse the tube can withstand by a factor as high as 5–7.

Acknowledgements Much of the work presented in this chapter was supported by the Office of Naval Research DOD MURI on Mechanics and Mechanisms of Impulse Loading, Damage and Failure of Marine Structures and Materials (ONR Grant No. N00014-06-1-0730), under program manager Dr. Y. D. S. Rajapakse. Mr. Tomohiro Nishiyama, currently at the Japan Patent Office, and Prof. Kazuaki Inaba, currently at the Tokyo Institute of Technology, executed the initial design and supervised the fabrication of the annular tube implosion fixture while working at Caltech. Mr. Jason Damazo and Dr. Rafal Porowski, currently at the research Center for Fire Protection in Poland, carried out preliminary experiments on buckling using the initial fixture design. Prof. Ravichandran of Caltech provided essential technical advice and encouragement as well as leadership of the Caltech MURI.

References

Abrahamson GR, Goodier JN (1962) Dynamic plastic flow buckling of a cylindrical shell from uniform radial impulse. In: Proceedings of the fourth national congress of applied mechanics, vol 2, pp 939–950

Abrahamson GR, Florence AL, Lindberg HE (1966) Investigation of response of simplified ICBM-type structures to impulsive loading. Technical Report AFWL-TR-65-136, Stanford Research Institute, 1966

Anderson DL, Lindberg HE (1968) Dynamic pulse buckling of cylindrical shells under transient lateral pressures. AIAA J 6(4):589–598

Batdorf SB (1947) A simplified method of elastic-stability analysis for thin cylindrical shells. Technical Report TR 847, NACA, 1947

Beltman W, Shepherd JE (2002) Linear elastic response of tubes to internal detonation loading. J Sound Vib 252(4):617–655

Budiansky B, Hutchinson JW (1965) Dynamic buckling of imperfection-sensitive structures. In: Gortler H (ed) Proceedings of the eleventh international congress of applied mechanics. Springer, New York, pp 636–651

Chonan S (1977) Response of fluid-filled cylindrical shell to a moving load. J Sound Vib 55(3):419–430

Damazo JS, Porowski R, Shepherd JE, Inaba K (2010) Fluid-structure interaction of submerged tubes subjected to impact generated stress waves. In: Proceedings of the 16th US national congress of theoretical and applied mechanics. ASME, 2010. USNCTAM2010-1279, June 27-July 2, 2010, State College, Pennsylvania, USA

Galletly GD, Bart R (1956) Effects of boundary conditions and initial out-of-roundness on the strength of thin-walled cylinders subject to external hydrostatic pressure. J Appl Mech 23:351–358

Goodier JN, McIvor IK (1964) The elastic cylindrical shell under nearly uniform radial impulse. J Appl Mech 31:259–266

Hegemier GA (1966) Instability of cylindrical shells subjected to axisymmetric moving loads. J Appl Mech 33(2):289–296

Hegemier GA (1967) Stability of cylindrical shells under moving loads by the direct method of Liapunov. J Appl Mech 34:991–998

Hutchinson JW, Budiansky B (1966) Dynamic buckling estimates. AIAA J 4(2):525–530
Hutchinson JW, Koiter WT (1970) Postbuckling theory. Appl Mech Rev 23(12):1353–1366
Inaba K, Shepherd JE (2010) Flexural waves in fluid-filled tubes subject to axial impact. J Pressure Vessel Technol 132:021302
Jones JP, Bhuta PG (1964) Response of cylindrical shells to moving loads. J Appl Mech 31(1):105–111
Joukowsky N (1900) Über den hydraulishen stoss in wasserleitungsröhren (on the hydraulic hammer in water supply pipes). Mémoires de l'Académie Impériale des Sciences de St. Péterbourg, 9(5). Series 8
Junger MC, Feit D (1972) Sound, structures, and their interaction, 2nd edn. MIT, Cambridge, MA
Kempner J, Pandalai KA, Patel SA, Crouzet-Pascal J (1957) Postbuckling behavior of circular cylindrical shells under hydrostatic pressure. J Aeronaut Sci 24:253–264
Koiter WT (1945) On the stability of elastic equilibrium. PhD thesis, Polytechnic Institute Delft
Korteweg D (1878) Über die fortphlanzungsgeschwindigkeit des schalles in elastisches röhren (on the velocity of propagation of sound in elastic pipes). Annalen der Physik und Chemie 9(5):525–542
Leissa AW (1973) Vibration of shells. Technical Report NASA SP-288, National Aeronautics and Space Administration, 1973
Lighthill J (1978) Waves in fluids. Cambridge University Press, Cambridge
Lin CT, Morgan CW (1956) A study of axisymmetric vibrations of cylindrical shells as affected by rotatory inertia and transverse shear. J Appl Mech 23(2):255–261
Lindberg HE, Firth RD (1967) Structural response of spine vehicles, volume II: Simulation of transient surface loads by explosive blast waves. Technical Report AFWL-TR-66-163, vol II, Stanford Research Institute, 1967
Lindberg HE (1988) Random imperfections for dynamic pulse buckling. J Eng Mech 114(7):1144–1165
Lindberg HE, Florence AL (1987) Dynamic pulse buckling. Martinus Nijhoff Publishers, Dordrecht, Netherlands
Lindberg HE, Anderson DL, Firth RD, Parker LV (1965) Response of reentry vehicle-type shells to blast loads. Technical Report LMSC-B130200-VOL-4-C, Stanford Research Institute, 1965
Lindberg HE (1964) Buckling of a very thin cylindrical shell due to an impulsive pressure. J Appl Mech 31:267–272
Lindberg HE (1974) Stress amplification in a ring caused by dynamic instability. J Appl Mech 41:392–400
Lindberg HE, Sliter GE (1969) Response of reentry-vehicle-type shells to transient surface pressures. Technical Report AFWL-TR-68-56, Stanford Research Institute, 1969
Lockhart D, Amigazo JC (1975) Dynamic buckling of externally pressurized imperfect cylindrical shells. J Appl Mech 42:316–320
Mann-Nachbar P (1962) On the role of bending in the dynamic response of thin shells to moving discontinuous loads. J Aerospace Sci 29:648–657
McIvor IK, Lovell EG (1968) Dynamic response of finite length cylindrical shells to nearly uniform radial impulse. AIAA J 6(12):2346–2351. The paper number is 64–144
McLachlan NW (1964) Theory and application of Mathieu functions, 1st edn. Dover Publications, New York
Mushtari KhM, Galimov KZ (1961) Nonlinear theory of thin elastic shells. Technical Report NASA-TT-F62, 1961. Translated from Tatknigoizdat, Kazan' 1957
Naghdi PM, Cooper RM (1956) Propagation of elastic waves in cylindrical shells, including effects of transverse shear and rotary inertia. J Acoust Soc Am 28(1):56–63
Schiffner K, Steele CR (1971) Cylindrical shell with an axisymmetric moving load. AIAA J 9(1):37–47
Shepherd JE, Inaba K (2010) Shock loading and failure of fluid-filled tubular structures. In: Shukla A, Ravichandran G, Rajapakse Y (eds) Dynamic failure of materials and structures, pp 153–190. Springer, New York

Shepherd JE (2009) Structural response of piping to internal gas detonation. J Pressure Vessel Technol 131(3):031204

Simitses GJ (1986) Buckling and postbuckling of imperfect cylindrical shells: A review. Appl Mech Rev 39(10):1517–1524

Simitses GJ, Aswani M (1974) Buckling of thin cylinders under uniform lateral loading. J Appl Mech 41(3):827–829

Simkins TE, Pflegl GA, Stilson EG (1993) Dynamic strains in a 60 mm gun tube: An experimental study. J Sound Vib 168(3):549–557

Sobel LH (1964) Effects of boundary conditions on the stability of cylinders subject to lateral and axial pressures. AIAA J 2(8):1437–1440

Stein M (1964) The influence of prebuckling deformations and stresses on the buckling of perfect cylinders. Technical Report TR R-190, NASA, 1964

Tang S-C (1965) Dynamic response of a tube under moving pressure. In: Proceedings of the American Society of Civil Engineers. Engineering mechanics division, vol 5, pp 97–122

Teng JG (1996) Buckling of thin shells: Recent advances and trends. Appl Mech Rev 49(4):263–274

Tijsseling AS (1996) Fluid-structure interactions in liquid-filled pipe systems: A review. J Fluids Struct 10:109–146

Timoshenko SP, Gere JM (1961) Theory of elastic stability, 2nd edn. McGraw-Hill, New York, NY

Wiggert DC, Tijsseling AS (2001) Fluid transients and fluid-structure interaction in flexible liquid-filled piping. Appl Mech Rev 54(5):455–481

Wylie EB, Streeter VL (1993) Fluid transients in systems. Prentice Hall, Upper Saddle River, NJ

Yamaki N (1969) Influence of prebuckling deformations on the buckling of circular cylindrical shells under external pressure. AIAA J 7(4):753–755

Yamaki N (1984) Elastic stability of circular cylindrical shells. Elsevier Science Publishing, Amsterdam, Netherlands

Chapter 8
The Use of Tubular Structures as Cores for Sandwich Panels Subjected to Dynamic and Blast Loading: A Current "State of the Art"

S. Chung Kim Yuen and G.N. Nurick

Abstract Blast loads from explosion (either accidental or intentional) can be mitigated by using thick armor systems that are often heavy and significantly affect the payload of the structure. In the design of structures for blast protection, sacrificial claddings which consist of high-energy-absorbing components are often used. Sandwich-type cladding structures with lightweight cores are now becoming more popular for blast mitigation applications, including armor systems, because of their capability to carry transverse loads with minimal weight penalty and absorb large amount of plastic energy. The choice of the sandwich core has significant influence on the performance of the sandwich panels. Typical cores are wood, different foam materials, or tubular structures. Because of its high energy absorption capabilities, tubular structures are commonly used. This chapter presents a "state of the art" on the use of tubular structures as cores for sandwich panels subjected to blast and dynamic loading.

8.1 Introduction

With the ever-increasing threats to impact and blast load, the demand for better protective structures is on the rise. While thick solid monolithic structures are currently in use, there is a drive to steer away from them as these types of structures may be cumbersome and heavy adding to existing structural load. As such there is an effort for developing and using sandwich panels for impact and blast protection.

Sandwich panels, usually lighter in weight and consisting of two plates referred to as face sheets separated by a core, have emerged as possible solutions to absorb energies from impact and impulsive loads. The core may consist of materials

S.C.K. Yuen • G.N. Nurick (✉)
Blast Impact and Survivability Research Unit (BISRU),
Department of Mechanical Engineering, University of Cape Town,
Private Bag, 7701 Rondebosch, Cape Town, South Africa
e-mail: gerald.nurick@uct.ac.za

A. Shukla et al. (eds.), *Blast Mitigation: Experimental and Numerical Studies*,
DOI 10.1007/978-1-4614-7267-4_8,

categorized as cellular (foam), micro-architectural (small-scale lattice type), and macro-architectural (larger scale plastic-deforming elements). The face sheets are often made of thin metal plates or composite laminates. The sandwich panels can be tailored, subject to ease of manufacturing, cost, and weight, to allow for potentially better performance than monolithic plates. Zhu and Lu (2007) and Chung Kim Yuen et al. (2009) presented overviews on the response of sandwich structures to impact and blast loading. Xue and Hutchinson (2004) compared the performance of three different core geometries of metal sandwich plates to that of solid plates of the same material and same mass. These studies have indicated that advanced sandwich structures can potentially have significant advantages over monolithic plates in absorbing the blast energy whether in air or underwater.

There are two significantly different types of sandwich paradigms: (1) sandwich panels and (2) sacrificial cladding. In a sandwich panel the back-face is free to deform. The performance of such sandwich panels is generally defined by the maximum back-face deformation for a given load. A sacrificial cladding structure, on the other hand, is a type of sandwich panel that is fixed or retrofitted to an existing structure. In this configuration, the back-face of the panel is assumed to be rigid. The panel is designed to deform in such a way that front face distributes the load more evenly across the core that is responsible to absorb the impact energies and transmit minimal/controlled forces to the main structure. A sacrificial cladding structure in the form of an efficient energy absorber should also be light in weight with ease of installation.

One of the frequently used means of energy dissipation to mitigate impact damage is the use of tubular structures. The tubular structures dissipate impact energies by various mechanisms such as plastic deformation and material fracture or tearing under either lateral loading or axial loading. Tubular structures with their extensive range of deformations are crucial in crashworthiness design. Crashworthiness relates to the capability of a structure/material to absorb high energy during impact in a progressive, controlled, and irreversible manner (Lu and Yu 2003). In the transport industry, tubular structures are one of the commonly used energy-absorbing devices because of the ability to absorb impact energy with minimal attenuation of survivable space. The thin-walled tubular structures can also deform plastically in several different modes such as inversion, folding, and splitting, each with a different energy dissipation capacity. In the last few decades, several studies have been carried out to investigate the energy absorbing capabilities of the tubular structures subjected to load in the lateral or the axial direction, for instance Lu and Yu (2003), Reid and Reddy (1978), Reddy and Reid (1979, 1980), Shim and Stronge (1986a), Gupta and Khullar (1995), Gupta and Velmurugan (1997), Gupta et al. (2001, 2005), Jones (2011, 2003), Reid (1993), Alghamdi (2001), and Chung Kim Yuen and Nurick (2008).

In view of the energy absorbing capabilities and characteristics, several authors have carried out research with a view to use tubular structures in cladding structures. Because of the high stiffness, tubular structures are used in sacrificial cladding panels and not in sandwich panels. To maximize energy absorption the tubular structures can be arranged either in the lateral (single or multilayers) or in the axial direction, as depicted in Fig. 8.1, to deform a large volume of material plastically.

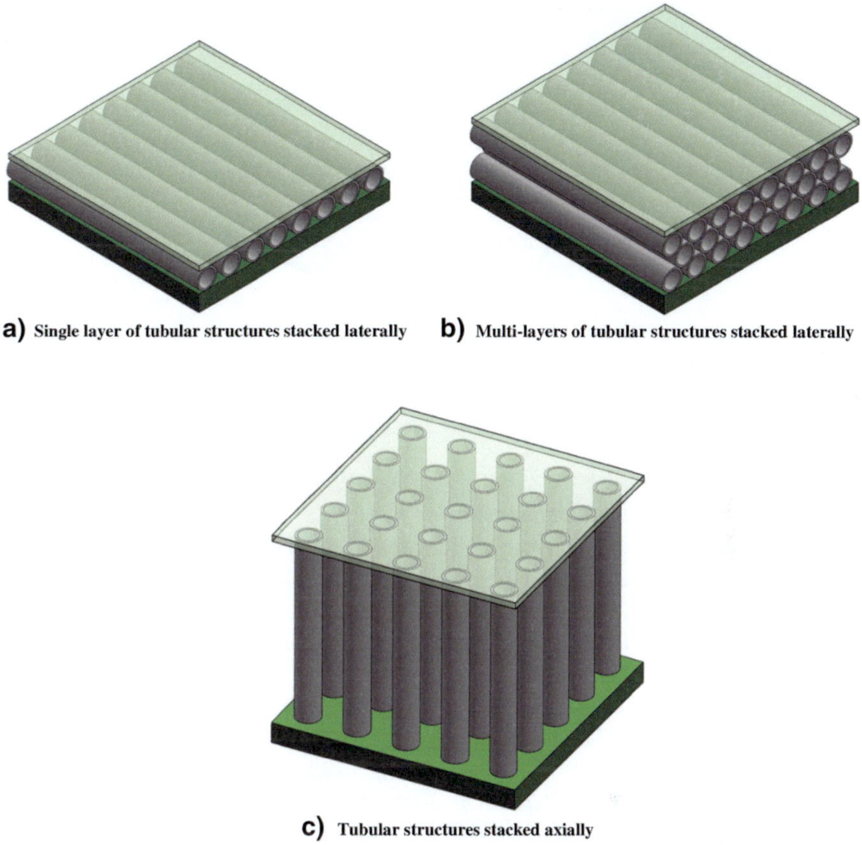

Fig. 8.1 Possible configurations of cladding panels using tubular structures. (**a**) Single layer of tubular structures stacked laterally. (**b**) Multilayers of tubular structures stacked laterally. (**c**) Tubular structures stacked axially

While there are numerous published studies on the use of tubular structures as energy absorbers this chapter presents:

- Studies on the tubular structures to be used in sacrificial cladding structures
- "State of the art" on the use of tubular structures as cores for sacrificial cladding structures subjected to impact (dynamic and blast) loading

8.2 Cores with Laterally Stacked Tubular Structures

Hitherto, no studies on the response of sandwich panels consisting of laterally stacked tubular structures as core have been presented. Palanivelu et al. (2011b, c) suggested the use of recyclable metal beverage cans stacked in the lateral direction,

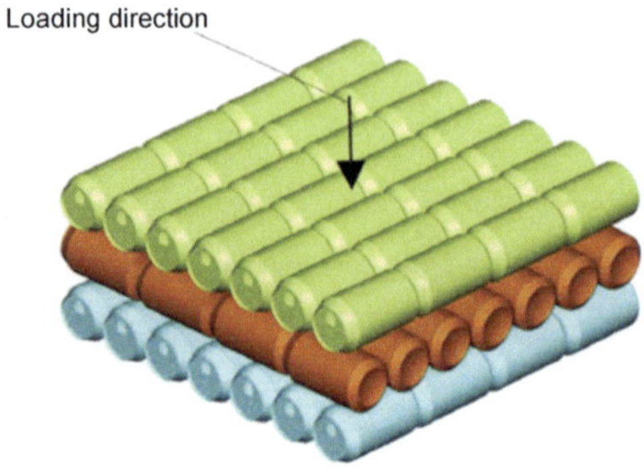

Fig. 8.2 Conceptual arrangement of empty recyclable metal beverage cans as core for sacrificial structure (Palanivelu et al. 2011b, c)

shown in Fig. 8.2, as core for sacrificial cladding but presented no results of the performance of "such" cladding structure.

Nonetheless, the lateral collapse of thin-walled tubular structures made of either metals or composites has received considerable attention for energy-absorbing system and has been studied by many researchers. The tubular structures, widely utilized in engineering, have earlier been examined through experiments and mathematical/numerical modelling.

Several studies have been carried out to characterize the behavior of single circular and quadrangular thin-walled tube subjected to a lateral load (Reid and Reddy 1978; Reddy and Reid 1979, 1980; Shim and Stronge 1986a; Gupta and Khullar 1995; Gupta and Velmurugan 1997; Gupta et al. 2001, 2005). The crushing process is driven by formation of plastic hinges. It was also reported that constraining the tubular structures so that the horizontal diameter of the tube cannot increase caused plastic hinges to develop during collapse and significantly increased energy absorption (Shim and Stronge 1986a; Reid et al. 1983). In view of the possibilities to stack tubular structures in multilayers as possible core for cladding, numerous authors have also conducted studies on a multitude of tubes stacked on top of one another (Shim and Stronge 1986b; Stronge and Shim 1987; Gupta and Khullar 1994). The localized crushing behavior of the multilayers of tubes (Fig. 8.3) showed similar characteristics of some ductile cellular solids with instability occurring as a result of tube buckling.

Other ways of improving the energy absorption of the tubular structures to lateral impact load include the use of a filler as reported by Hall et al. (2002). The specific energy absorption in filled tubes was found to increase. In transversely tested tubes the foam deformed laterally showing a capability of spreading the deformation. Fan et al. (2011) also reported on the crushing of foam-filled sandwich tubes. Typical progressive collapse of the sandwich tubes is shown in Fig. 8.4. It was found that the deformation patterns showed progressive buckling, initiated locally at the impact end, and gradually developed to the distal end during the crushing process.

Based on the various studies, it is clear that tubular structures loaded in the lateral direction can be used as efficient energy-absorbing components as core in cladding structures.

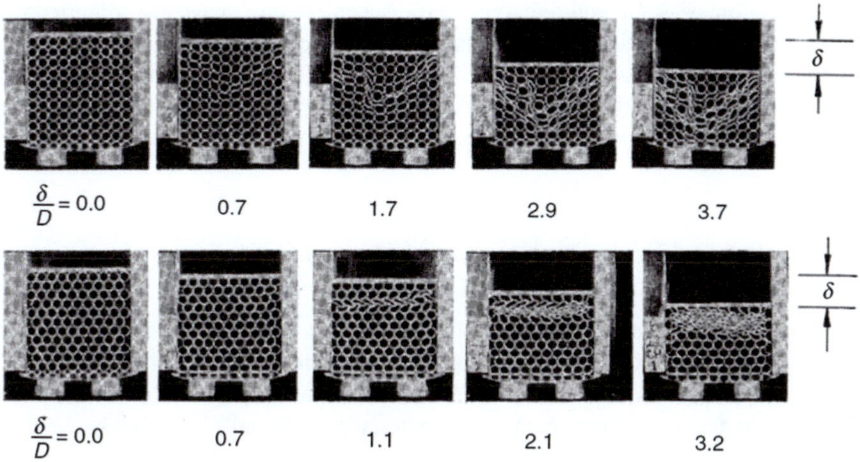

Fig. 8.3 Compression of arrays of tube (*top*: square packed; *bottom*: hexagonally packed) (Shim and Stronge 1986b)

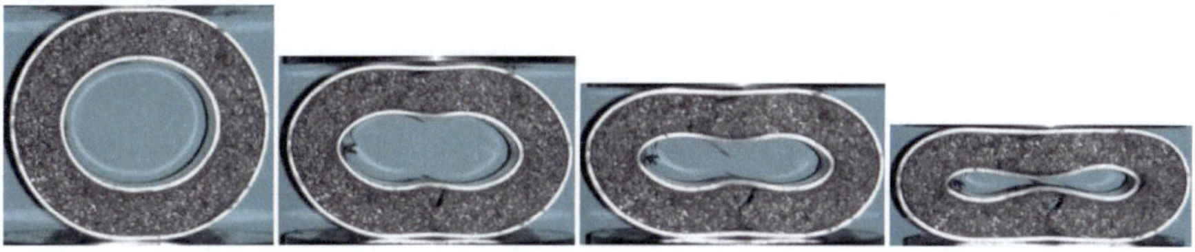

Fig. 8.4 Progressive collapse of foam-filled sandwich tubes (Fan et al. 2011)

8.3 Cores with Axially Stacked Tubular Structures

Tubular structural members irrespective of geometries and materials have been widely established as energy absorbers of choice, especially in the transport industry. These extruded thin-walled structures have the ability to absorb and convert large amounts of kinetic energy into plastic strain energy in high-impact situations. Thin-walled tubular structures can sustain large geometric deformations, strain hardening effects, strain-rate effects, temperature effects, and various interactions between different deformation modes (such as bending and stretching) when deforming under compression in the axial direction.

Studies on the behavior of thin-walled structures were pioneered in the 1960s by Pugsley and Macaulay (1960) and Alexander (1960). Subsequently, there has been ongoing interest on the axial crushing behavior of these structures. Numerous authors, for example Lu and Yu (2003), Jones (2011, 2003), Reid (1993), Alghamdi (2001), and Chung Kim Yuen and Nurick (2008), have presented overviews on the response of single tubular structures with or without triggers to different types of axial loads (quasi-static and dynamic load). Several authors have also reported on the use of multiple cell profiles (multiple tubes placed concentrically inside one another); see Seitzberger et al. (1997, 2000) and Chung Kim Yuen et al. (2008). Nevertheless, none of these studies were carried out with a view to use the tubular structures in cladding panels to absorb or redistribute impact energies.

8.3.1 Tubes Made with Composite Materials

With the increased demand of lightweight structures and the need for retrofitting old structures, many researches have begun to implement the use of tubular structures as cores in sacrificial cladding panels for better energy absorbers. Several authors have focused on the crushing characteristics of "small" tubes with a view to use them as cores in cladding structures. Kakogiannis et al. (2009) discussed the failure of a pultruded glass fiber-reinforced composite tube subjected to axial impact load transmitted by a small attached mass exposed to blast load. The composite tubes, made of pultruded glass fiber reinforced with epoxy vinyl ester matrix, were loaded by detonating plastic explosive PE4 onto a free striking mild steel disc (108 g). After detonation the free striking mass was accelerated onto the tubular structure in an axial direction causing the tube to crush. Typical failed specimens are shown in Fig. 8.5. Debonding and delamination mechanism were observed where the tubes failed. In the axial direction the tubes "petalled."

Palanivelu et al. (2010a, b, c, d, e, 2011c) carried out experiments and numerical investigations to evaluate the performance of hollow and foam-filled composite tubes with different geometrical shapes, shown in Fig. 8.6, for sacrificial cladding applications. Similar failure modes were observed for composite tubes, made of E-glass fabric and polyester resin, subjected to quasi-static and dynamic impact load. The tubes exhibited uniform and progressive crushing failure modes with circumferential delamination, axial cracks, lamina bending, and fiber fracturing. The inclusion of polyurethane foam provided additional wall strength and stability

Fig. 8.5 Photograph showing failed specimens and corresponding striking masses accelerated by the detonation of 5, 9, and 13 g of PE4. Kakogiannis et al. (2009)

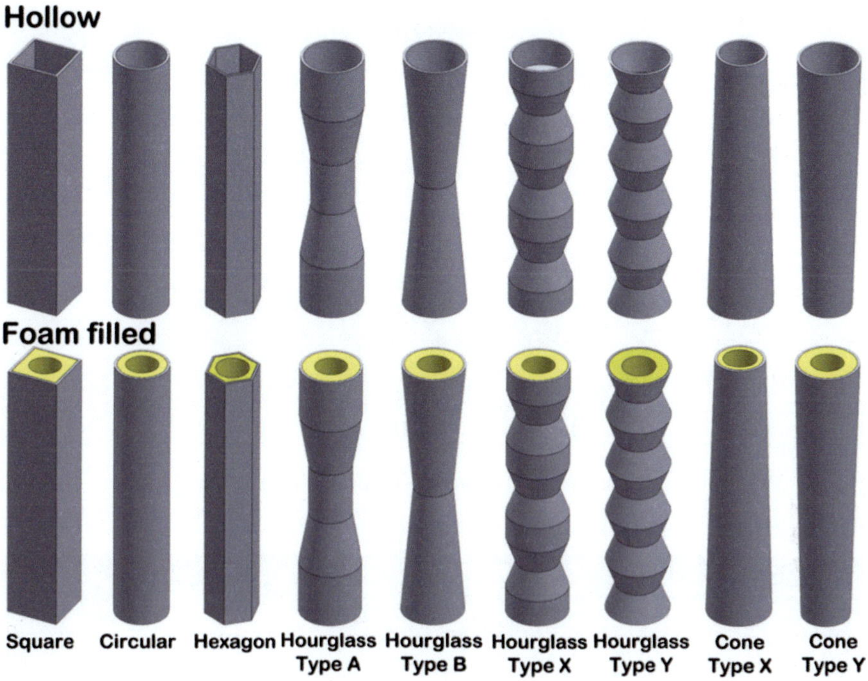

Fig. 8.6 Schematic showing different geometrical shape of composite tubes evaluated by Palanivelu et al. (2010a, b, c, d, e, 2011c)

for uniform and progressive crushing failure mode suppressing any circumferential delamination process and fiber fracturing and reduced the specific energy absorption significantly.

Palanivelu (2011) also presented blast testing on a representative sacrificial cladding structure made of either 25 or 37 polyurethane foam-filled pultruded glass polyester composite tubes and sandwich composite skin panels. The skin was made from composite panels consisting of glass fiber, polyester resin, and divinycell foam. Soft foam holders were used to assemble the composite tubes on the rear skin of the panels. To prevent penetration of the glass fiber into the blast-facing skin, thin aluminum plates (0.5 mm in thickness) were fixed on crushing side of the skin. The blast loads were generated by detonating either 100 or 150 g of C4 at one end of the two connecting sewage concrete pipes (standoff distance of 4.2 m). The results from the blast tests showed that the stiffness of the rear skin panel was not adequate. Consequently, local and global bending of the skin was observed because of the relatively high stiffness of the composite tubes. Nevertheless, the composite tubes exhibited uniform and progressive failure modes such as delamination, axial cracks, lamina bending, fiber fracturing, and compression of the polyurethane foam.

Tarlochan et al. (2012) also presented the results of polymer composite tubular structures subjected to quasi-static loading with a view to design and fabricate tubular

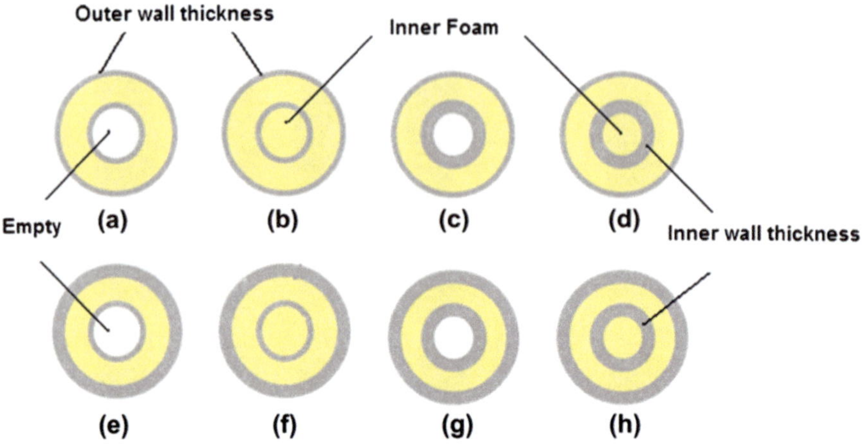

Fig. 8.7 Various specimen geometry configurations (Tarlochan et al. 2012)

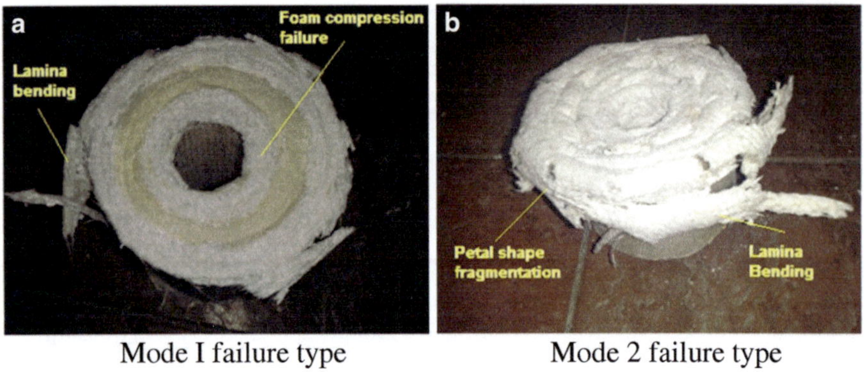

Fig. 8.8 Photograph showing failure of the polymer composite tube (Tarlochan et al. 2012). (**a**) Mode I failure type. (**b**) Mode 2 failure type

sandwich panels that have potential as energy-absorber devices. The composite sandwich specimens were fabricated from glass fiber, polystyrene foam, and epoxy resin. The various specimen geometry configurations tested are shown in Fig. 8.7. In general, the specimen collapsed progressively in folding patterns with the composites exhibiting high energy absorption capabilities and high crush force efficiency. Figure 8.8 depicts the two different modes of failure observed. Mode I is associated with the progressive end crushing with the formation of fronds spreading outwards and inwards of the tube. Mode II is associated with fronds spreading mostly outwards. These collapse modes show the "splaying" and "lamina-bending" type of stable brittle fracture. Energy dissipated was associated with bending of the fronds and composite lamina bundle, splaying and fragmentation of the composite skins, crack growth, and sliding of the external and internal fronds.

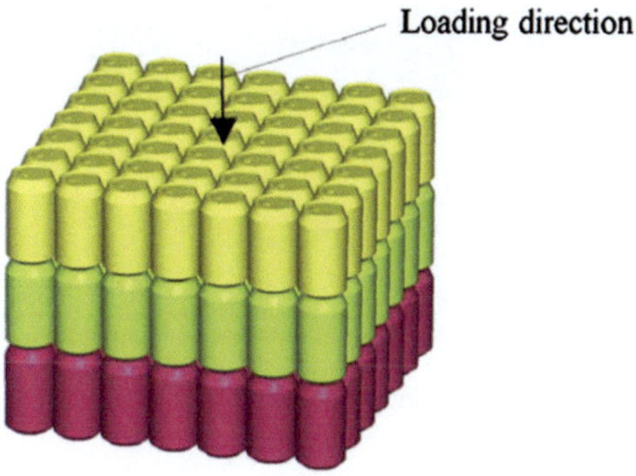

Fig. 8.9 Suggested arrangement of empty recyclable metal beverage cans as core for sacrificial structure (Palanivelu et al. 2011a)

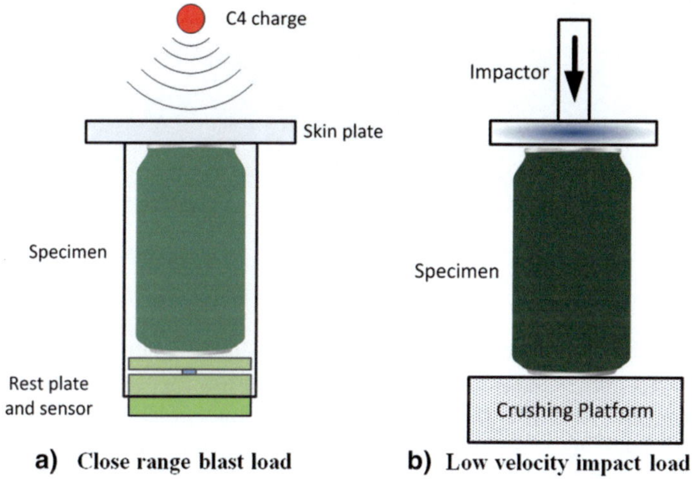

Fig. 8.10 Schematic representation of the experimental arrangements. (a) Close-range blast load. (b) Low-velocity impact load

To create "greener" sacrificial claddings, Palanivelu et al. (2011b, c) suggested the use of readily available wasted metal beverage cans that are environmentally friendly and noncorrosive. Before performing any tests on the conceptual sacrificial cladding structures, shown in Fig. 8.9, Palanivelu et al. (2011b, c) carried out experimental and numerical blast study and low-velocity axial impact tests on single empty recyclable metal beverage can to understand the crushing mechanisms, deformation patterns, and corresponding energy absorption of the structure. The beverage cans with an average mass of 26 g consisted of steel for the body and aluminum for the cover. A schematic representing the experimental arrangements is shown in Fig. 8.10a for the close-range blast load tests and Fig. 8.10b for the low-velocity impact tests. The close-range free air blast tests were carried out using 20 g spherical charge of C4 explosive detonated at a standoff distance of 300 mm. The low-velocity impact tests were carried out using drop mass. The beverage cans

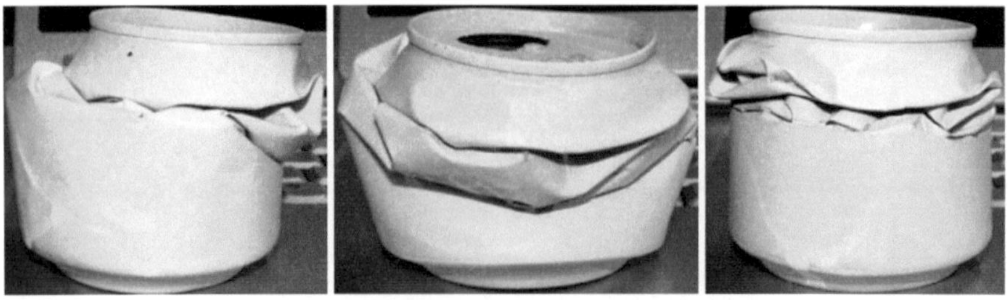

Fig. 8.11 Photograph showing crushed empty beverage cans as a result of the close-range free air blast (Palanivelu et al. 2011a)

Fig. 8.12 Photograph showing progressive crushing stages of empty beverage can for an initial impact velocity of 4.9 m/s (Palanivelu et al. 2011b)

exhibited similar crushing characteristics of circular tubes. Controlled and progressive failure pattern were observed for both loading scenarios. Figure 8.11 shows crushed empty beverage cans loaded using explosive. Figure 8.12 shows the different progressive buckling stages of a beverage can subjected to an impact velocity of 4.9 m/s. Buckling was initiated in the mid-wall section of the can with triangular folds as a result of the smaller wall thickness in this region. The inclusion of four 3.5 mm holes drilled in the bottom section of the structure to evacuate air inside the beverage can during the crushing process was found to exhibit higher deformation length without any changes in the deformation pattern.

Palanivelu (2011) and Palanivelu et al. (2011d) later manufactured two configurations of the cladding structure with the core made up of either 25 or 37 empty beverage cans (stacked axially) and the skin made from composite panels consisting of glass fiber, polyester resin, and divinycell foam. To prevent delamination of the skin, the outer face sheets were sewn through the foam with aramid fibers. Figure 8.13 shows the experimental setup. Plastic explosive, C4, detonated at one end of the two connecting sewage concrete pipes (standoff distance of 4.2 m) generated a planar shock wave to impact the cladding structure located at the other end of the pipe against a solid concrete block.

The empty beverage cans crushed progressively during the explosive loading and generally showed an asymmetric failure pattern (diamond failure mode), shown

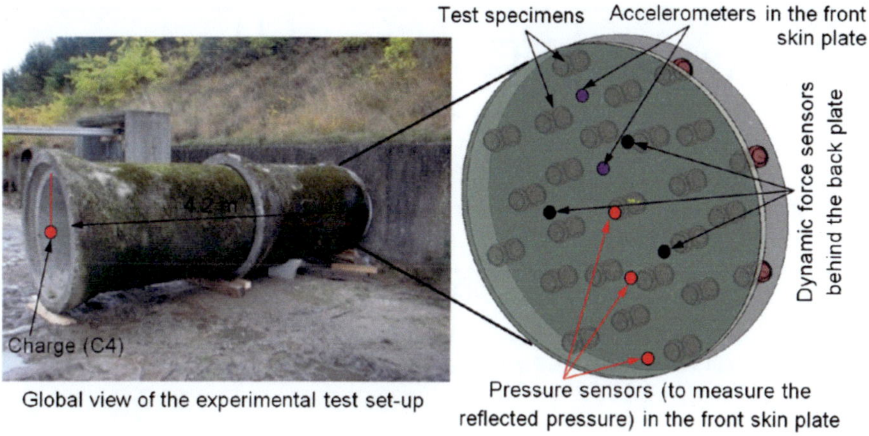

Fig. 8.13 Schematic of the experimental setup (Palanivelu et al. 2011d)

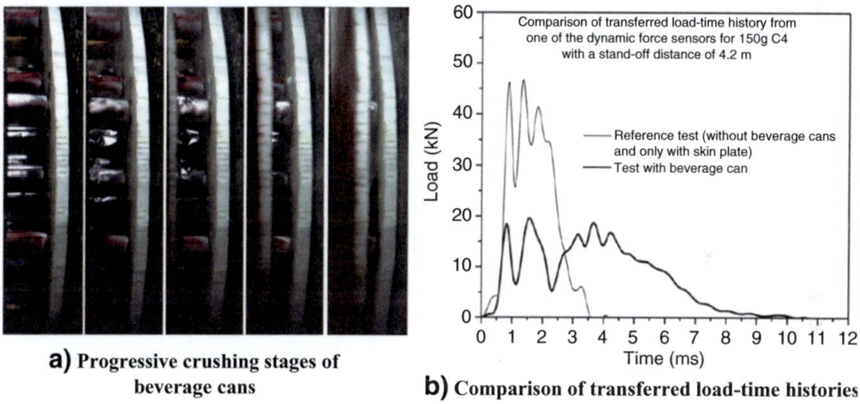

Fig. 8.14 Results from the blast tests (Palanivelu et al. 2011d). (**a**) Progressive crushing stages of beverage cans. (**b**) Comparison of transferred load-time histories

in Fig. 8.14a. The effectiveness of the proposed sacrificial cladding structure was evaluated by comparing load-time histories of the structure to the sandwich composite panels without any beverage cans. The corresponding transferred load to the concrete structure from one of the load cells is shown in Fig. 8.14b. It was observed that the use of the beverage cans reduced the peak crush load and extended the total duration of the blast loading event considerably.

Depending on the size of the cladding structures there are numerous factors that dictate the energy absorption performance of the structure to any impact load, for instance, the size, geometry, and material of the tubular structure. Other important factors include the quantity and placement of tubes as core material within the cladding panel. If there are too many tubular structures the core may become too rigid

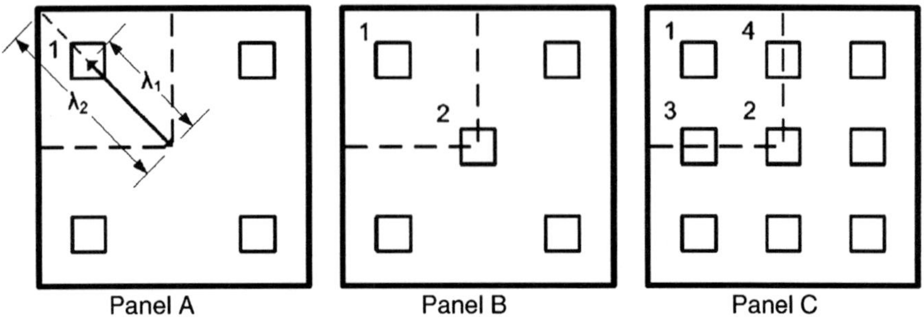

Fig. 8.15 Schematic of tube layouts used by Theobald and Nurick (Theobald and Nurick 2007). *Dashed lines* indicate symmetry used in FEM model

Table 8.1 Plate design variables used by Theobald and Nurick (2007)

Design variable	Values used in numerical study
Panel layout	A, B, C (Fig. 8.15)
λ	0.443; 0.471; 0.500; 0.528; 0.556, 0.585; 0.613; 0.641; 0.669; 0.698
Tube thickness (mm)	0.5; 0.55; 0.6; 0.65; 0.7
Top plate thickness (mm)	2.0; 2.25; 2.5; 2.75; 3.0
Tube aspect ratio (R)	4; 4.5; 5; 5.5; 6

and does not behave like appropriate energy absorbers. If the number of tubular structures is too small the sacrificial cladding structure may be ineffective as energy absorbers. The placement of the tubular structure is also important as load bearers of the cladding structure for the desirable buckling mode under impact load. In essence, the tubular core must be strong enough to bear any structural load but also weak enough to plastically deform and absorb any impact energy in a predictable and controllable manner.

Theobald and Nurick (2007, 2010) investigated the response of sandwich cladding panels with thin-walled square tubes as the core material to blast loads. Theobald and Nurick (2007) reported on a parametric study to examine the effects and interaction between several design variables for three different tube layouts, shown schematically in Fig. 8.15. The design variables for buckling stability and best energy absorption performance used are listed in Table 8.1. While the tube aspect ratio showed only small effect on core buckling stability and energy absorption, the thickness of the tube highly influenced the onset of buckling and hence the stability of the core. Buckling stability and absorption performance were also shown to be highly sensitive to tube placement because of the interaction between the top plates and the tubes.

To ensure efficient energy absorption, Theobald and Nurick (2007) controlled the formation of the plastic lobes during the crushing process in the parametric study using optimization technique. The typical overall crushed shape of the tubes obtained in the study is shown in Fig. 8.16. Mode I crush behavior was described as "perfect"

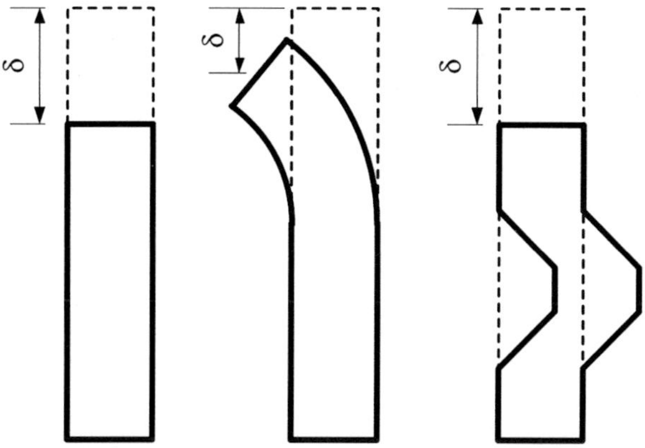

Fig. 8.16 Tube crushing modes observed in numerical simulations (Theobald and Nurick 2007)

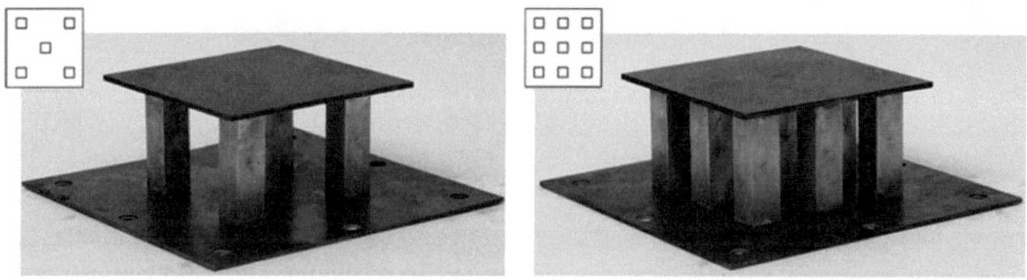

Fig. 8.17 Photographs of typical aluminum tube panels after construction (Theobald and Nurick 2010)

dynamic plastic or progressive symmetric buckling. Mode J behavior was related to global buckling, or a combination of progressive buckling and global buckling. Mode K referred to crush behavior where large non-axial deformations occurred primarily away from the proximal end of the tube. Modes J and K were further subdivided based upon the direction in which the non-axial displacements occur. For best performance at the maximum stroke length, modes J and K were undesirable.

Based on the outcome of the parametric study (Theobald and Nurick 2007), Theobald and Nurick (2010) manufactured and tested three different tube-core cladding panels for blast energy absorbers. The cores were made from either annealed mild steel or 6063-T3 aluminum alloy tubes of nominal dimensions 21.4 × 21.4 × 0.6 mm (machined down from 25.4 × 25.4 × 2.0 mm tubes). Hemispherical indentation triggers were implemented in the tubes to induce progressive and symmetric buckling. The tubular cores were attached to the top and bottom plates by means of soft solders. Conventional welding was deemed unfit because of the high heat required that may cause either warping in the top and bottom plate or changes in the material properties of the panel. Figure 8.17 shows panels made with five and nine aluminum tubular

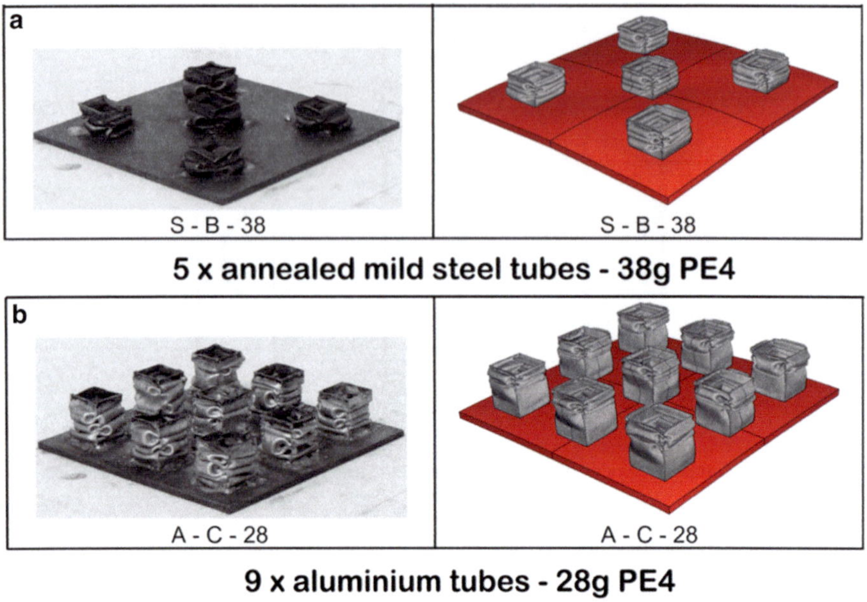

Fig. 8.18 Experimental and numerical deformation of tubular core to blast load (Theobald and Nurick 2010). (**a**) 5× annealed mild steel tubes—38 g PE4. (**b**) 9× aluminum tubes—28 g PE4

cores. The panels, 156×156 mm in cross section, were "uniformly" blast loaded by detonating a circular disc of plastic explosive (PE4) of diameter 75 mm at a standoff distance of 190 mm from one end of a blast tube. Varying masses of explosive were used to provide a range of panel responses.

Typical response of the tubular cores to blast load is shown in Fig. 8.18. At smaller blast loads, irregular buckling modes occurred possibly due to variable trigger performance. In cases where the full stroke of the tubes was used, symmetric buckling modes were observed for higher charge masses. In general, the panel crush distance increased with increasing masses of explosives and decreased with an increasing number of tubes in the panel core. For the same charge mass, the aluminum tube cores exhibited significantly higher crushed distance than the annealed mild steel tube cores. The results of the numerical simulation of the experiments indicated that the energy absorption was primarily confined to the core tubes with minimal plastic strains developed in the top plate.

Chung Kim Yuen et al. (2011) extended the study by Theobald and Nurick (2010) to investigate the potential of sandwich panels to low-velocity impact. While being similar in configurations, the panels used by Chung Kim Yuen et al. (2011) did not contain any welds. The tubular cores were 108 mm in length of which 8 mm was located within the outer skin via groove cutouts. In this case, the tubular cores were manufactured from mild steel and were 20×20×0.9 mm in cross section. Three tube configurations were investigated: as-received tubes, tubes with circular cutouts on opposite sides, and tubes with dents on opposite sides. The imperfections were located at the midsection of the tube. The circular cutouts were 6 mm in diameter.

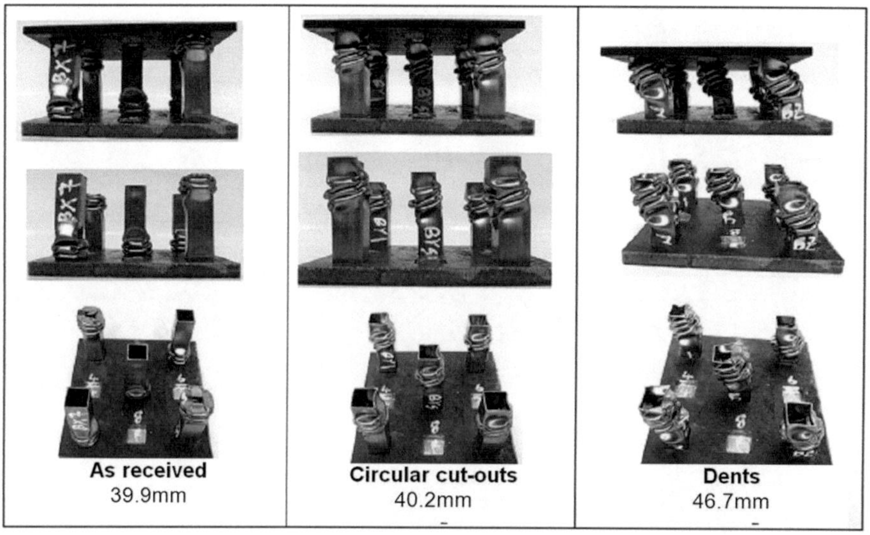

Fig. 8.19 Photographs showing different views of dynamically crushed sandwich panels consisting of five tubes with different triggers (drop height: 2 m) (Chung Kim Yuen et al. 2011)

The dents were 5 mm deep and generated using a 6 mm hemispherical indenter. The low-velocity impact tests were performed in a drop hammer rig with a constant drop mass of 240 kg and drop heights ranging from 1.2 to 3 m to result in range impact velocities between 4.9 and 7.7 m/s. The majority of the tubular specimens failed in the progressive buckling mode with the exception of one or two tubes. Figure 8.19 shows different views of dynamically crushed sandwich panels consisting of five tubes with different triggers subjected to an axial load of 240 kg dropped from a height of 2 m. As expected, sandwich panels made up of lower number of tubes crushed more than sandwich panels with more tubes and tubes with triggers generally had higher crushed distance than as-received tubes.

High-speed camera footage (captured at 10,000 frames per second) showed that for sandwich panels made up with as-received tubes, the impacted end moved progressively towards the non-impacted end. In some instances, some tubes started to buckle at the non-impacted end progressing towards the impacted end. For tubes with triggers, buckling was generally initiated at the "weaken" region. Figure 8.20 shows the transient response of a sandwich panel consisting of nine tubes with dent imperfections. In this case, all nine tubes within the sandwich panel buckled in the same manner.

Finite element simulations were also carried out and validated using high-speed photography footage obtained from experiments. In general, the finite element results showed very good correlation for crush initiation, final crushed shape, and mean crushed force. Figure 8.21 shows typical comparison of the final crushed response to experiments for sandwich panel consisting of nine "as-received" tubes.

The results from these investigations suggest that a sandwich panel made with an array of tubular structures can be an effective energy-absorbing structure.

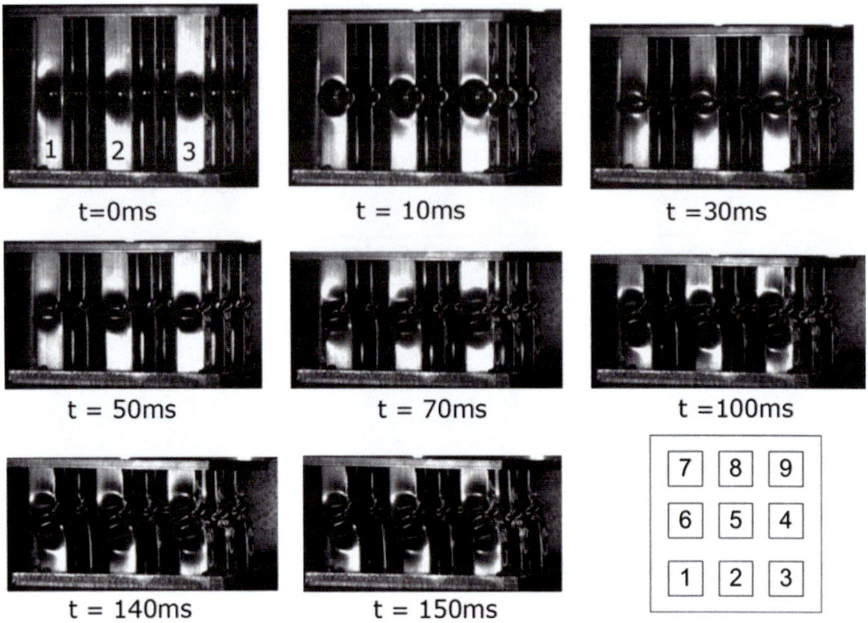

Fig. 8.20 Transient collapse of sandwich panel consisting of nine tubes with dent triggers (drop height: 3 m) (Chung Kim Yuen et al. 2011)

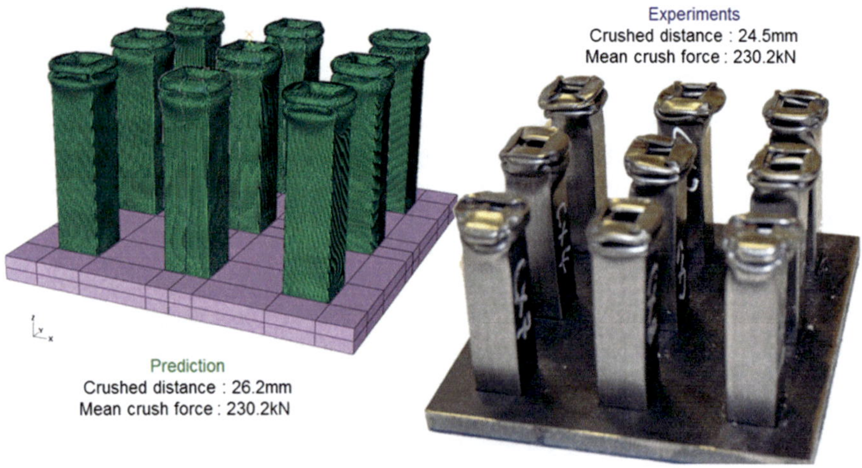

Fig. 8.21 Comparison of simulation and experiments for final crush mode for panel made up with nine "as-received" tubes (drop height: 3 m) (Chung Kim Yuen et al. 2011)

The placement of the tubes within such a panel and choices of the tubes can be further optimized to better dissipate energy while collapsing depending on the application and loading conditions.

Xiong et al. (2011) developed a new lightweight sandwich panel made up with vertically aligned hollow Al–Si alloy tubes as core construction, shown in Fig. 8.22.

Fig. 8.22 Photograph showing a sandwich panel with hollow Al–Si alloy tube core (Xiong et al. 2011)

Fig. 8.23 Failure of hollow tubes under compressive load (Xiong et al. 2011)

The cast tubes were bonded to carbon fiber composite face sheets using epoxy adhesive. These panels were, however, not assessed under any impact tests. When subjected to in-plane compression and three-point bending, debonding of the cores to the face sheets was observed. Under compressive loading (Fig. 8.23), the Al–Si alloy tube core configuration showed superior specific strength under crushing compared to common metallic and stochastic foam cores. Consequently, Xiong et al. (2011) suggested that these Al–Si hollow tube cores can be considered to design lightweight multifunctional structures once further tests at higher strain rate loading are carried out.

Using a different approach but with a similar goal of developing novel design for cost-effective crashworthy composite structures, Pitarresi et al. (2007) quantified the energy absorption characteristics of hollow rectangular sections fabricated with tubes and five other different cores (single corrugation, double corrugation, dimpled, perpendicular webs, and diagonal and perpendicular webs). The manufacturing process of the tube was not provided because of proprietary design details. Based on quasi-static axial compression experiments, the best energy absorption performance, by a considerable margin, was exhibited by the samples with the tubular ties. The measured specific energy absorptions were in excess of 40 kJ/kg. Figure 8.24 shows the specimen failed by extensive, progressive, local brittle fracture of both the facings and the tubular core reinforcement. The core with the smaller diameter tubes (10 mm) was found to provide higher specific energy absorptions than the larger diameter tube core (20 mm).

8.4 Conclusions

Sandwich structures, due to their extremely high flexural stiffness-to-weight ratio and high strength-to-weight ratio, offer potential as energy absorbers. Currently, there are limited studies available in the open literature on the performance of

Fig. 8.24 Photograph showing crushed specimen with tubular ties (Pitarresi et al. 2007)

sandwich panels with tubular cores. The few experiments conducted thus far show that it is likely that the different panel types will offer improvement in different loading regimes (different loading intensities, duration, and area). Irrespective of the stacking sequences, there exists significant potential for these sandwich panels to improve protection against impact and blast loading. The sandwich panels with tubular cores can be designed and optimized using analytical and numerical techniques in various panel configurations that will allow best performance for the required application. One of the major drawbacks to evaluate the sandwich panels with tubular cores is the manufacturing of such panels. With improvements and advances in manufacturing techniques, these difficulties may be overcome in the future. This should be beneficial for impact and blast resistance.

References

Alexander JM (1960) An approximate analysis of the collapse of thin cylindrical columns. Quart J Mech Appl Math 13(1):10–15

Alghamdi AAA (2001) Collapsible impact energy absorbers: an overview. Thin Wall Struct 39(2):189–213

Chung Kim Yuen S, Nurick GN (2008) The energy absorbing characteristics of tubular structures with geometric and material modifications: an overview. Appl Mech Rev 61(2):020802-1–020802-15. doi:10.1115/1.2885138

Chung Kim Yuen S, Nurick GN, Starke RA (2008) The energy absorption characteristics of double-cell tubular profiles. Lat Am J Solid Struct 5(4):289–317

Chung Kim Yuen S, Nurick GN, Theobald MD, Langdon GS (2009) Chapter 10, Sandwich panels subjected to blast loading. In: Shukla A, Ravichandran Y, Rajapakse Y (eds) Dynamic failure of materials and structures. Springer, New York, pp 297–325. ISBN ISBN: 197814419 04454

Chung Kim Yuen S, Nurick GN, Witbeen HL (2011) The response of sandwich panels made of thin-walled tubes subjected to axial load. Int J Protect Struct 2(4):477–498

Fan Z, Shen J, Lu G (2011) Investigation of lateral crushing of sandwich tubes. Procedia Eng 14:442–449. doi:10.1016/j.proeng.2011.07.055

Gupta NK, Khullar A (1994) Lateral collapse of orthogonal and non-orthogonal cross-layered arrays of square and rectangular tubes. Int J Mech Sci 36(5):449–467. doi:10.1016/0020-7403(94)90048-5

Gupta NK, Khullar A (1995) Collapse load analysis of square and rectangular tubes subjected to transverse in-plane loading. Thin Wall Struct 21(4):345–358. doi:10.1016/0263-8231(95)93619-W

Gupta NK, Sekhon GS, Gupta PK (2001) A study of lateral collapse of square and rectangular metallic tubes. Thin Wall Struct 39(9):745–772. doi:10.1016/S0263-8231(01)00033-7

Gupta NK, Sekhon GS, Gupta PK (2005) Study of lateral compression of round metallic tubes. Thin Wall Struct 43(6):895–922. doi:10.1016/j.tws.2004.12.002

Gupta NK, Velmurugan R (1997) An analysis of axial crushing of composite tubes. J Comp Mater 31(13):1262–1286

Hall IW, Guden M, Claar TD (2002) Transverse and longitudinal crushing of aluminum-foam filled tubes. Scripta Mater 46(7):513–518. doi:10.1016/S1359-6462(02)00024-6

Jones N (2003) Several phenomena in structural impact and structural crashworthiness. Eur J Mech A Solid 22(5):693–707

Jones N (2011) Structural impact, 2nd edn. Cambridge University Press, Cambridge, pp 1–604. ISBN ISBN: 9781107010963

Kakogiannis D, Van Hemelrijck D, Wastiels J, Van Ackeren J, Palanivelu S, Van Paepegem W, Nurick GN, Chung Kim Yuen S (2009) Experimental and numerical study of pultruded composite tubes under blast loading. In: Dyckmans G (ed) Proceedings of the 9th international DYMAT conference on the mechanical and physical behaviour of materials under dynamic loading (DYMAT 2009), Brussels, Belgium, pp. 1677–1683. (Vol 2) EDP Sciences, France

Lu G, Yu T (2003) Energy absorption of structures and materials. Woodhead Publishing Limited, Cambridge, pp 1–403. ISBN ISBN: 1 85573 688 8

Palanivelu S (2011) Energy absorption of crushable tubes for protective structures under static impact and blast loading. Universiteit Gent, Belgium, pp 1–590. ISBN ISBN: 978-90-8578-438-8

Palanivelu S, Paepegem WV, Degrieck J, Vantomme J, Kakogiannis D, Ackeren JV, Hemelrijck DV, Wastiels J (2010a) Comparison of the crushing performance of hollow and foam-filled small-scale composite tubes with different geometrical shapes for use in sacrificial cladding structures. Compos B Eng 41(6):434–445. doi:10.1016/j.compositesb.2010.05.009

Palanivelu S, Paepegem WV, Degrieck J, Vantomme J, Kakogiannis D, Ackeren JV, Hemelrijck DV, Wastiels J (2011a) Crushing and energy absorption performance of different geometrical shapes of small-scale glass/polyester composite tubes under quasi-static loading conditions. Compos Struct 93(2):992–1007. doi:10.1016/j.compstruct.2010.06.021

Palanivelu S, Van Paepegem W, Degrieck J, De Pauw S, Vantomme J, Wastiels J, Kakogiannis D, Van Hemelrijck D (2011b) Low velocity axial impact crushing performance of empty recyclable metal beverage cans. Int J Impact Eng 38(7):622–636. doi:10.1016/j.ijimpeng.2011.02.008

Palanivelu S, Van Paepegem W, Degrieck J, Kakogiannis D, Van Ackeren J, Van Hemelrijck D, Wastiels J, Vantomme J (2010b) Comparative study of the quasi-static energy absorption of small-scale composite tubes with different geometrical shapes for use in sacrificial cladding structures. Polymer Test 29(3):381–396. doi:10.1016/j.polymertesting.2010.01.003

Palanivelu S, Van Paepegem W, Degrieck J, Kakogiannis D, Van Ackeren J, Van Hemelrijck D, Wastiels J, Vantomme J (2010c) Parametric study of crushing parameters and failure patterns of pultruded composite tubes using cohesive elements and seam, Part I: central delamination and triggering modelling. Polymer Test 29(6):729–741. doi:10.1016/j.polymertesting.2010.05.010

Palanivelu S, Van Paepegem W, Degrieck J, Reymen B, Ndambi JM, Vantomme J, Kakogiannis D, Wastiels J, Van Hemelrijck D (2011c) Close-range blast loading on empty recyclable metal beverage cans for use in sacrificial cladding structure. Eng Struct 33(6):1966–1987. doi:10.1016/j.engstruct.2011.02.034

Palanivelu S, Van Paepegem W, Degrieck J, Reymen B, Segers E, Ndambi JM, Vantomme J, Van Ackeren J, Wastiels J, Kakogiannis D, Van Hemelrijck D (2011d) Performance of sacrificial cladding structure made of empty recyclable metal beverage cans under large-scale air blast load. Appl Mech Mater 82:416–421. doi:10.4028/www.scientific.net/AMM.82.416

Palanivelu S, Van Paepegem W, Degrieck J, Van Ackeren J, Kakogiannis D, Van Hemelrijck D, Wastiels J, Vantomme J (2010d) Experimental study on the axial crushing behaviour of pultruded composite tubes. Polymer Test 29(2):224–234. doi:10.1016/j.polymertesting.2009.11.005

Palanivelu S, Van Paepegem W, Degrieck J, Van Ackeren J, Kakogiannis D, Wastiels J, Van Hemelrijck D, Vantomme J (2010e) Parametric study of crushing parameters and failure patterns of pultruded composite tubes using cohesive elements and seam: part II. Multiple delaminations and initial geometric imperfections. Polymer Test 29(7):803–814. doi:10.1016/j.polymertesting.2010.07.005

Pitarresi G, Carruthers JJ, Robinson AM, Torre G, Kenny JM, Ingleton S, Velecela O, Found MS (2007) A comparative evaluation of crashworthy composite sandwich structures. Compos Struct 78(1):34–44. doi:10.1016/j.compstruct.2005.08.008

Pugsley AG, Macaulay M (1960) The large scale crumpling of thin cylindrical columns. Quart J Mech Appl Math 13(1):1–9

Reddy TY, Reid SR (1979) Lateral compression of tubes and tube-systems with side constraints. Int J Mech Sci 21(3):187–199. doi:10.1016/0020-7403(79)90023-7

Reddy TY, Reid SR (1980) Phenomena associated with the crushing of metal tubes between rigid plates. Int J Solid Struct 16(6):545–562. doi:10.1016/0020-7683(80)90005-0

Reid SR (1993) Plastic deformation mechanisms in axial compressed metal tubes used as impact energy absorbers. Int J Mech Sci 35(12):1035–1052

Reid SR, Drew SLK, Carney JF III (1983) Energy absorbing capacities of braced metal tubes. Int J Mech Sci 25(9/10):649–667. doi:10.1016/0020-7403(83)90074-7

Reid SR, Reddy TY (1978) Effect of strain hardening on the lateral compression of tubes between rigid plates. Int J Solid Struct 14(3):213–225. doi:10.1016/0020-7683(78)90026-4

Seitzberger M, Rammerstorfer RF, Degischer HP, Gradinger R (1997) Crushing of axially compressed steel tubes filled with aluminium foam. Acta Mech 125:93–105

Seitzberger M, Rammerstorfer FG, Gradinger R, Degischer HP, Blaimschein M, Walch C (2000) Experimental studies on the quasi-static axial crushing of steel columns filled with aluminium foam. Int J Solid Struct 37(30):4125–4147

Shim VPW, Stronge WJ (1986a) Lateral crushing of thin-walled tubes between cylindrical indenters. Int J Mech Sci 28(10):683–707. doi:10.1016/0020-7403(86)90013-5

Shim VPW, Stronge WJ (1986b) Lateral crushing in tightly packed arrays of thin-walled metal tubes. Int J Mech Sci 28(10):709–728. doi:10.1016/0020-7403(86)90014-7

Stronge WJ, Shim VPW (1987) Dynamic crushing of a ductile cellular array. Int J Mech Sci 29(6):381–406. doi:10.1016/0020-7403(87)90001-4

Tarlochan F, Ramesh S, Harpreet S (2012) Advanced composite sandwich structure design for energy absorption applications: blast protection and crashworthiness. Compos B Eng 43(5): 2198–2208. doi:10.1016/j.compositesb.2012.02.025

Theobald MD, Nurick GN (2007) Numerical investigation of the response of sandwich-type panels using thin-walled tubes subject to blast loads. Int J Impact Eng 34(1):134–156. doi:10.1016/j.ijimpeng.2006.04.003

Theobald MD, Nurick GN (2010) Experimental and numerical analysis of tube-core claddings under blast loads. Int J Impact Eng 37(3):333–348. doi:10.1016/j.ijimpeng.2009.10.003

Xiong J, Ma L, Wu L, Li M, Vaziri A (2011) Mechanical behavior of sandwich panels with hollow Al-Si tubes core construction. Mater Design 32(2):592–597. doi:10.1016/j.matdes.2010.08.016

Xue Z, Hutchinson JW (2004) A comparative study of impulse-resistant metal sandwich plates. Int J Impact Eng 30(10):1283–1305

Zhu F, Lu G (2007) A review of blast and impact of metallic and sandwich structures, EJSE 92–101

Chapter 9
Mitigation of Loading on Personnel in Light-Armored Vehicles Using Small Model Testing

Thomas Brodrick, Ryan Hurley, and W.L. Fourney

Abstract Death and injury are a major problem when military personnel are riding in a vehicle which is struck by the detonation forces from a buried mine. This chapter presents the results from small-scale testing aimed at reducing the forces on vehicle occupants due to the detonation of a buried mine. With the recent deployment of mine-resistant ambush protected (MRAP) vehicles by the United States Department of Defense (DOD) the mortality rates have dropped dramatically. The very large acceleration forces on the body, however, still result in major injury to vehicle occupants—especially brain injury. The aim of this research is to find ways to reduce the forces on the occupants such that major injury can be avoided. Various methods of mitigating forces are examined and the results are compared from the standpoint of acceleration, jerk, and head injury criterion (HIC).

9.1 Introduction and Background

The Dynamic Effects Laboratory has been investigating the loading caused by underbelly blasts on simulated vehicles over the past several years. This is a very serious problem as the type of warfare for our military has changed dramatically after the 9/11 attacks on the United States. The rapidly induced vehicle acceleration that results from the detonation of a buried mine can have dramatic effects on vehicle occupants. Accelerations can cause loss of blood flow to vital parts of the brain, concussion, TBI, or death (Blackman n.d.). With a significant number of troop casualties and injuries resulting from explosive attacks, it is of great interest for the military to have vehicles designed to minimize the risk posed by these attacks.

T. Brodrick• R. Hurley • W.L. Fourney (✉)
Dynamic Effects Laboratory, Department of Mechanical Engineering,
University of Maryland, 1131 Martin Hall, College Park, MD 20742, USA
e-mail: four@umd.edu

A. Shukla et al. (eds.), *Blast Mitigation: Experimental and Numerical Studies*,
DOI 10.1007/978-1-4614-7267-4_9, © Springer Science+Business Media New York 2014

As of July 31, 2010, explosives were responsible for 2,688 of the 4,384 US troops killed and 25,652 of the 39,187 US troops wounded during the global war on terrorism (Blackman n.d. 2007). Screenings performed between 2006 and 2009 showed that approximately 20 % of all combat troops suffer from some degree of traumatic brain injury (TBI). These injuries result in an approximate treatment cost of $2 billion per year (Blackman n.d.). The explosive devices causing these injuries are typically improvised explosive devices (IEDs) or land mines detonated below personnel transport vehicles.

Until 2007, the United States was not heavily invested in MRAP vehicles, instead favoring the up-armored Humvee, which offered very little protection in the event of explosive detonation beneath the vehicle (Hellman 2007). The inherent benefit of the MRAP arises from its V-shaped hull, which deflects blast energy, such as ejected soil, outwards. Small-scale testing for vehicles with angled bottoms to our knowledge started within the Dynamic Effects Lab in 2004 and 2005. A master's thesis was conducted by Kevin Genson (Genson 2006) during that time frame which showed that the loading on a vehicle with an angled bottom of 13° reduced the impulse delivered to the vehicle to approximately 50 % of that for a flat-bottomed vehicle. In July of 2006 a complete series of small-scale testing was conducted on two MRAP configurations and the results again showed that there was a significant decrease in the loading on the vehicle based upon shaped bottoms. In 2007, the DOD approved the purchase of 23,000 MRAP vehicles for use in the global war on terrorism (Hellman 2007). By November 2008, the percentage of casualties in the war attributed to IEDs was only 5 %, down from 60 % 1 year earlier (Lamb et al. 2009).

The effectiveness of IEDs has clearly been severely reduced by the implementation of MRAPs, as shown in Fig. 9.1 (Lamb et al. 2009). Congress approved $3 billion for MRAP purchase in 2007, and $16.8 billion in 2008 (Hellman et al. 2008). In 2010, the total MRAP program cost reached $22 billion (Hellman 2009).

This chapter describes research aimed at protecting the health and safety of personnel in dangerous regions around the world. The need for increasing MRAP safety will be of interest to all parties involved with MRAP design and purchase. In particular, the US Government, including the US Army and DOD, and private MRAP manufacturers will be interested in the results of this research. The research will hopefully be the beginning of a larger effort to optimize the design of structures that provide safety from rapidly applied accelerations. It will also demonstrate the need for additional research describing physiological human response to acceleration.

9.1.1 History of the Human Acceleration Problem

Putting very large number of MRAPs into service during 2007 and 2008 has resulted in an amazing reduction in the fatality rate of involved troops but soldiers returning from service are nonetheless subjected to large acceleration forces that can result in brain damage.

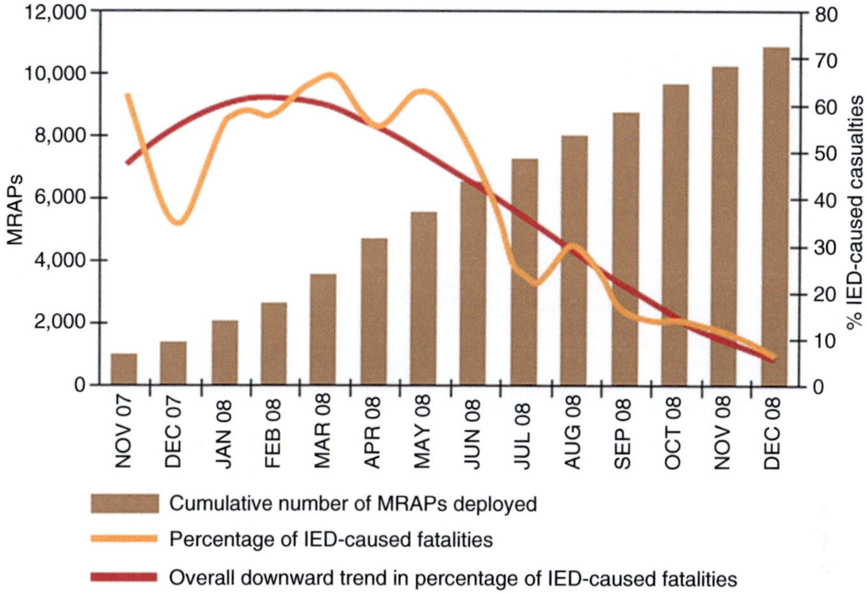

Fig. 9.1 IED-caused fatalities and MRAP deployment (Lamb et al. 2009)

The adverse physiological consequences of vertical acceleration were first observed in the early 1900s. In 1919, doctors began noting a phenomenon known as "fainting in the air" caused by sustained airplane accelerations of 4.5 g or higher (Head 1919; Brown and Lechner 1956. Similar phenomena such as blackout and gray out were also observed around the same time. Over the coming decades, these effects became attributed to cessation of blood flow in the eyes and brain (Duane 1953).

Early research on the response of humans to vertical accelerations focused on nonlife-threatening acceleration levels because technology did not yet exist to allow accelerations large enough to cause dramatic injuries. Research began to evolve in the 1950s to incorporate a range of acceleration thresholds that could cause severe injury and death. Investigators began to understand that rapid vertical acceleration could cause, in addition to blackout and unconsciousness, fracture of the spinal cord, and brain contact with the skull (Blackman n.d.).

Stoll (Stoll 1956) determined that the severity of human response to vertical acceleration depends on not only maximum acceleration but also the rate that acceleration is applied (the jerk). Eiband (Eiband 1959) published data showing the effects of acceleration on scales smaller than previously investigated. Eiband (Eiband 1959) also developed one of the first graphs illustrating the vertical acceleration tolerance thresholds of human beings. This graph, shown in Fig. 9.2, has been reproduced in other studies and is the basis for early acceleration criteria for vehicle seat crashworthiness. In this graph, a threshold value of 23 g for 5.5 ms is suggested in order to avoid the "area of moderate injury."

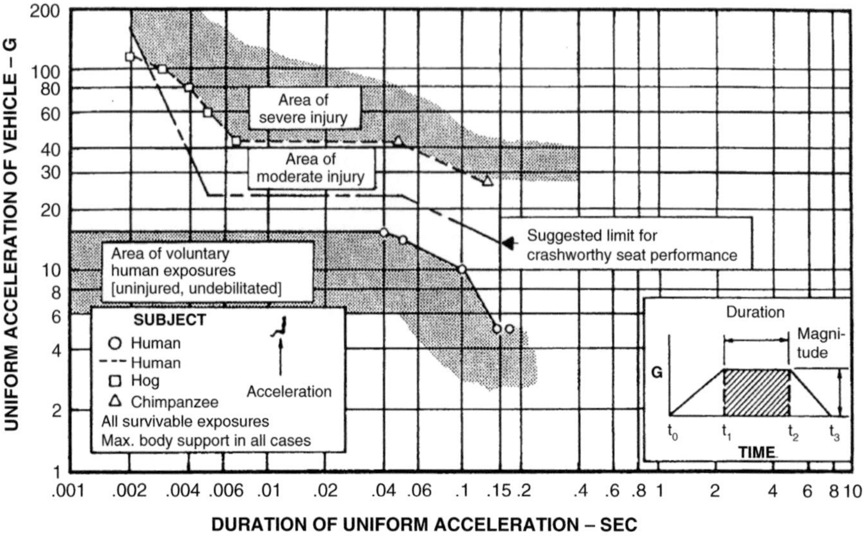

Fig. 9.2 Vertical acceleration tolerance thresholds (Eiband 1959)

Another milestone in the study of human response to vertical acceleration was the development of the dynamic response index (DRI) (Stech and Payne 1969) . The DRI is a model that assesses the physical response of the human body to upward acceleration and is based on a differential equation describing the stiffness of the spinal cord. The DOD and US Army have developed limiting thresholds for human vertical acceleration tolerance based on the DRI and Eiband's curves. The DOD has suggested limiting human acceleration to 25 G for 100 ms (Department of Defense 1998). The US Army has suggested a limit of 23 G for 25 ms (US Army Aviation Systems Command 1989).

An assessment of potential human injury can also be made with the HIC (McHenry 2004). HIC is a widely used quantitative analysis of the head injury risk posed by a particular incident. It is most commonly used in assessing injury risk in vehicle crash tests (Henn 1998). HIC provides a good quantitative analysis of risk because it captures elements of both jerk and duration of acceleration. The equation for HIC is shown below:

$$\mathrm{HIC} = \left[\frac{1}{t_2 - t_1} \int_{t_1}^{t_2} a\,\mathrm{d}t \right]^{2.5} (t_2 - t_1)$$

In this chapter, the full-scale HIC values obtained from testing will be compared to threshold values, and to one another. Acceleration envelopes will be compared to the DOD, US Army, and Eiband's recommendations. In addition, values of jerk will be analyzed and discussed. The study uses small-scale testing which can provide very accurate results at very little cost compared to full-size testing.

9.1.2 Energy Absorbers and Acceleration Mitigation

An energy absorber is a device that dissipates kinetic energy in a non-recoverable manner. The absorber may dissipate energy through plastic deformation, viscous damping, friction, or fracture (Olabi et al. 2007). Any structure that deforms plastically due to kinetic energy input dissipates energy from the system and is therefore an energy absorber. Energy absorbers have a wide range of applications in many fields of engineering, but have been used extensively in the study of mechanical response to extreme loading.

The majority of literature available on the topic of energy absorbers and acceleration mitigation comes from the last half of the twentieth century. During this time, an increased interest in structural mechanics and dynamics birthed a number of publications dedicated to structural plasticity and impact loading. The International Journal of Mechanical Science, which began in 1960, and the International Journal of Impact Engineering, which began in 1983, became major sources of the early research on energy absorbers.

In the 1980s, researchers began to focus on practical applications for energy absorbers. Large-scale implementation began in cars, aircraft, crash barriers, road bridges, nuclear reactors, offshore structures, oil tankers, and many other machines and devices (Alghamdi 2000). Energy absorbers owe their prominence to their ability to take a variety of geometries and sizes depending on the application.

Only a small class of energy absorbers has become prevalent in acceleration mitigation applications. Tube shapes, because of their efficiency, have become the dominant choice in applications involving acceleration mitigation. Tubes have been investigated for performance in the following modes of deformation: inversion, splitting, lateral indentation, lateral flattening, and axial crushing (Alghamdi 2000). While the inversion and splitting deformation modes have not yet found any practical applications, lateral indentation, lateral flattening, and axial crushing modes have been widely used in energy mitigation problems. Applications with these deformation modes can be seen in everything from automobile bumpers to energy-absorbing seat mechanisms (EASMs) (Alghamdi 2000).

Research in the static and quasi-static energy absorption capacity of tubes has revealed that the axially symmetric buckling failure mode has the potential to dissipate ten times more energy than the lateral flattening failure mode (Alghamdi 2000). Nevertheless, lateral flattening has been extensively researched and has proven to be more effective than lateral indentation and easier to implement than axial crushing (Alghamdi 2000). Lateral indentation, the failure mode utilized by car bumpers, is the least effective.

Alem and Strawn (Alem and Strawn 1996) first investigated the use of tube-type energy absorbers in EASMs in order to quantify the energy dissipation from land mine blasts. The primary purpose of using EASMs with energy absorbers was to attenuate acceleration to survivable levels. In 2002, a study proposed the use of axially oriented tubes in an EASM for agricultural aircraft (Kellas 2002). The seat mechanism relied on the symmetric axial crushing of these tubes for maximum energy dissipation. The study had favorable results, significantly reducing both acceleration values on the seat and loads transferred to human occupants. Several subsequent studies supported the effectiveness of axially crushed thin-walled tubes in EASMs (Tabiei and Nilakantan 2009).

Polyurea is an elastomer with an effective modulus that increases as its strain rate increases. As a result, literature has suggested that polyurea is able to absorb vast amounts of energy when stretched over a plate subject to rapid bending (Xue and Hutchinson 2007). Ballistic projectile testing has shown that coating on one side of a metal plate can change the response of the plate from full penetration to full fracture mitigation (Amini et al. 2006). Furthermore, computational testing of polyurea used as a helmet suspension-pad material has shown a significant reduction in the loading experienced by the brain (Grujicic et al. 2010).

The favorable properties of polyurea found through experimentation and computation suggest that it is an ideal candidate for energy dissipation. Research at the University of Maryland with a particular type of polyurea manufactured by Line-X has shown that the elastomer is able to reduce frame acceleration when used as a coating on a vehicle hull (Brodrick 2010). The polyurea is also useful in mitigating vibrations associated with energy transfer between the hull and the frame (Brodrick 2010). Other experimental tests illustrate the immense benefit of coating blast-threatened structures with polyurea (www.paxcon.com).

9.2 Test Program

Earlier testing in the lab investigating the accelerations imparted to the floorboard of a simulated vehicle indicated that those accelerations are very large compared to those imparted to the vehicle's frame. The decision was made to concentrate on reducing accelerations on the frame and plan on attaching seats to the frame rather than to the floorboard. This chapter examines two separate series of tests. The first series was composed of 15 tests and mainly used "V" hulls (with the exception of the last three tests). The second series was composed of 10 tests and used inverted "V" hulls. The first 15 tests were designated T1–T15 and the last 10 tests were designated Tests 1–10.

9.2.1 Tests T1–T15

These tests were conducted on simulated vehicles and various means of attenuating the accelerations were tried. Accelerometers from PCB Piezotronics (Model 350C02) were used for acceleration measurements (three accelerometers were used on each model). These accelerometers were chosen because of dynamic range, their ability to withstand a large g level, a long cable length, and size of the accelerometers. Three accelerometer holes were tapped in each vehicle's frame, recording

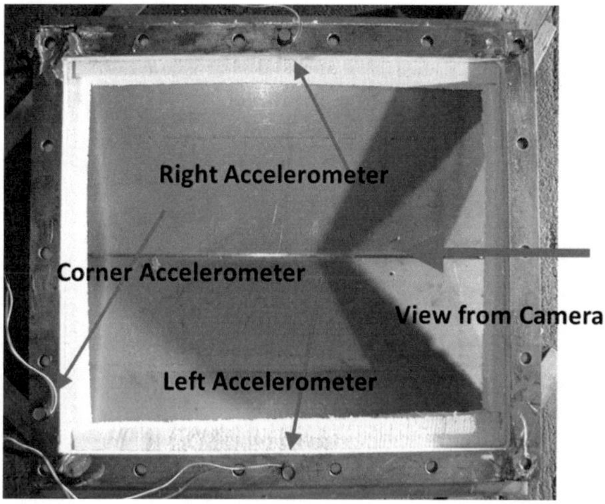

Fig. 9.3 Accelerometer placement (*top view*)

Table 9.1 Calibration factors

Serial number	Calibration factor (g's/mV)
19445	8.84956
19446	10.7527
30279	9.7087

accelerations of the left, right, and corner of the frame (see Fig. 9.3). Kenlube grease was used to improve the bond between accelerometers and the metal frame. A thin layer of grease was applied to the thread of the accelerometer and around the holes on the frame where accelerometers were screwed in. Accelerometers were calibrated up to 10,000 Hz, and required calibration factors were used to convert their output in voltage to acceleration in g-forces. Table 9.1 shows the calibration factors for each accelerometer used in the tests. Accelerometer placement can be seen below in Fig. 9.3 from overtop the test specimen. Left, right, and corner accelerometers are labeled as viewed by the camera.

All tests were performed in the sandpit located in the basement of the Dynamic Effects Lab. A diagram of the test setup can be seen below in Fig. 9.4. Accelerometers were connected to a signal amplifier and then split between two oscilloscopes. The firing box trigger was connected to the camera and the oscilloscopes, while the firing wire was connected to the charge. When triggered, the oscilloscopes and camera recorded data (camera data was sent to a laptop for viewing purposes).

Figure 9.5 shows one of the simulated vehicles ready for testing. This configuration will be later referred to as the control frame for the aluminum series. All frames were 40.64 cm along the longer side and 35.56 cm along the shorter side. Notice the "V"-shaped hull. The connections between the hull and the frame were 16.2 cm bolts along both of the longer sides. In the figure, the vehicle is sitting on the sand with the buried charge beneath it; the tank is filled with water up to the point of the

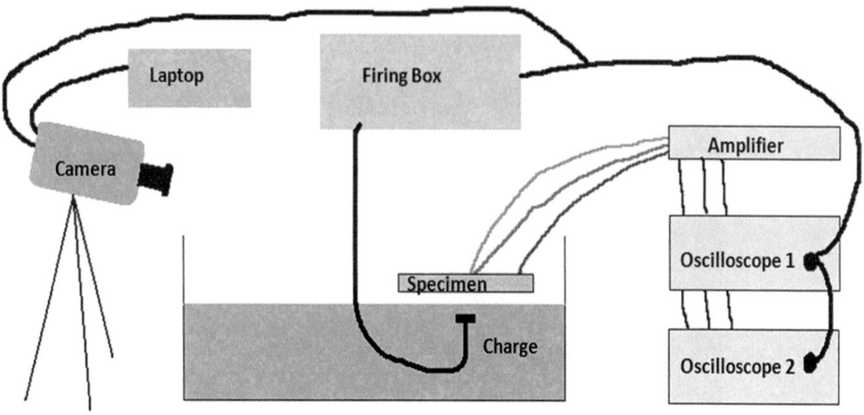

Fig. 9.4 Test setup

Fig. 9.5 Test setup

sand/air interface. The steel tank is 1.524 m by 1.524 m square by 61 cm deep. It is equipped with a water saturation control system that fills the tank. The system consists of a column that fills up with water, which is connected to the bottom of the box. This provides an even saturation of water in the sand. The bottom of the pit contains a mesh over a layer of coarse gravel, which prevents sand from contaminating the water saturation system. On top of the mesh is sand purchased from Home Depot.

In the first series of 6 tests using an aluminum frame, the model was scaled such that an experimental charge weight of 1 g would have the same net effect as

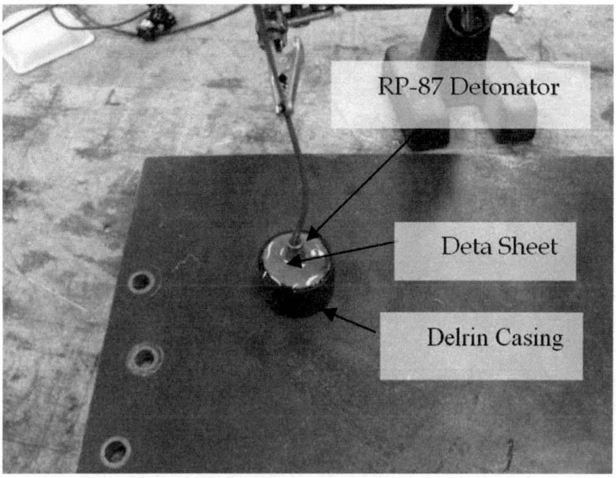

Fig. 9.6 Schematic of explosive charge

would 2.27 kg of TNT. For the remaining 19 tests, the model was scaled such that an experimental charge weight of 4.4 g would have the same net effect as would 4.54 kg of TNT.

All tests discussed used explosive charges made from an RP-87 Exploding Bridge Wire detonator, Deta Sheet, and a Delrin plastic charge casing, held together by heavy duty epoxy manufactured by Loctite. Deta Sheet was the main explosive used and consisted of 63 % pentaerythritol tetranitrate (PETN) and 37 % plasticizer. An explosive charge with labeled components is shown in Fig. 9.6. The RP-87 detonator is manufactured by Teledyne Technologies Incorporated and contains 26 mg of PETN and 43 mg of cyclotrimethylenetrinitramine, a common military explosive referred to as RDX [http://www.teledyne.com]. The charge shown in Fig. 9.6 is the 4.4 g charge. The 1 g charge used the same detonator but less Deta Sheet.

The first 12 tests used a "V"-shaped bottom while the last 13 tests used an inverted "V"-shaped bottom (a pocket plate). The angle used on the vehicle bottom was 13°. In earlier testing using flat bottoms, 7° bottoms, 13° bottoms, and 20° bottoms it was determined that the biggest reduction in impulse delivered to the vehicle was obtained with the 13° angle. Although continued increase in angle reduced the impulse, it was small compared to the decrease obtained at 13°.

In terms of a full-scale vehicle the first six tests equated to results for a 2.27 kg TNT charge buried 10.16 cm deep with a standoff distance to the closest point on the vehicle of 106.7 cm. For the last 19 tests the results would equate to a 4.54 kg TNT charge buried 10.16 cm deep with a standoff distance of 81.3 cm for the full-size vehicle. For a normal 4.54 kg charge, the expected depth of burial would be 10.16 cm and the normal standoff distance would be 40.64 cm. Therefore the loadings on all of these vehicles were much less than what would be expected in the field. For the first series, the charge size was small and the standoff distance was very large. For the remaining tests, the standoff distance was about twice the norm. Unfortunately, it was not possible to equate the results of the first 6 tests to those of the last 19.

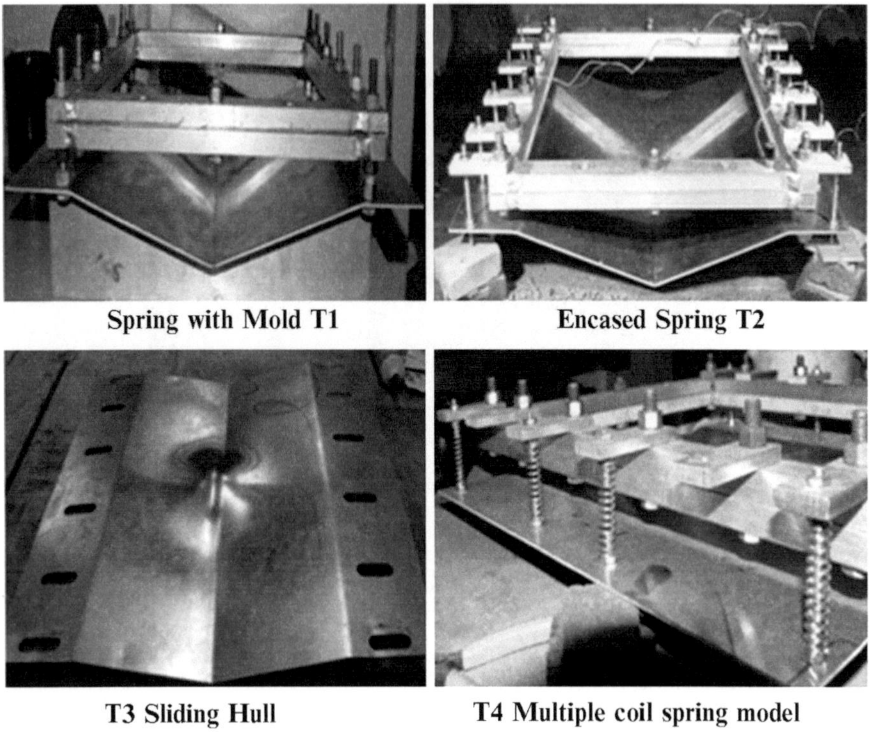

Fig. 9.7 Models used in Tests T1 through T4

The procedure followed was that a control frame was tested for the aluminum frame, for the steel frame, and for the steel frame with the inverted "V" bottom. Then various mitigation techniques were employed to see if the accelerations on the frame could be reduced.

9.2.2 Aluminum Frame Tests

The models used in the six tests in the aluminum frame series are shown below in Figs. 9.7 and 9.8. The charge size, the standoff distance, and the depth of burial were kept constant. The control frame configuration (Test T1) is shown in Fig. 9.7 and consisted of 12 straight bolts (6 equally spaced along the longer edges of the model) attaching the hull to the frame.

The second test configuration (Test T2) used aluminum "beams" that were 0.7 cm × 2.54 cm × 6.3 cm long and connected the straight bolts to the aluminum frame. The third configuration (T3) used a hull that could slide in the sidewise direction. Instead of holes in the hull for attaching to the vertical bolts (without the beams) the hull had slots and the bolts were not torqued but rather were tightened such that the hull could slide sidewise as the upward force impacted on the hull.

Fig. 9.8 Model which used the softer, less stiff white foam

In the fourth configuration, the hull was mounted to the frame via studs located concentrically within springs, as shown in Fig. 9.7. This arrangement allows for compression of the springs when a load is applied. Note in the figure that only four bolts were utilized on each side (with springs). The spring constant for each of the springs was 8,231 N/m and the four springs were equally spaced along the longer sides of the frame.

The last two tests with the aluminum frame (T5 and T6) used the control frame but a sheet of closed cell foam was placed between the hull and frame. In the first foam test, a very stiff pink foam was used while in the second a less stiff white foam was used. A picture of the model with the white foam is shown in Fig. 9.8. The stiffness of the pink foam was 651,822 N/m and the stiffness of the white foam was 48,685 N/m.

A comparison of the resulting accelerations obtained for the first six tests using the aluminum frame is shown in Fig. 9.9.

In all tests the displacement–time curves for the points on the frame where the accelerometers were attached were recorded with a high-speed video camera. The accelerations measured at those points were integrated twice to obtain displacement curves. If the two curves agreed it was assumed that the accelerometer data was valid. If there was disagreement between the two displacement–time curves the acceleration values were deemed to be incorrect. In the results above, notice that in only three of the six tests were all accelerations felt to be valid. In the first test, two of the three values were felt to be incorrect and in the other two cases two of the three values were felt to be correct. From the results, the four tests (the beam-type attachment, the sliding hull, the spring connections, and the white foam) all gave similar results and these accelerations were about two-thirds of the accelerations measured in the control frame and the frame with the pink foam. It was felt that the pink foam was too stiff and really did not provide much, if any, mitigation.

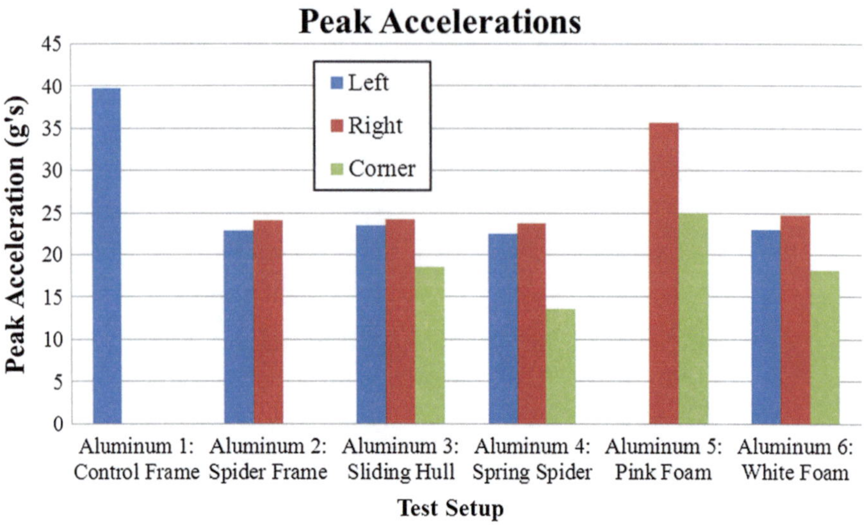

Fig. 9.9 Peak accelerations from aluminum frame test series

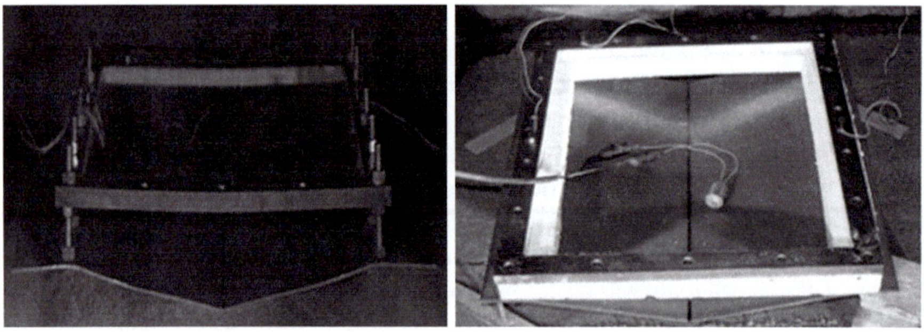

T7 Steel Control Frame **T8 Third Foam Test Model**

Fig. 9.10 Models used in T7 and T8

9.2.3 Steel Frame Tests

The next series of tests were with the steel frame. These tests used a greater amount of explosive with a smaller standoff distance, and were therefore subjected to greater loads. Six tests were conducted, including a test with a control frame. The control frame is shown in Fig. 9.10. It is identical to the aluminum control frame, except that the frame is made of steel and therefore stiffer.

The next test (T8) again used the white foam, but in this case just a strip of the foam was placed between the hull and frame, with no foam in the center. The model for T8 is also shown in Fig. 9.10. The electrical device shown is used to ensure that the camera and the scopes trigger simultaneously and is used prior to wiring up the explosive. Recall that the white foam was less stiff than that which was used in the

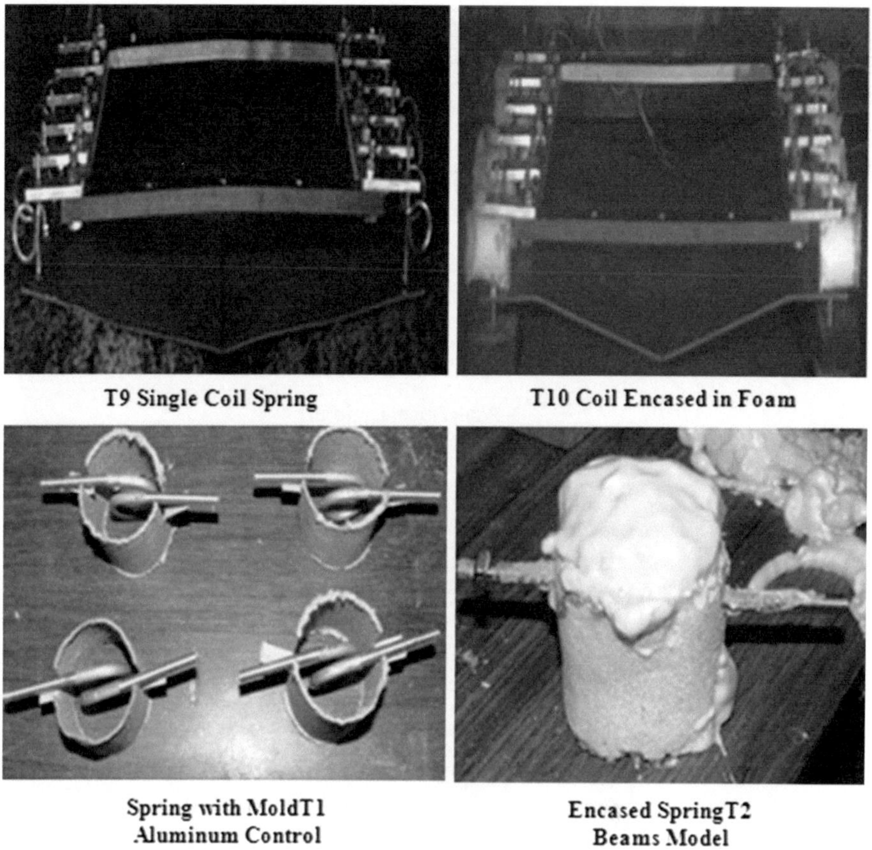

Fig. 9.11 Models for Tests T9 and T10

sixth test in the aluminum series and appeared to do a better job of mitigating the acceleration than did the stiffer pink foam.

The next model (Test T9) used a 0.48 cm diameter bolt that was bent to form a single coil between the hull and frame in conjunction with the "beam" attachment as used before. To make the springs, a 0.48 cm diameter aluminum straight bolt was wrapped around a 1.9 cm mandrel to form the single-coil spring. The spring constant for one of these springs was 105,076 N/m. The model for T9 is shown in Fig. 9.11.

The next model (T10) used the same attachment as Test 9 but foam was placed around the springs to assist in carrying the load and damping out any vibrations. The spring constant for each of these foam-encased springs was 112,781 N/m, only slightly higher than the uncased springs. The model is shown in Fig. 9.11. The next test (T11) used the control frame but had a polyurea/polyurethane spray coating on the hull (both top and bottom). A photograph of that model is shown in Fig. 9.12. The final test (T12) in this series used commercially available isolator cables to connect the hull to the frame. The spring constant for the isolators, which used 0.56 cm braided cable, was 16,287 N/m in shock loading. This model is also pictured in Fig. 9.12.

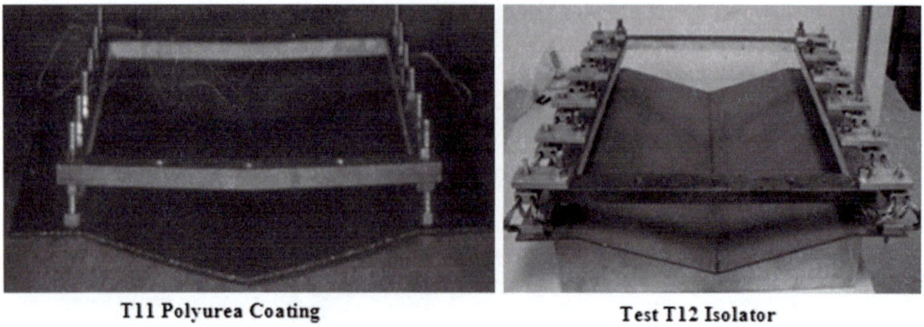

Fig. 9.12 Models for Tests T11 and T12

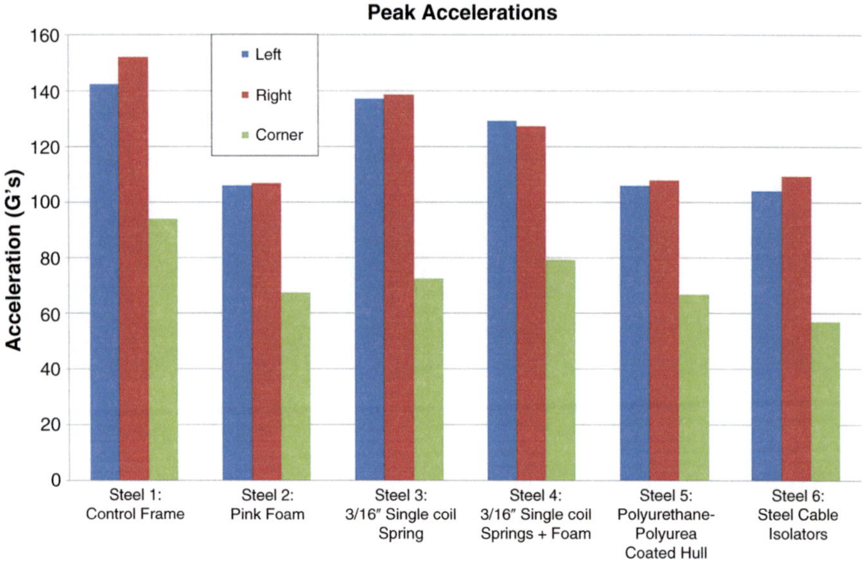

Fig. 9.13 Acceleration results for steel frame test series

A comparison of the accelerations measured in these six tests is shown in Fig. 9.13. The single-coil spring and the foam-enclosed single-coil spring (T8 and T9) helped very little. The other three techniques (foam, polyurethane/polyurea coating, and commercial cable isolator) seemed to do the best and reduced the accelerations by about 30 %.

9.2.4 Inverted "V" Hull Tests

The next three tests (T13, T14, and T15) used the inverted "V" hull. Again, the first test employed the control frame configuration, shown in Fig. 9.14, and used straight

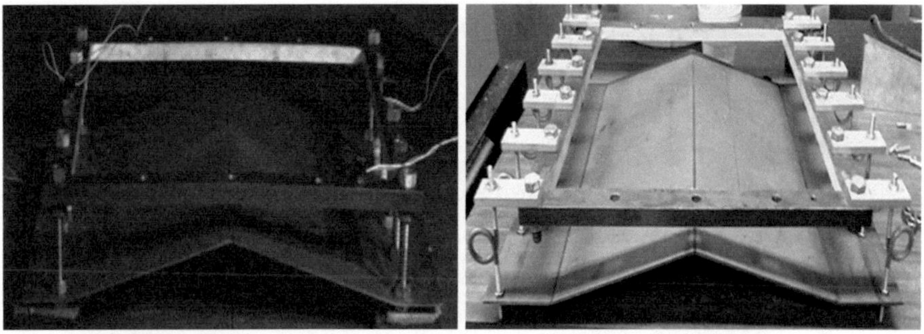

T13 Control for Inverted T14 Inverted with Spring

Fig. 9.14 Models for Tests 13 and 14 with the inverted hull

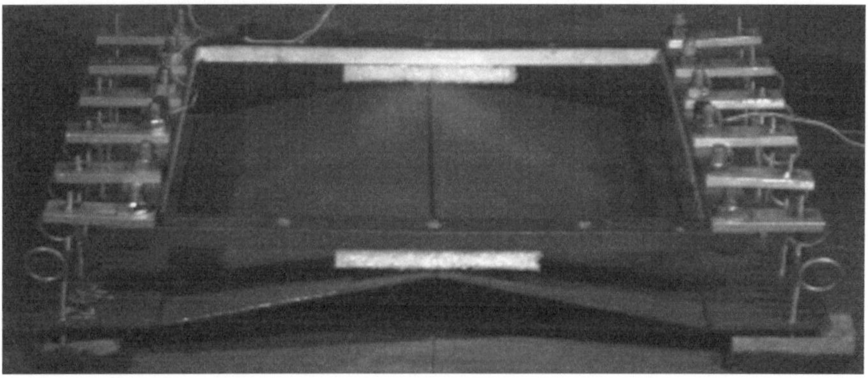

Fig. 9.15 Model used in Test 15

bolts to attach the frame to the hull. The second model used the "beam" arrangement along with 0.48 cm single spring made from straight bolts bent around a 1.9 cm diameter mandrel, and used 6 springs along each side (spring constant of 105,076 N/m). The models for these two tests are shown in Fig. 9.14.

The final model in this series (T15) used 0.32 cm diameter single coils, made by wrapping straight aluminum bolts around the same-size mandrel (spring constant of 26,269 N/m). The model is shown below in Fig. 9.15. In addition, a small strip of white foam was used just below the frame such that, if the hull deformation is too great, the hull would hit the foam rather than the frame itself.

A comparison of the results of these three tests is shown in Fig. 9.16. Notice first that going from the "V" hull to the inverted "V" hull reduced the accelerations by about 1/3 (see Fig. 9.13/Control Frame and Fig. 9.16/Control Frame). The 0.48 cm single-coil spring again did not appear to help very much. However, the 0.32 cm single-coil spring (less stiffness) with the foam strip reduced the acceleration by about 20 % over the control frame.

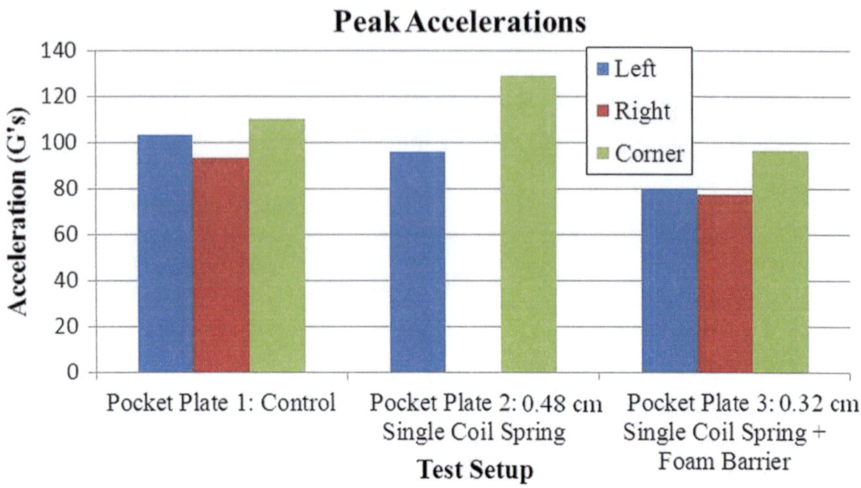

Fig. 9.16 Comparison of peak accelerations from inverted hull series

Table 9.2 Small- and full-scale dimensions

	Small-scale tests performed in this chapter	Corresponding full-scale values
SoD to hull (cm)	2.54	25.66
SoD to frame (cm)	8.05	81.33
DoB (cm)	0.99	10.01
Vehicle length (cm)	40.64	410.46
Vehicle width (cm)	35.56	359.16

9.2.5 Final Mitigation Test Series

Having found out from the test series described above that the inverted "V" hull appears to reduce accelerations when compared to the "V" hulls, we next proceeded to investigate mitigation techniques in conjunction with inverted "V" hulls. In this test series we conducted ten tests. These tests will be labeled Tests 1–10. All tests utilized a 13° inverted "V" hull. All of the ten tests used 4.4 g of explosive charge and the model used was scaled to a charge of 4.54 kg of TNT. When model dimensions, depth of burial, and standoff distance (to the bottom of the floorboard) are scaled the full-size values are listed in Table 9.2 above. Notice that all of these values correspond closely to what might be found in the field—except for the standoff distance. As in the previous tests the value of 81.28 cm is about twice what would be found for a current-day vehicle.

9.3 Vehicle Fames

Three vehicle frames were used for the ten tests performed for this last series. The vehicle frames were changed in order to optimize the properties of the vehicle being tested.

Fig. 9.17 Photos of the original frame and the pocket frame

9.3.1 Original Steel Frame

The first frame tested was a rectangular frame constructed of 1,018 steel with outer dimensions 35.56 cm by 40.64 cm, as shown in Fig. 9.17. The steel bars used to construct the frame were 2.54 cm wide and 1.27 cm tall. The weight of the frame was 3.54 kg. Bolt holes of 0.95 cm diameter were drilled around the perimeter of the frame at 7.62 cm center to center spacing along the 40.64 cm spans. Threaded holes were tapped 0.762 cm into the frame at the center of each 40.64 cm span and at one corner with a ¼″-28 male tap. These tapped holes served as mounting locations for the accelerometers.

9.3.2 Pocket Frame

The second frame tested was a modified rectangular frame which will be referred to as a pocket frame. The pocket frame was constructed of 1,018 steel with outer dimensions 35.56 cm by 40.64 cm and a height of 3.18 cm. The weight of the pocket frame was 3.99 kg. Bolt holes of 0.95 cm diameter were drilled through the 40.64 cm spans at 7.62 cm center to center spacing. Threaded holes were tapped into the frame in the same size and location as in the first steel frame in order to provide identical mounting locations for accelerometers. The pocket frame can be seen in Fig. 9.17.

9.3.3 Reinforced Pocket Frame

A third frame was created by modifying the pocket frame and will be referred to as the reinforced pocket frame. The original pocket frame was modified by welding pieces 2.54 cm wide, 12 gage strips of 1,018 sheet steel in several locations to the existing 2.54 cm × 1.27 cm steel bars, as shown in Fig. 9.18. The weight of the reinforced pocket frame was 4.47 kg.

Fig. 9.18 Reinforced pocket frame

9.4 Vehicle Hulls

Two inverted "V"-shaped vehicle hulls were used for these tests. The vehicle hulls were all formed from 40.64 cm × 40.64 cm square sections of 1,018 sheet steel with 12 gage thicknesses.

9.4.1 Hull V 1

The first hull used for testing was an inverted "V"-shaped hull, bent to 13° angles with a hand sheet metal bender. This hull is referred to as IV-1.

9.4.2 Hull V 2

The second hull used for testing was constructed of the same material and method as IV-1. However, after bending the plate, sections were cut out of the ends of the hull using a metal grinder. The motivation for this alteration was to provide increased space for permanent hull deflection before impact with the vehicle frame. The length of cutouts across the hull depended on the amount of additional hull deformation that was expected. This hull will be referred to as IV-2 and can be seen in Fig. 9.19.

The test series contained the following 10 tests shown in Table 9.3.

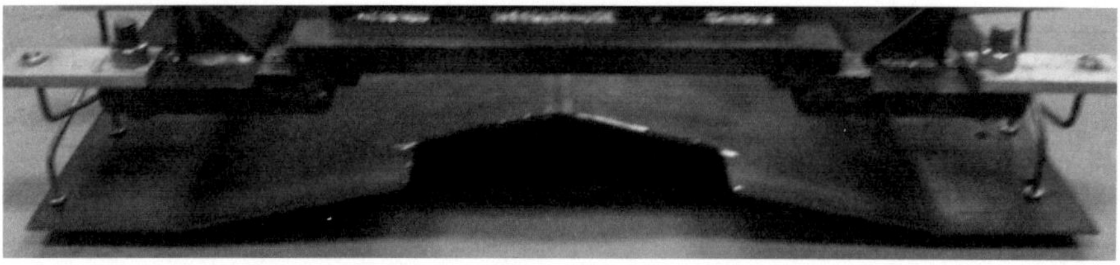

Fig. 9.19 IV-2 Hull features cutout ends

Table 9.3 Test series vehicle hulls

Test no.	Fame	Hull type	Mitigation technique	Specimen weight (kg)	Specimen mass (g)	SoD to frame (cm)	SoD to hull (cm)	DoB (cm)
1	Control	IV-1	None	7.84	7,839	8.05	2.54	0.99
2	Control	IV-1	0.48 cm coils	7.49	7,491	8.05	2.54	0.99
3	Control	IV-1	0.32 cm coils	7.35	7,351	8.05	2.54	0.99
4	Pocket	IV-1	None	8.31	8,311	8.05	2.54	0.99
5	Pocket	IV-1	Isolators	9.04	9,039	8.05	2.54	0.99
6	Reinforced pocket	IV-2	Compression strut truss, hull cuts	8.75	8,750	8.05	2.54	0.99
7	Reinforced pocket	IV-2	Lateral tubes	8.56	8,557	9.14	2.54	0.99
8	Reinforced pocket	IV-2	Lateral tubes, compression struts	8.71	8,713	9.14	2.54	0.99
9	Reinforced pocket	IV-2	Lateral tubes, hull cuts, hull end plates	8.57	8,570	9.14	2.54	0.99
10	Reinforced pocket	IV-2	Polyurea coat, lateral tubes, hull cuts and plates	11.65	11,653	9.14	2.54	0.99

Figure 9.20 shows the model used in the first test. It is the control frame against which all further results are compared. From Table 9.3 above, it is seen that it used the control frame and the hull IV-1.

The frame was connected to the hull with 5 straight 0.95 cm bolts equally spaced along the long side (40.64 cm long). The three positions of the accelerometers are also shown.

The model used in Test 2 is shown in Fig. 9.20. In this model 0.48 cm diameter straight bolts were wrapped around a 1.91 cm mandrel to form a single-coil spring and then ten such springs were used to attach the hull to the frame in conjunction with 0.7 cm × 2.54 cm × 6.3 cm thick 6,061 aluminum "beams." The third model was almost identical to the second but it used 0.32 cm diameter straight bolts wrapped around the 1.91 cm mandrel to form the single-coil spring. The spring constant for the 0.48 cm diameter spring was 105,076 N/m while for the 0.32 cm diameter one it was 26,269 N/m. The fourth model (Fig. 9.20) used the pocket frame described above but no form of mitigation and once again is attached to the hull by 0.95 cm

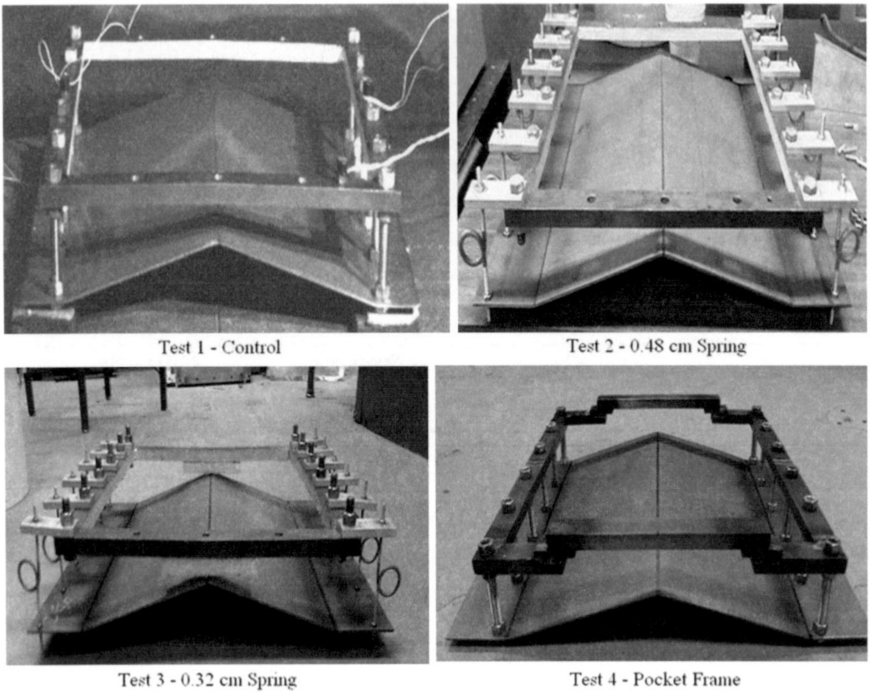

Fig. 9.20 Models used in Tests 1, 2, 3, and 4

bolts. The attempt here was to prevent the hull from striking the frame once it was loaded by the upward traveling ejecta. The connections on the frame were wielded.

The fifth model, shown in Fig. 9.21, used the pocket frame and commercially available cable isolation devices at each of the attachment points for the hull to the frame. The isolation device is made of twisted cables. It was model SM7-165A and was sold by IDC and had a spring constant of 16,287 N/m in shock loading. Also note that the hull is also braced with a V-shaped metal sheet to prevent the hull from striking the frame when loaded by the explosive detonation.

The sixth model tested is shown in Fig. 9.21. After 2 tests were conducted with the pocket frame, additional frame stiffness was still desired. The reinforced pocket frame was created to provide additional stiffness to help reduce large vibrations visible in high-speed photography and accelerometer data. In addition, the IV-2 hull was introduced to allow more hull deformation before frame impact, resulting in more energy dissipation and a less violent impact. That is to say part of the hull was cut away to prevent it from hitting the frame. The installation of a compression member was the main mitigation technique of this test. The compression member was created with 0.32 cm diameter aluminum rods arranged in three "X" patterns on each side of the vehicle, as shown in Fig. 9.21.

The seventh model tested is shown in Fig. 9.21 and used the reinforced pocket frame in conjunction with lateral crushing of tube-like elements. Aluminum

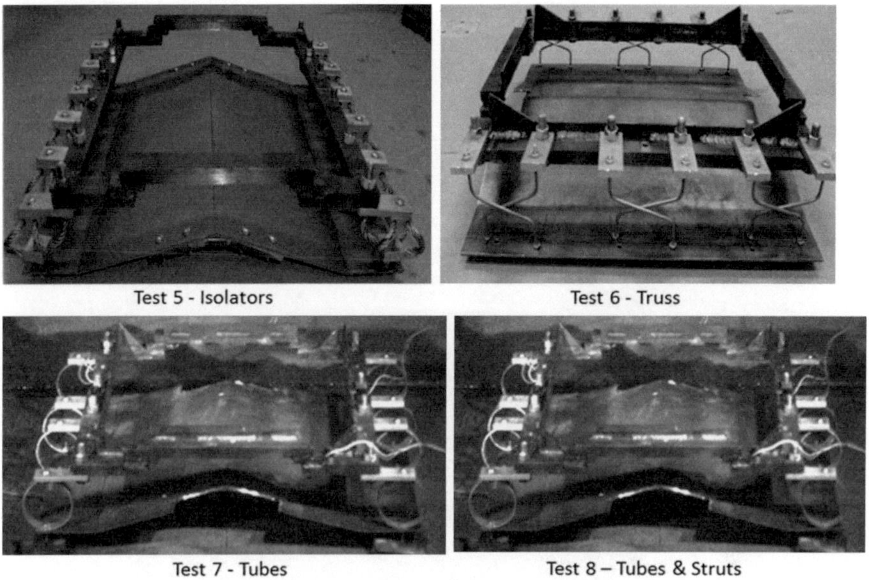

Fig. 9.21 Models used in Tests 5, 6, 7, and 8

6061-T6 tubes were installed and configured to crush in this specimen. The intent was to utilize one of the more powerful energy dissipation techniques of lateral tube flattening. The tubes were 5.72 cm in outer diameter and 5.47 cm in inner diameter, had a thickness of 0.125 cm, and a cut length of 1.59 cm. Notice also that the model used the stiffened pocket plate frame as well as a cutout for the hull to prevent the hull from contacting the frame.

The eighth model tested is shown in Fig. 9.21. Because impact of the hull and the frame was still deemed to be a problem with the IV-2 hulls and a major source of energy transfer, several techniques were investigated to prevent violent contact. The first technique was to install three aluminum struts across the width of the specimen in the back, middle, and front. The struts were made of 0.48 cm diameter aluminum rod, threaded on each end by a 10–32 die, and attached to each side of the hull with small tapped aluminum blocks. The blocks were attached to the top surface of the sides of the hull. Unfortunately the struts are difficult to see in the photo.

Test 9 was similar to the other test specimens with laterally installed aluminum tubes. Rather than using compression rods spanning across the width of the specimen to minimize violent impact, "end plates" were installed on the front and back of the hull where the hull was cut out. These end plates, shown in Fig. 9.22 on the vehicle hull, were created from 2.03 mm 1,018 sheet steel, bent with the assistance of a tabletop clamp.

During hull deformation, the end plate will deform laterally due to in-plane movements, creating another means of energy dissipation. The end plates were attached to the hull with rivets. The overall test specimen can be seen in Fig. 9.22.

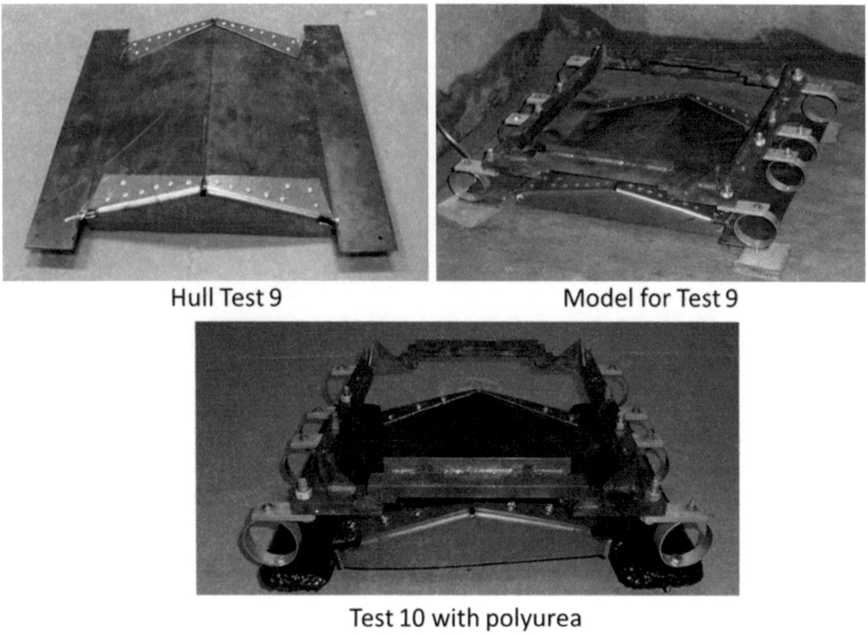

Fig. 9.22 Models used in Tests 9 and 10

The final test performed in this series utilized the most effective acceleration mitigation strategies. In addition, the test involved a thick double coating of polyurea on the IV-2 hull. Polyurea or similar elastomers have been tested in large-scale tests with coatings of several inches on vehicle hulls. This test was intended to simulate an extreme condition of hull coating in which vehicle cost and weight is not an issue. The test was performed in order to determine what effects polyurea may have in large thicknesses, since previous testing by Brodrick (2010) has already shown that it has desirable properties in thin coatings.

Each side of the specimen hull was coated with 0.76 cm of polyurea. Aluminum tubes were attached to the hull and frame, as before. Similarly, end plates were machined and bolted through the hull. The specimen can be seen in Fig. 9.22.

9.5 Results

The values obtained in the model testing were scaled up to a 4.54 kg charge. Tabulated results from all ten tests are presented in Table 9.4 below. Again, this table shows scaled up values and gives an indication of whether or not the criteria thought to be important was passed or failed. For Eiband if the acceleration exceeds 23 g's for more than 23 ms the test failed. For the Army it is exceeding 23 g's for 25 ms. For the DOD it is exceeding 25 g's for more than 100 ms. All tests but Test

Table 9.4 Test result series vehicle hulls

Test no.	Eiband—23 g's—5.5 ms	Army—23 g's—25 ms	DoD—25 g's—100 ms	HIC	Peak jerk (m/s^3)	Impulse (kg m/s)
1	Fail	Fail	Fail	327.4	481.7	4,882
2	Fail	Fail	Fail	637.7	504.8	5,205
3	Fail	Fail	Fail	541.1	379.7	5,725
4	Fail	Fail	Pass	44.1	574.6	4,518
5	Fail	Fail	Fail	592.2	358.8	6,302
6	Fail	Fail	Fail	106.5	83.15	4,928
7	Fail	Fail	Fail	91.3	96.6	4,215
8	Fail	Fail	Fail	49.6	72.6	4,787
9	Fail	Fail	Pass	107.3	78.8	5,087
10	Pass	Pass	Pass	7.5	31.9	5,189

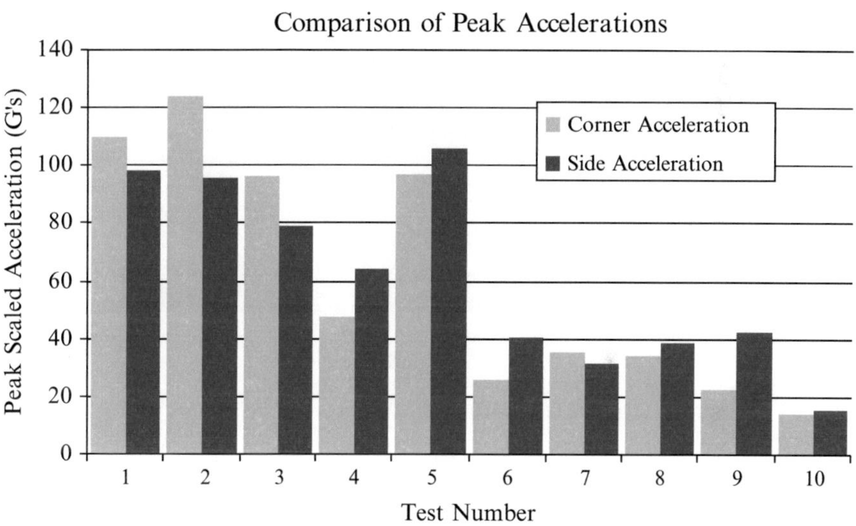

Fig. 9.23 Peak accelerations in Tests 1–10

No. 10 failed the Eiband and Army acceleration envelope criteria. All tests but Test Numbers 3, 9, and 10 failed the DOD criteria. The HIC, peak jerk, and impulse are also presented in the table and are discussed later.

Figure 9.23 shows the peak acceleration values obtained for the 10 tests. Recall that Test 1 is the control test and all subsequent values should be compared to that result. Note also that the acceleration values obtained in this series of tests for the control frame agree very well with results obtained for the control frame as presented above for the inverted "V" frame test (approximately 100 g's in both cases). Values are presented for both the accelerometers at the corner location and for the location at the center of the frame. Tests 2, 3, and 5 did not perform much better than the control frame when g level was the only criteria. However, the other mitigation

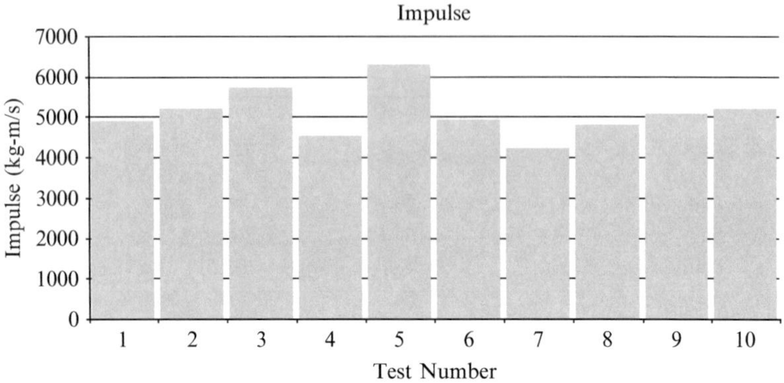

Fig. 9.24 Impulse values for Tests 1–10

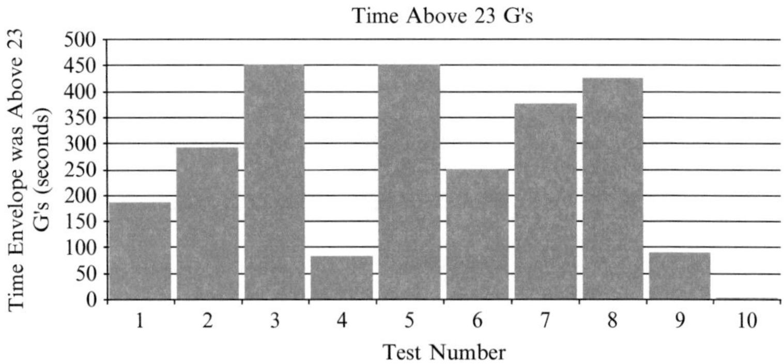

Fig. 9.25 Time above 23 g's for Tests 1–10

techniques appear to be quite effective in reducing peak g levels. Test 4 was fairly effective and that involved going to the pocket frame and no other means of mitigation. The very best results shown were for Model 10 which used all of the best forms of mitigation discovered to be good (lateral tubes, hull end plates, and polyurea coating). These techniques knocked the peak acceleration down to 10 % of what it was for the control test. Notice from Table 9.4 that as various mitigation techniques were used the weight went up and also notice that for Tests 7, 8, 9, and 10 the standoff distance also increased slightly. The increased mass would normally result in higher impulse delivered to the target while the increased standoff distance would result in reduced impulse. Figure 9.24 shows the impulse delivered to the simulated model for each of the 10 tests. There was not a lot of variation in the impulse delivered. Test 5 with the isolators resulted in the largest impulse.

Figure 9.25 shows the time above 23 g's for all of the tests. With this criterion, only Test 10 passed and Tests 4 and 9 were the next best.

Keep in mind that the results presented above are for vehicles with a standoff distance nearly twice as much as could be expected in the field with a normal vehicle.

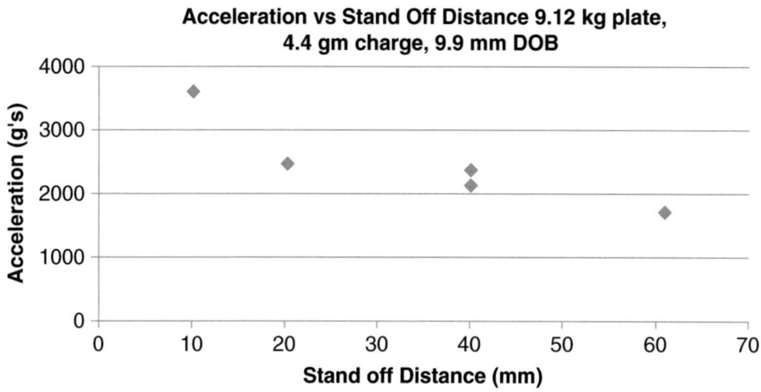

Fig. 9.26 Acceleration as a function of standoff distance

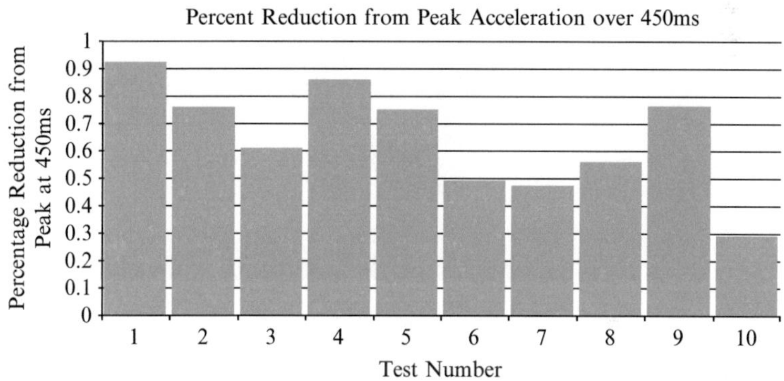

Fig. 9.27 Peak reduction in accelerations after 450 ms for Tests 1–10

Figure 9.26 presents work conducted on the investigation of accelerations felt on a small flat plate and shows acceleration as a function of standoff distance. Keep in mind that the standoff distance of interest is 40.64 cm and a charge size of 4.54 kg. For the case shown in Fig. 9.26, the two points (at 40 mm) show what would be expected for the small flat plate at a standoff distance (in full scale) of 40.64 cm and a charge size of 4.54 kg. The standoff distances used in these tests would be closer to 81 cm. Hence the accelerations experienced in this study are less than those actually to be expected in the field.

Figure 9.27 shows the percent reduction in peak acceleration over 450 ms and gives an idea of the damping in the system. Unlike what might be expected, the polyurea coating performs poorly. This is likely due to the fact that the accelerations for the polyurea were so low to begin with.

Figures 9.28 and 9.29 present the peak jerk values as well as HIC values calculated for all 10 of the models. The HIC values for Tests 4, 6, 7, 8, 9, and 10 are quite

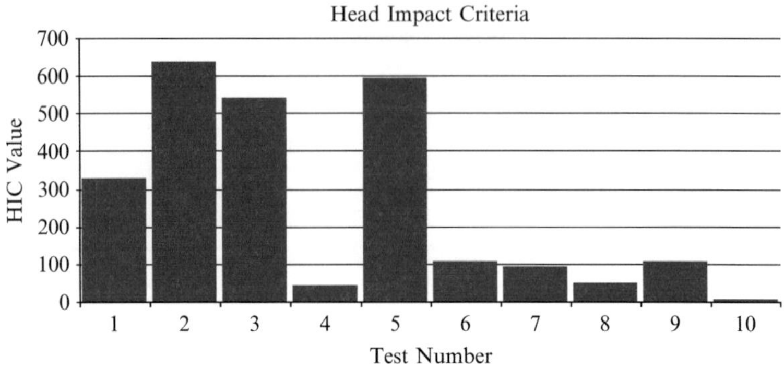

Fig. 9.28 HIC values for Tests 1–10

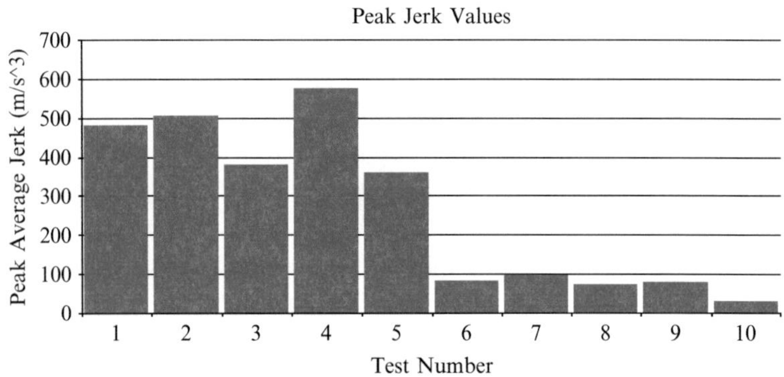

Fig. 9.29 Peak jerk values for Tests 1–10

low compared to the other tests. Again HIC is mainly used for automobile crash testing and is a combination of acceleration and jerk. It is unclear what these results imply other than 6 of the 10 tests gave values much lower than the other four. An HIC value for an automotive crash test is deemed to be good if it is below 1,000. There is currently not a good understanding in the medical community as to what the mechanism of brain injury is. It could be the soft tissue of the brain being injured when it contacts the hard skull during rapid movement. It could be damage to the small axions due to differential movement of the more massive parts of the brain. It could be destruction of blood vessels due to differential movement of the parts of the brain. It has also been speculated that it could be the result of increased blood pressure due to contraction of the abdomen when subjected to compression as a result of the detonation of the explosive. There is currently an increase in research within the medical and engineering communities to better understand exactly what the mechanism for brain injury is.

9.6 Conclusions

Acceleration data suggests that the reinforced pocket frame with aluminum tubes demonstrated the greatest proficiency to mitigate peak acceleration values. Tests 6–10, all of which utilized the IV-2 hull, had the lowest peak accelerations of all ten tests. Additionally, Tests 7–10 used laterally crushed aluminum tubes as the primary mitigation technique, which performed well when compared to other tests.

Vehicle 10 demonstrated the advantage of a large coating of polyurea: peak accelerations, HIC, and jerk are significantly reduced but damping properties for Test 10 were much worse. There would likely, however, be a high fabrication price. In reality, a full-scale model of vehicle 10 would be cost prohibitive due to the 0.76 cm top and bottom coating of polyurea, which would scale to approximately 7.6 cm.

While peak accelerations with any of the vehicles in Tests 6–10 were at least 50 % lower than those using the original control frame, a balance must be reached in order to minimize both the peak accelerations and the long-term acceleration envelope. A vehicle that exploits the benefits of the rigid connection in Tests 1 and 4 to maximum hull deformation, but also incorporates the energy-mitigating capabilities of IV-2 hull, aluminum tubes, and hull plates, can likely perform better than all previously tested vehicles in acceleration mitigation and damping.

9.7 Future Considerations

In the future, the best performing vehicles in this chapter should be altered and retested. Thinner coatings of polyurea should be used to determine its optimal thickness in hull applications and its ultimate contribution to peak accelerations and damping properties. In addition, other energy-absorbing collapsible devices, such as axial buckling of tube-like structures, should be investigated to determine their efficiency. An effort should be made to understand the relationship between the static stiffness properties of hull–frame connections and energy dissipation capacity over short and long durations of time.

Research on the medical effects of rapid acceleration should also continue in order to determine appropriate analysis techniques for scaled vehicle testing. This research should focus on the contribution of acceleration envelopes, peak jerk, and jerk curves on human response.

Eventually, effects of secondary acceleration mitigation, such as the installation of EASM, should be combined with the research presented in this chapter. The interaction between the two dissipation devices will provide more true insight into the acceleration response felt by human vehicle occupants.

Acknowledgments This chapter is based on work supported, in part, by the US Army Research Laboratory, Aberdeen, MD, NSWC Indianhead Division, TARDEC, and the Center for Energetic Concepts Development. In particular, the authors would like to acknowledge the support over the years from Leslie Taylor (NSWC) and David Fox(TARDEC & ARL). The authors particularly wish to thank the undergraduates and graduate students in the Dynamic Effects Laboratory, who actually carried out most of the tests which provided the data used in this chapter, and Ulrich Leiste, Lab Manager of the Dynamic Effects Laboratory.

References

Alghamdi A (2000) Collapsible impact energy absorbers: an overview. Thin-Walled Struct 39:189–213

Alem N, Strawn G (1996) Evaluation of an energy absorbing truck seat for increased protection from landmine blasts. (USAARL Report No. 96-06)

Amini M, Isaacs J, Nemat-Nasser S (2006) Effect of polyurea on the dynamic response of steel plates. Proceedings of the 2006 SEM annual conference and exposition on experimental and applied mechanics. St. Louis, MO

Blackman E (2008) Helmet protection against traumatic brain injury: a physics perspective. Retrieved from: http://siadapp.dmdc.osd.mil

Brodrick T (2010) Mitigation of frame acceleration induced by a buried charge. University of Maryland, College Park

Brown JL, Lechner M (1956) Acceleration and human performance; a survey of research. J Aviat Med 27(1):32–49

Department of Defense (1998) Crew systems crash protection handbook (JSSG-2010-7). Washington DC

Duane T (1953) Preliminary Investigation into the Study of the Fundus Oculi of Human Subjects under Positive Acceleration. (Report No. NADC-MA-5303). Naval Air Development Center, Johnsville, PA

Eiband M (1959) Human tolerance to rapidly applied accelerations: a summary of the literature. Lewis Research Center, Cleveland, OH

Genson K (2006) Vehicle shaping for mine blast damage reduction. University of Maryland, College Park

Grujicic M, Bell W, Pandurangan T (2010) Blast-wave impact-mitigation capability of polurea when used as helmet suspension-pad material. Mater Des 31:4050–4065

Head H (1919) The sense of stability and balance in the air. The medical problems of flying. Oxford

Hellman C (2007) The MRAP: A case study in military planning (and congressional response). Retrieved from The Center for Arms Control and Non-Proliferation website: www.armscontrolcenter.org

Hellman C, Sharp T (2008) FY 2009 Pentagon (DOD) defense budget spending request. Retrieved from The Center for Arms Control and Non-Proliferation. Website: www.armscontrolcenter.org

Hellman C (2009) Analysis of the fiscal year 2010 Pentagon spending request. Retrieved from The Center for Arms Control and Non-Proliferation website: www.armscontrolcenter.org

Henn HW (1998) Crash tests and the head injury criterion. Teach Math Appl 17(4):162–170

Kellas S (2002) Energy absorbing seat for agricultural aircraft (NASA/CR-2002-212132). Retrieved from http://ntrs.nasa.gov

Lamb C, Schmidt M, Fitzsimmons B (2009) MRAPS, irregular warfare, and Pentagon reform. Joint Force Q 55(4):76–85

McHenry B (2004) Head injury criterion and the ATB. ATB Users' Group 2004

More Attacks, Mounting Casualties (2007) The Washington Post. Retrieved from: http://www.washingtonpost.com/. Accessed 30 Sept 2007

Olabi A, Morris E, Hashmi M (2007) Metallic tube type energy absorbers: a synopsis. Thin-Walled Struct 45:706–726

Stoll A (1956) Human tolerance to positive G as determined by the physiological end points. J Aviat Med 27(4):356–367

Stech E, Payne P (1969) Dynamics of the human body (AD701383). Froster Engineering Development Corporation, Englewood, CO
US Army Aviation Systems Command (1989) Aircraft crash survival design guide (Publications No. USAAVSCOM TR89-22A). Fort Eustis, VA
Tabiei A, Nilakantan G (2009) Axial crushing of tubes as an energy dissipating mechanism for reduction of acceleration induced injuries from mine blasts underneath infantry vehicles. Int J Imp Eng 39:729–736
Xue Z, Hutchinson J (2007) Neck retardation and enhanced energy absorption in metal-elastomer bilayers. Mech Mater 39:473–487
www.paxcon.com
http://www.teledyne.com

Chapter 10
Blast Response of Sandwich Composites: Effect of Core Gradation, Pre-loading, and Temperature

N. Gardner, S. Gupta, E. Wang, and Arun Shukla

Abstract The dynamic behavior of various sandwich composites made of E-glass vinyl-ester facesheets, Corecell™ foam, and polyurea have been studied using a shock tube apparatus. The effects of core gradation, pre-loading, and temperature have been investigated. The overall dimensions of the specimens were held constant, with the only differences arising in the overall thickness of the core. During the shock tube testing, high-speed photography coupled with the nonintrusive optical technique of 3-D DIC was utilized to capture the real-time deformation process, as well as failure mechanisms. Postmortem analysis was carried out to evaluate the overall blast performance of these sandwich composites. Results indicated that (1) increasing the number of monotonically graded core layers reduces the impedance mismatch between successive layers, allowing for a stepwise compression of the core, and thus reduces the strength of the incoming stress wave. (2) The location of the polyurea interlayer has a significant positive effect on the response of composite panels to shock wave loading, both in terms of failure mitigation and energy absorption, if it is placed opposite the blast-receiving side. (3) In-plane compressive loading coupled with the transverse blast loading induces local buckling in the front facesheet, with the amount of damage proportional to the level of compressive loading. (4) The blast performance of sandwich composites at room temperature (22 °C) is superior to its performance at high temperatures (80 °C) and low temperatures (–40 °C).

N. Gardner • S. Gupta • A. Shukla (✉)
Dynamic Photomechanics Laboratory, Department of Mechanical, Industrial and Systems Engineering, University of Rhode Island, 92 Upper College Road, Kingston, RI 02881, USA
e-mail: shuklaa@egr.uri.edu

E. Wang
Department of Aerospace Engineering, University of Illinois at Urbana-Champaign, Urbana, IL 61801, USA

A. Shukla et al. (eds.), *Blast Mitigation: Experimental and Numerical Studies*,
DOI 10.1007/978-1-4614-7267-4_10,

10.1 Introduction

Today, terrorist attacks in the form of blast loadings and explosions are occurring much more frequently (Perl and O'Rourke 2001; Perl 1998). Due to the increased threat of damage to civilian and defense structures, conventional structural materials need to be replaced with more blast-resistant materials. This has lead to the development of sandwich structures, which have important applications in the naval and aerospace industry. Due to their construction, they have many advantages including high strength/weight ratio, high stiffness/weight ratio, and energy absorption capabilities. During normal service, these sandwich structures are sometimes subjected to dynamic loads such as air blasts, underwater blasts, and impact loadings. These dynamic loadings can occur when these structures are exposed to pre-loadings, and varying environmental conditions (high and low temperatures). Understanding the overall behaviors and failure mechanisms is pivotal, and will help lead to optimally designed lightweight structures that can mitigate energy and maintain structural integrity when subjected to blast loadings.

Sandwich structures consist of two thin, stiff facesheets, usually of the same thickness, separated by a lightweight, thicker core. The facesheets carry almost all of the bending and in-plane loads, while the core helps to stabilize the facesheets and defines the flexural stiffness and out-of-plane shear and compressive behavior. When sandwich structures are subjected to high-intensity impulse loadings, such as air blasts, the core materials play a crucial role in the dynamic behavior and overall structural response. Their properties assist in dispersing the mechanical impulse that is transmitted into the structure, and thus protect anything located behind it (Xue and Hutchinson 2003; Fleck and Deshpande 2004; Dharmasena et al. 2008).

Common cores are made of metallic and nonmetallic honeycombs, cellular foams, balsa wood, PVC, truss, and lattice structures. Extensive research exists in the literature regarding the dynamic response of sandwich structures consisting of the various core materials and geometric structures (Dharmasena et al. 2008; Zhu et al. 2008; Nurick et al. 2009; Radford et al. 2004; Wang et al. 2011; Tagarielli et al. 2008; Atas and Sevim 2010; McShane et al. 2006, 2007). Dharmasena et al. (2008), Zhu et al. (2008), and Nurick et al. (2009) have tested sandwich structures with a metallic honeycomb core material. Their results indicated that the parameters of core materials can effectively reduce the transmitted impulse and damage of the back facesheet. Radford et al. (2004) and Wang et al. (2011) have conducted metal foam projectile impact experiments to simulate a blast loading on sandwich structures with metal foam cores. Tagarielli et al. (2008) and Atas and Sevim (2010) have investigated the dynamic response of sandwich beams with PVC and balsa wood cores. McShane et al. (2006), (2007) have investigated the underwater blast response of sandwich composites with a prismatic lattice (Y-frame, corrugated), as well as simulated an air blast, using metal foam projectiles, on sandwich composites with pyramidal lattice cores. These studies have indicated that advanced sandwich structures can potentially have significant advantages over monolithic plates of equivalent mass in absorbing the blast energy, whether in air or underwater.

In recent years, with its ability to improve structural performance and damage resistance of structures, as well as effectively dissipate blast energy, the application of polyurea to sandwich structures has become a new area of interest. Although the behavior of polyurea has been investigated (Yi et al. 2005; Amirkhizi et al. 2006; Hoo Fatt et al. 2004; Roland et al. 2006), there have been no studies regarding the dynamic behavior of sandwich structures with a functionally graded core and polyurea interlayer. Tekalur et al. (2008) experimentally studied the blast resistance and response of polyurea-based layered composite materials subjected to blast loading. Results indicated that sandwich materials prepared by sandwiching the polyurea between two composite skins had the best blast resistance compared to the EVE composite and polyurea-layered plates. Bahei-El-Din et al. (2006) numerically investigated the blast resistance of sandwich plates with a polyurea interlayer under blast loading. Their results suggest that separating the composite facesheet from the foam core by a thin interlayer of polyurea can be very beneficial in comparison to the conventional sandwich plate design. Amini et al. (2010a, b, c, d) experimentally and numerically studied the dynamic response of circular monolithic steel and steel–polyurea bilayer plates to impulsive loads. More importantly they focused on the significance of the relative position of the polyurea layer with respect to the loading direction. Results indicated that the polyurea layer can have a significant effect on the response of the steel plate to dynamic impulsive loads, both in terms of failure mitigation and energy absorption, if it is placed on the back face of the plate. On the contrary, they also found that polyurea can enhance the destructive effect of the blast, promoting (rather than mitigating) the failure of the steel plate if applied on the impact side of the plate.

Aside from using polyurea for blast mitigation, functionally graded materials, where the material properties vary gradually or layer by layer within the material itself, have gained much attention (Hossain et al. 2007; Kiernan et al. 2009; Li et al. 2001; Apetre et al. 2006; Wang et al. 2009). Since the properties of the layered/graded material can be designed and controlled, they show great potential to be an effective core material for energy absorption and blast mitigation. Hossain et al. (2007) have experimentally studied the compressive behavior of a functionally graded foam material (FGFM) system and energy absorption under quasi-static loading conditions. The results indicated stepwise crushing from the lower density to the higher density foams. Kiernan et al. (2009) numerically investigated the propagation of a stress wave through a virtual FGFM. It was concluded that the amplitude of a stress wave propagating through a FGFM can be shaped by the gradient functions according to which the foam density varies through the direction of wave propagation. Li et al. (2001) numerically examined the response of layered and graded metal–ceramic structures under impulsive loadings. It was concluded that the choice of gradation has a great significance on the impact applications and the particular design can exhibit better energy dissipation properties. Apetre et al. (2006) numerically investigated the impact response of sandwich beams with a functionally graded core. Their results showed that a reasonable core design can effectively reduce the shear forces and strains within the structure. Consequently, they can mitigate or completely prevent impact damage on sandwich composites. In the previous work by the authors

(Wang et al. 2009), they investigated the blast resistance of sandwich composites with a discretely layered foam core. They concluded that monotonically increasing the density of the foam core allows for a stepwise compression of the core layers and thus reduces the dynamic pressures imparted on the back facesheet.

However, these investigations mainly focus on the transverse compression of the sandwich composites under blast loading, without pre-loading and/or varying temperatures, both of which will most likely affect the blast mitigation performance. Some investigations on the dynamic response of prestressed composite structures under low-velocity transverse impact loadings can be found in literature (Robb et al. 1995; Whittingham et al. 2004; Heimbs et al. 2009; Sun and Chen 1985; Choi 2008). Robb et al. (1995) carried out the first experimental investigation on the low-velocity impact response of E-glass-reinforced/polyester laminated plates under different in-plane uniaxial and biaxial prestress. They found that the effects of the prestress were significant only at the highest level of prestress (equivalent to 6,000 με) and the shear loading produced the greatest impact damage area on the specimen. Whittingham et al. (2004) performed similar experiments on carbon fiber/epoxy laminated plates under tensile and shear prestress. They also found that low prestress levels (<1,500 με) had no significant effect on the peak force of impact, absorbed energies, and penetration behavior. Heimbs et al. (2009) tested carbon fiber/epoxy laminated plates under an in-plane compressive preload. An increase in deflection and energy absorption was observed under a preload of 80 % of the buckling load. Sun and Chen (1985) and Choi (2008) analytically investigated the effects of prestress on the dynamic response of composite laminates. They found that the initial in-plane tensile load increased the peak contact force while reducing the total contact duration and deflection. The compressive load reacted oppositely. However, contact loading, such as impact loading, will induce localized damage, which is different from air or underwater blast loading. Thus, these results cannot be extended to the blast response of composite structures.

There are a few studies reported in literature on the impact response of composite laminates and sandwich composites at low and high temperatures (Jang et al. 1987; Dutta 1994; Bibo et al. 1994; Salehi-Khojin et al. 2007; Erickson et al. 2005). Jang et al. (1987) studied the effect of temperature on the failure behavior of composites and demonstrated that the composites show a great degree of delamination and cracking arising from thermal expansion mismatch between the fibers and matrix at extreme temperatures. Thus, the environment in which a sandwich composite operates plays an important role in determining its failure mechanisms. Dutta (1994) tested the energy absorption and brittleness of graphite/epoxy composites at room and high temperature under low-velocity impact and concluded that the properties of these composites slightly depend upon the temperature. Bibo et al. (1994) studied the impact performance of carbon fiber composites at 22, 80, and 150 °C. Their results indicated that high-temperature testing increases the spread of delaminations in carbon fiber composite laminates. Salehi-Khojin et al. (2007) studied hybrid kevlar and carbon fiber composite sandwiches under low-velocity impact at ambient temperatures ranging from −50 to 120 °C. They found that the back surface damage and fiber breakage decreased considerably with increasing temperature. Erickson et al. (2005)

investigated the effect of temperature on the impact behavior of composite sandwich panels at −25, 25, and 75 °C and showed that the temperature significantly affects the absorbed energy and the maximum impact force. All of the above-mentioned studies indicated that temperature had a significant effect on the dynamic properties and failure mechanisms of sandwich composites and composite laminates subjected to impact loading. However, no studies regarding the effect of temperature on the blast response of sandwich composite have been reported.

The current experimental investigation will focus on the blast resistance and mitigation behaviors of novel sandwich structures as well as provide insight into the effect of in-plane pre-loading and temperature on their overall blast response (Gardner et al. 2012a; b; Wang and Shukla 2012; Gupta and Shukla 2012). Understanding the overall behaviors and failure mechanisms will help to better understand the material systems and sandwich structures during normal service and operations, as well as aid in the development of optimally designed lightweight structures to mitigate energy and maintain structural integrity when subjected to blast loadings. Due to the increased threat of damage to civilian and defense structures in the form of terrorist attacks and blast loading, as well as the loss of human lives, a comprehensive understanding on the blast response of composites and sandwich structures is essential. The sandwich composites were fabricated and subjected to shock wave loading generated by a shock tube. The shock pressure profiles and real-time deformation images were carefully analyzed to reveal the failure mechanisms of these sandwich composites. DIC analysis was implemented to investigate the real-time deflection, in-plane strain, and velocity of the back face of the specimens. Postmortem analysis was also carried out to evaluate the overall blast performance of these sandwich structures.

10.2 Material Systems and Fabrication

10.2.1 Facesheet (Skin) and Core Materials

The facesheet materials utilized in this study are E-glass vinyl-ester (EVE) composites. The woven roving E-glass fibers of the skin material consisted of 8 layers of plain weave fabric placed in a quasi-isotropic layout $[0/45/90/-45]_s$. The fibers were made of the 0.61 kg/m^2 areal density plain weave. The resin system used was Ashland Derakane Momentum 8084, with the front and back skins (facesheets) consisting of identical layups and materials. The EVE composite consisted of a 55 %/45 % volume fraction of glass (fiber) to resin, as determined using proper ASTM standard D 2584. Figure 10.1 shows a schematic of the sandwich composite with skin and core materials.

The core materials used in the present study consisted of styrene acrylonitrile (SAN) foams (A-series, M-series, and P-series), which are manufactured by Gurit SP Technologies specifically for marine sandwich composite applications, as well as Dragonshield-HT polyurea (PU), which is manufactured by Specialty Products

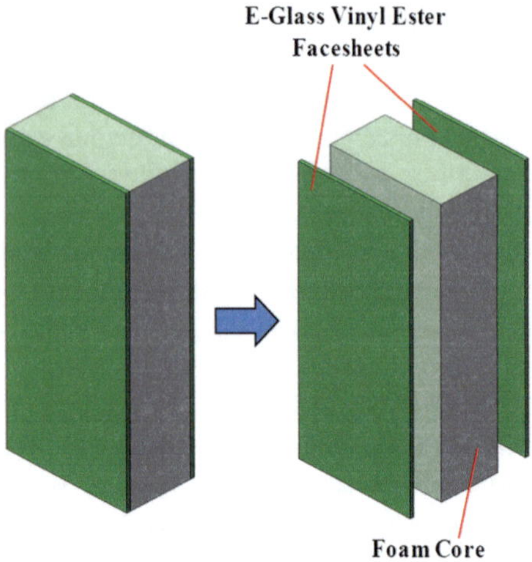

Fig. 10.1 Schematic of sandwich composite with skin and core

Table 10.1 Quasi-static material properties of the EVE facesheet, foam cores, and PU

	Nominal density, ρ (kg/m^3)	Compressive modulus, E (MPa)	Compressive strength, σ_y (MPa)	Acoustic wave impedance (kg/m^{2}s)
E-glass vinyl-ester composite	1,800	13,600 (longitudinal) 3,900 (transverse)	220	4.9×10^6 (longitudinal) 2.6×10^6 (transverse)
A300	58.5	32	0.5	4.3×10^4
A400	69	41	0.6	5.3×10^4
A500	92	64	0.9	7.7×10^4
A800	150	117	2.1	13.2×10^4
Dragonshield—HT polyurea	1,000	–	–	–
P600	122	125	1.8	–
M100	108	107	1.6	–

Inc. (SPI) specifically for energy absorption and blast mitigation. For the various investigations, approximately six different types of Corecell™ foam were used, A300, A400, A500, and A800 (core gradation), P600 (in-plane compressive loading), and M100 (temperature effects). Table 10.1 lists important material properties of the core materials, foam and polyurea, as provided by the manufacturer's data (http://www.gurit.com; http://www.specialty-products.com), as well as the material properties of the facesheet (Gardner 2012) material. The material properties of the facesheet and the core materials were determined using proper ASTM standards, D 3410 and D 1621, respectively.

10.2.2 Sandwich Composites

The vacuum-assisted resin transfer molding (VARTM) process was utilized to fabricate the sandwich specimens. During the VARTM process, the sandwich specimens

Table 10.2 Various material systems, dimensions, and areal densities for the different investigations

Investigation	Width × length (mm)	Facesheet thickness	Core materials	Core thickness	Areal density (kg/m²)
Core gradation effects	102 × 254	~4 mm	A500	38 mm	19.0
			A300, A800	19 mm each	18.5
			A300, A500, A800	12.7 mm each	19.0
			A300, A400, A500, A800	9.5 mm each	20
Polyurea effects			A300, A500, A800, PU	12.7 mm, 12.7 mm, 6.35 mm, 6.35 mm	26.0
Pre-loading effects			P600	25.4 mm	17.2
Temperature effects			M100	25.4 mm	16.8

Fig. 10.2 Cell microstructure of A-series foam core layers

were infused under the same conditions, i.e., temperature, humidity, and vacuum pressure (760 mmHg (1 atm)), with the same volume of resin. For the various investigations studied, the core consisted of different materials and layer thickness. Table 10.2 lists the different material systems, dimensions, and areal densities for the various material systems studied and their respective investigations.

For the studies involving core gradation (functionally graded and polyurea), the sequence of the foam core material was designed based on increasing the one-dimensional acoustic wave impedance of the layers. It can be seen in Table 10.1 that for the A-series foam, the A300 foam has the lowest nominal density (ρ) and compressive modulus (E) of the four foams, followed by the A400, A500, and A800 foams, respectively. Since both the nominal density and the compressive modulus are increasing from the A300 foam to the A800 foam, the one-dimensional acoustic wave impedance (Z) is also increasing, and shows the following relationship:

$$Z = \rho C = \rho \sqrt{E / \rho} \tag{10.1}$$

$$Z_{A300} < Z_{400} < Z_{A500} < Z_{A800} \tag{10.2}$$

Note that the cell structures of the four A-series foams are very similar and the only difference appears in the cell wall thickness and node sizes, which accounts for the different densities of the foams. The SEM images of the cell microstructures can be seen in Fig. 10.2.

10.3 Experimental Setup and Procedure

10.3.1 Shock Tube

The shock tube apparatus used to obtain the controlled dynamic loading is shown in Fig. 10.3a. Shock tubes offer the advantages of plane wave fronts, wave parameters that are easily controllable and repeatable, and uniform loading over the shock tube muzzle diameter (Kumar et al. 2012). A complete description of the shock tube and its calibration can be found in (LeBlanc et al. 2007). In principle, the shock tube consists of a long rigid cylinder, divided into a high-pressure driver section and a low-pressure driven section, which are separated by a diaphragm. By pressurizing the high-pressure driver section, a pressure difference across the diaphragm is created. When this pressure differential reaches a critical value, the diaphragm ruptures. The subsequent rapid release of gas creates a shock wave, which travels down the shock tube to impart shock loading on the specimen at the muzzle end.

When the shock wave impacts the specimen located at the end of the muzzle, the wave is reflected at a higher pressure than that of the incident shock pressure. The theoretical details on the equations for shock tubes have been previously established in the literature (Wright 1961). There are four basic theoretical assumptions which are used to describe the gas flow in shock tube:

1. The gas flow is one dimensional.
2. The gas is ideal and has constant specific heats.
3. Heat transfer and viscosity effects are neglected.
4. Diaphragm rupture is instantaneous and does not disturb the subsequent gas flow.

The shock tube apparatus utilized in the present study has an overall length of 8 m, consisting of a driver, driven, converging, and muzzle section. The diameter of the driver and driven section is 0.15 m. Figure 10.3b shows detailed dimensions and locations of the muzzle, specimen, supports, and pressure sensors (PCB102A). The sensors are mounted at the end of the muzzle section to measure the incident pressure and the reflected pressure during the experiment. The final muzzle

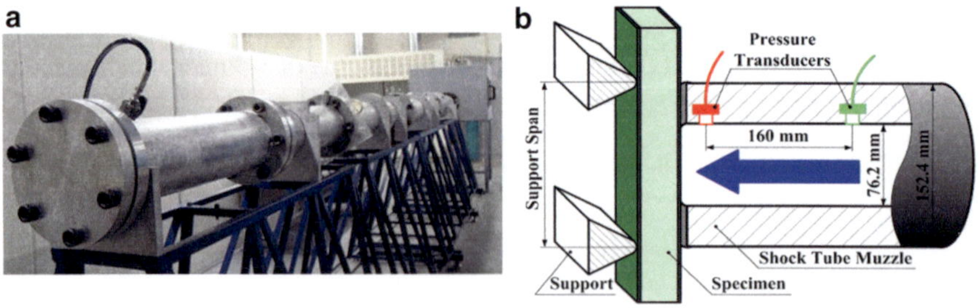

Shock tube
Detailed dimensions of the muzzle

Fig. 10.3 Shock tube apparatus. (**a**) Shock tube. (**b**) Detailed dimensions of the muzzle

Table 10.3 Various loading, boundary, and environmental conditions for the different investigations

Investigation	Boundary condition	Incident peak pressure	Reflected peak pressure	Incident wave velocity	Environmental conditions
Core gradation	Simply supported	1.0 MPa	5.0 MPa	1,050 m/s	Room temp
Polyurea	Simply supported	1.0 MPa, 1.5 MPa	5.0 MPa, 7.5 MPa	1,050 m/s, 1,300 m/s	Room temp
In-plane pre-loading	Simply supported	1.0 MPa	5.0 MPa	1,050 m/s	Room temp
Temperature	Simply supported	0.9 MPa	4.0 MPa	1,000 m/s	High temp (80 °C), low temp (–40 °C)

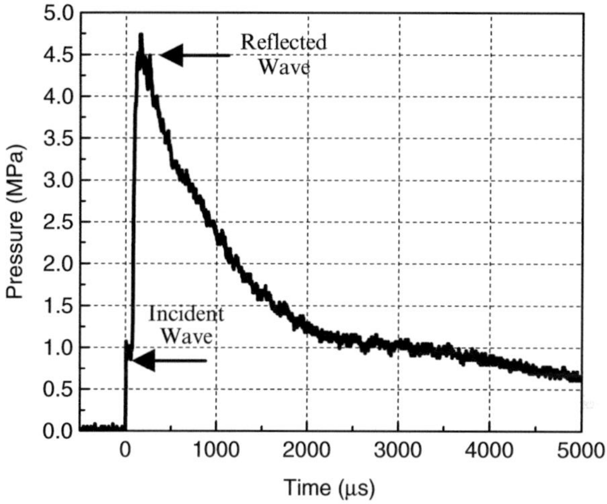

Fig. 10.4 Typical pressure profile from transducer closest to the specimen

diameter is 0.076 m. The distance between the two sensors is 0.160 m and the distance between the second sensor and the end of the muzzle is ~0.020 m.

10.3.1.1 Loading, Boundary, and Environmental Conditions

Various loading, boundary, and environmental conditions were utilized in the different experiments. Table 10.3 lists the various conditions for each investigation. For each configuration studied, at least three samples were tested. A typical pressure profile obtained from the transducer closest to the specimen (~0.020 m away) can be seen in Fig. 10.4. It should be noted that both pressure transducers were utilized to obtain the shock wave history, i.e., incident/reflected pressure and incident/reflected velocity. However, only the pressure transducer closest to the specimen was utilized to obtain the pressure applied on the specimen.

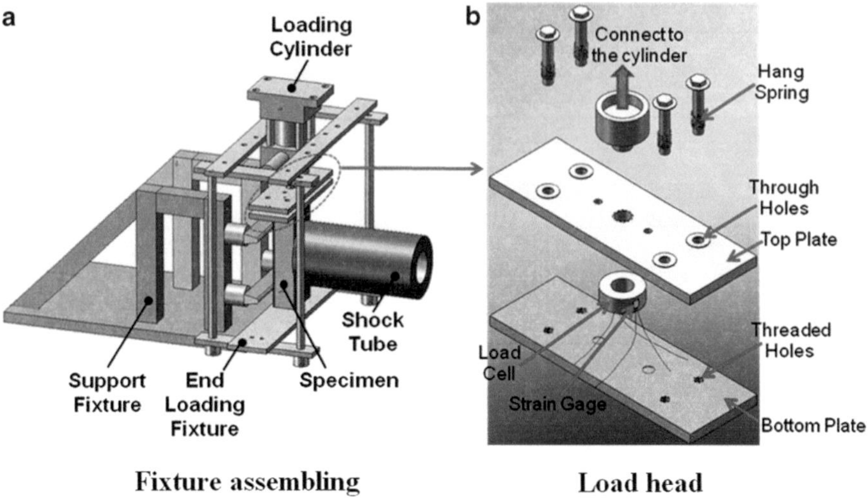

Fig. 10.5 In-plane compressive loading fixture. (**a**) Fixture assembling. (**b**) Load head

10.3.1.2 Experimental Fixtures and Procedure

To apply the in-plane compressive loading (static) on the sandwich composite panels, a fixture was designed and fabricated as shown in Fig. 10.5. The loading head is connected to a hydraulic loading cylinder, which is mounted on the frame. An aluminum cylinder with an outer diameter of 50.8 mm and an inner diameter of 38.1 mm is positioned between two plates as shown in Fig. 10.5b. Two strain gages, which are attached on this aluminum cylinder, measure the deformation of this cylinder and consequently provide the load applied on the specimen. The support fixture and in-plane compressive loading fixture are all securely fastened inside the dump tank of the shock tube. The in-plane compression loading was applied on the specimen and held at a constant level until the specimen was subjected to the transverse shock wave loading. The loading head is fixed during the entire experimental process. The load will be released when the vertical dimension of the specimen is reduced.

For testing the sandwich composite panels under high temperatures, a thermally insulated enclosure was designed around the support fixture, as shown in Fig. 10.6a. A heating element having a maximum output power of 1 kW was utilized to heat the internal environment and a circulation fan was used to ensure proper air-circulation. By adjusting the output power of the heating element, various temperatures in sandwich specimens (steady state) were achieved and then these temperatures were calibrated according to the power supplied to the heating element. Figure 10.6b shows a typical temperature–time plot for a sandwich specimen. To perform the experiment at a steady-state core temperature of 80 °C, the specimen was heated for approximately 180 min and then was subjected to shock wave loading.

For low-temperature testing, the sandwich composite panels were cooled down to −70 °C by placing the specimen between two panels of dry ice, as shown in

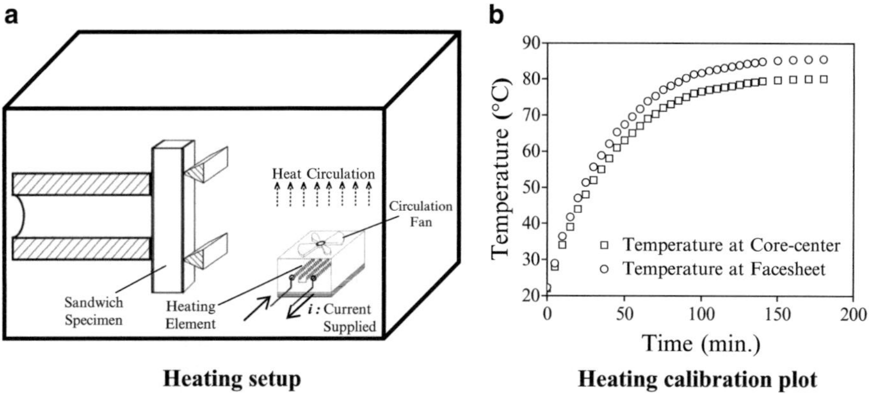

Fig. 10.6 Heating setup for shock tube. (**a**) Heating setup. (**b**) Heating calibration plot

Fig. 10.7a. Dry ice panels were placed approximately 15 mm away from the specimen. A temperature–time plot of the sandwich specimen under cooling is shown in Fig. 10.7b. After 75 min, the specimen reached a uniform steady-state temperature of −70 °C. Once this was achieved, the specimen was quickly removed and placed inside the support fixture located in the dump tank (within shock tube facility). Due to the surrounding environment (ambient temperature), the temperature of the sandwich panels increased non-uniformly as shown in Fig. 10.7b. The specimen was held in room temperature for approximately 8 min in order to achieve −40 °C within the core. At this time, the temperature of facesheet was −15 °C.

10.3.2 High-Speed Photography System

Two high-speed photography systems were used in the present study, as shown in Fig. 10.8. A high-speed 3-D DIC system, consisting of two high-speed digital cameras [Photron SA1], was placed facing the back side of the specimen to obtain the real-time full-field in-plane strain, along with out-of-plane deflection and velocity of the back facesheet. A randomized speckle pattern was placed directly on the back facesheet of the sandwich composite to ensure good contrast of the images. Another high-speed digital camera, [Photron SA1], was placed perpendicular to the side surface of the specimen to capture the side-view deformation images and mechanisms of failure. A framing rate of 20,000 fps was utilized which gives an interframe time of approximately 50 μs.

The DIC technique is utilized to capture the real-time full-field response of the back face of the sandwich composite panels. It is a nonintrusive, optical technique which allows for the capturing of the real-time dynamic response of sandwich composites through the use of high-speed photography and specialized software (Photron FASTCAM Viewer and VIC 3D). In order to capture the three-dimensional

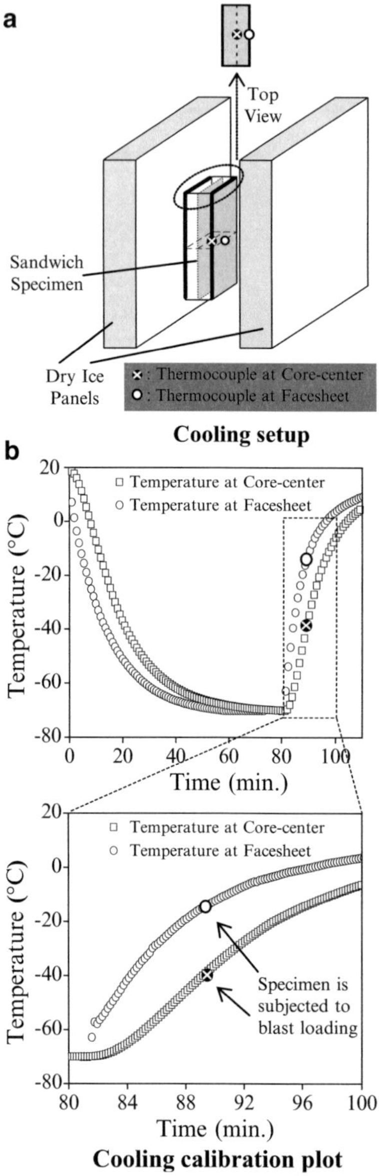

Fig. 10.7 Cooling setup for shock tube. (**a**) Cooling setup. (**b**) Cooling calibration plot

response of the specimens, two cameras must be used in stereo configuration. This means that the cameras must be calibrated and have synchronized image recording throughout the entire blast event. The calibration of the cameras is performed by placing a grid consisting of a known pattern of points (dots) in the exact location as to where the specimen will be placed during the experiment. This grid is then translated and rotated both in- and out-of-plane while manually recording a series of images. Due to the fact that the grid pattern has predetermined spacing, the

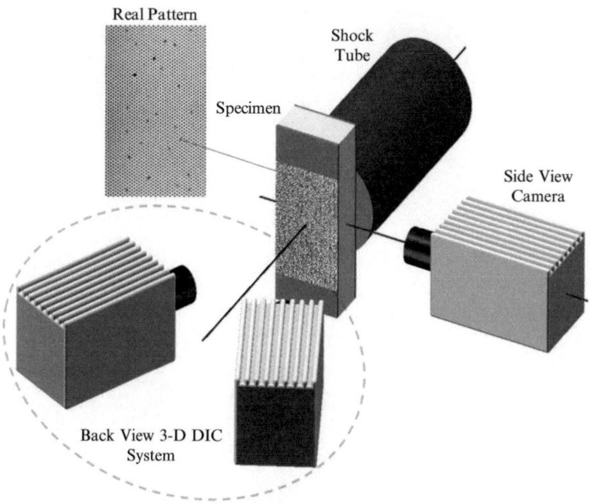

Fig. 10.8 High-speed photography setup (back-view DIC and side-view)

coordinates of the center of each point (dot) are extracted from each image. Since the coordinate location of each dot is extracted uniquely for each camera, this allows for a correspondence of the coordinate system between cameras (Sutton et al. 2009). DIC post-processing is performed utilizing the image pairs that are recorded during the blast loading event. Using the VIC-3D software package, as distributed by Correlated Solutions (http://www.correlatedsolutions.com), common pixel subsets of the randomized speckle pattern are matched between the deformed images and the undeformed image. By matching the pixel subsets of the random speckle pattern, the three-dimensional location of distinct points on the face of the panel throughout time is obtained. This technique has been applied as a full-field measurement technique in many applications, including shock loading (Tiwari et al. 2009).

10.4 Experimental Results and Discussion

10.4.1 Effects of a Functionally Graded Core

Functionally graded materials, where the material properties vary gradually or layer by layer within the material itself, have gained much attention in recent years (Hossain et al. 2007; Kiernan et al. 2009; Li et al. 2001; Apetre et al. 2006; Wang et al. 2009). Since the properties of the layered/graded material can be designed and controlled, they show great potential to be an effective core material for energy absorption and blast mitigation. The following sections elucidate the effects of a functionally graded SAN foam core on the blast response of sandwich composites.

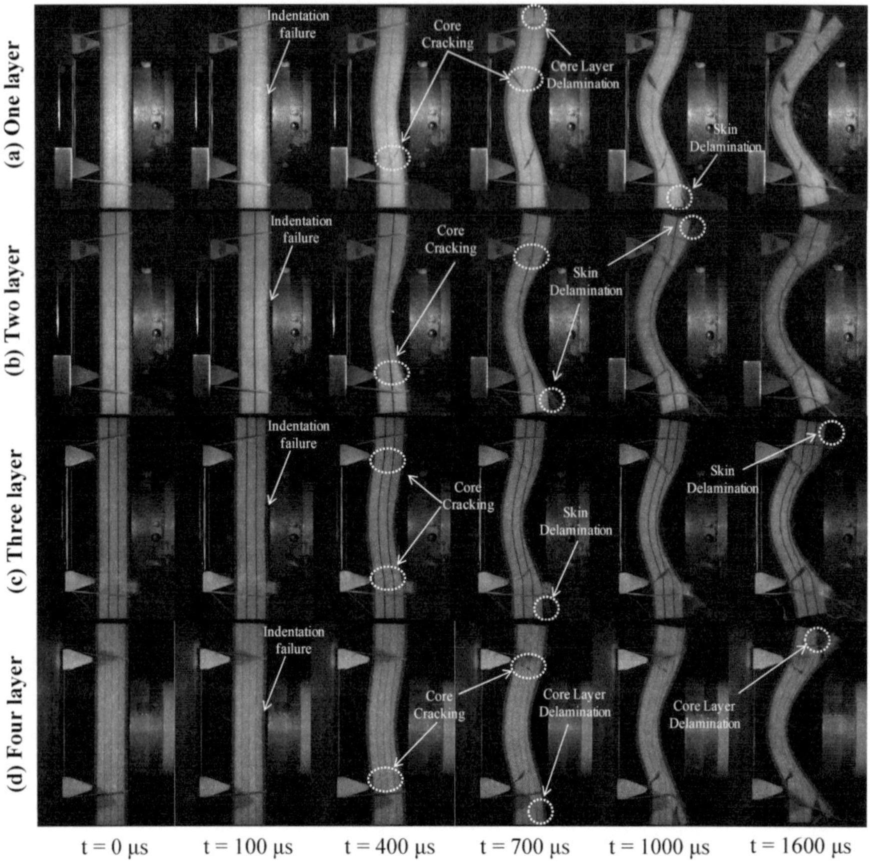

Fig. 10.9 High-speed images for (**a**) one, (**b**) two, (**c**) three, and (**d**) four layer of core gradation

10.4.1.1 Real-Time Deformation

The real-time observations of the transient behavior for the four different core configurations subjected to shock wave loading are shown in Fig. 10.9. The shock wave (pressure wave) propagates from the right side of the image to the left side and some detailed deformation mechanisms are pointed out in the figures. It should be noted that the time scheme used to represent the images in each configuration is identical. Therefore for each of the four configurations investigated, the images are correlated based on the same time per frame. This allows for a better comparison between the different configurations in regard to the failure mechanisms and extent of damage observed. Also, since each configuration is graded monotonically by increasing the acoustic wave impedance, the damage processes are identical. First, indentation failure (initiation of core compression) is observed, followed by core compression, core cracking, and then finally delamination, either between the skin and core or at the core layer interfaces.

Figure 10.9a shows the real-time blast loading response for the sandwich composites with one layer of core gradation (A500). From these images, it can be seen that the indentation failure begins at approximately $t = 100$ μs. At $t = 400$ μs core cracking is evident. The crack starts at the lower support and propagates from the back face towards the front face. By $t = 700$ μs more core cracking is observed and delamination between the core layers can be seen. Skin delamination is evident at $t = 1{,}000$ μs between the front facesheet and first core layer of foam, along with heavy first-layer core compression. By $t = 1{,}600$ μs the core cracks have propagated completely through the core from the back face to the front face, and the amount of delamination has increased. Also, the first-layer core compression has reached a maximum, approximately 30 % of its original thickness.

The real-time blast loading response for the sandwich composites with two layers of core gradation (A300/A800) can be seen in Fig. 10.9b. Indentation failure for this configuration is observed at approximately $t = 100$ μs. By $t = 400$ μs, first-layer core compression is very evident (A300). Also at this time, core cracking starting from the back facesheet where the lower support is located can be seen. By $t = 700$ μs more core cracking can be observed, as well as skin delamination between the front skin and the first core layer of foam (located at the bottom of the specimen). At $t = 1{,}000$ μs more skin delamination can be observed between the front facesheet and the first core layer of foam, but this time it is located at the top of the specimen. By $t = 1{,}600$ μs, the core cracks have propagated completely through the core from the back facesheet to the front facesheet. Also, the core compression in the first core layer has reached a maximum, approximately 75 % of its original layer thickness.

Figure 10.9c shows the real-time blast loading response for the sandwich composites with three layers of core gradation (A300/A500/A800). It can be seen from these images that the indentation failure begins at approximately $t = 100$ μs. Following indentation failure, heavy first-layer (A300) core compression is observed as well as core cracking by approximately $t = 400$ μs. By $t = 700$ μs, the first core layer has compressed to a maximum, reaching a critical strain level and initiating the onset of indentation and compression in the second core layer of foam (A500). Also at this time, skin delamination between the front skin and the first core layer of foam is evident at the bottom of the specimen. At $t = 1{,}000$ μs, more core compression is observed in the second core layer. Finally by $t = 1{,}600$ μs, skin delamination can be seen at the top of the specimen between the front facesheet and the first core layer of foam. The core cracking has propagated completely through the core from the back facesheet to the front facesheet. Also at this time, the core compression in the first and second layers has reached a maximum, approximately 70 and 50 %, respectively.

The real-time blast loading response for the sandwich composites with four layers of core gradation (A300/A400/A500/A800) is shown in Fig. 10.9d. From these images, indentation failure is observed at approximately $t = 100$ μs. At $t = 400$ μs, heavy core compression is observed in the first core layer (A300). Also at this time, core cracking has initiated and is located at the bottom of the specimen where the supports are. By $t = 700$ μs a second core crack is evident at the top of the specimen where the upper support is located. Delamination is evident at the bottom of the specimen, between the first and second core layers of foam. Also at this time, the first core layer of foam has compressed to a maximum, reaching a critical strain level and initiating the onset of indentation and compression in the second core layer of foam (A400). By $t = 1{,}000$ μs, the second core layer has compressed to a maximum, reaching a critical strain level, causing indentation and compression in the third core layer of foam (A500). Finally by $t = 1{,}600$ μs, the total core compression has reached a maximum, the first layer and second core layer have compressed approximately 65 %, while the third layer has compressed approximately 30 %. Also at this time, delamination between the first and second core layer of foam is evident, and occurs at the top of the specimen. The core cracks have stopped propagating towards the front facesheet, and unlike the other configurations studied, these core cracks never propagated completely through the core.

In order to better evaluate the blast performance of the four different core layer arrangements studied, a more in-depth look at the later time frames ($t = 1{,}000$ μs and onward) of the high-speed images (Fig. 10.9) must be investigated. For the sandwich composite with one layer of core gradation (as shown in Fig. 10.9a), it is evident at $t = 1{,}000$ μs that there are approximately four (4) major damage areas present. These damage areas consist of two heavy core cracks occurring in the central region of the specimen, one where the bottom support is located and the other just below the top support, as well as core delamination at the top of the specimen between the foam core layers, and skin delamination at the bottom of the specimen between the front facesheet and first core layer of foam.

Using two layers of core gradation, as shown in Fig. 10.9b, at $t = 1{,}000$ μs again four (4) major damage areas exist. These damage areas include two heavy core cracks in the central region of the specimen, where the supports are located, as well as skin delamination between the front facesheet and first core layer of foam, both at the bottom and top of the specimen.

Utilizing three layers of core gradation, as shown in Fig. 10.9c, at $t = 1{,}000$ μs only three (3) damage areas are present. These damage areas include two core cracks located in the central region of the specimen where the supports are located, as well as skin delamination, which occurs at the bottom of the specimen between the front facesheet and first layer of foam core. By $t = 1{,}600$ μs, the fourth damage area becomes present and consists of skin delamination between the front skin and foam core which occurred at the top of the specimen.

For the sandwich composite with four layers of core gradation, as shown in Fig. 10.9d, again it can be observed that at $t = 1{,}000$ μs only three (3) areas of damage are present. These damage areas consist of two core cracks occurring where the supports are located and propagating into the central region of the specimen. Also at this time core delamination can be observed, which occurs at the bottom of the specimen between the first and second core layer of foam. The fourth damage zone can be observed at $t = 1{,}600$ μs. This damage area consists of core delamination at the top of the specimen, which occurs between the first and second core layer of foam.

Therefore by looking at Fig. 10.9, it is clearly evident that when using sandwich composites with one and two layers of core gradation, four damage areas are present by $t = 1{,}000$ μs. When using sandwich composites with three and four layers of core

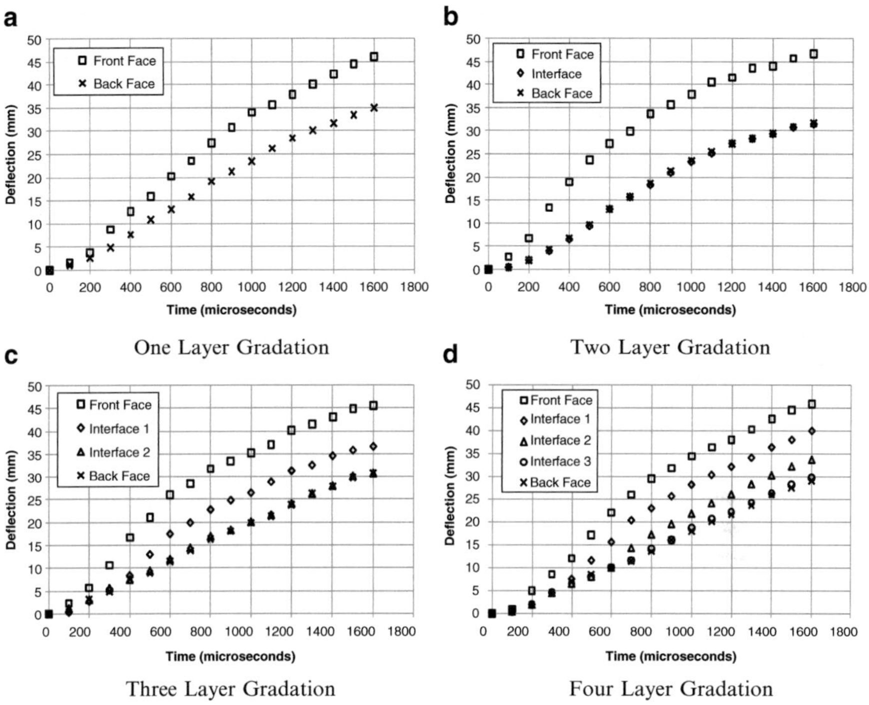

Fig. 10.10 Midpoint deflection of all configurations. (a) One-layer gradation. (b) Two-layer gradation. (c) Three-layer gradation. (d) Four-layer gradation

gradation, only three (3) damage areas are present at $t = 1{,}000$ μs. The fourth damage zone isn't observed until later in the panel deformation. Therefore, it can be concluded that higher levels of core layer gradation (i.e., 3 and 4 layers) allow for a delay in the arrival of the fourth damage zone.

10.4.1.2 Deflection

The midpoint deflections of each graded sandwich panel and all of its constituents were obtained from the high-speed images and a typical response can be seen in Fig. 10.10. For one-layer gradation, the midpoint deflection for the front face (front skin) and back face (back skin) of the specimen was plotted and can be seen in Fig. 10.10a. It is evident from the figure that at $t = 1{,}600$ μs the front face deflects to approximately 46 mm, while the back facesheet deflects approximately 35 mm. Therefore the difference between the deflection of the front face and deflection of the back face signifies the total amount of compression observed in the core. Therefore, it can be concluded that the core compressed approximately 11 mm, which is 30 % of its original thickness (38 mm).

When using two layers of core gradation, the midpoint deflection of the front face (front skin), interface (between first and second core layer), and back face of the specimen were plotted and are shown in Fig. 10.10b. It can be seen from the figure that at $t = 1{,}600$ μs, the front face has deflected to approximately 46 mm, while the interface and the back face deflect to approximately 33 mm. The difference between the deflection of the front face and deflection of the interface indicates the total amount of compression observed in the first core layer of foam (A300 layer). Therefore, the first core layer of foam (A300) compressed approximately 13 mm, or 75 % of its original thickness (19.05 mm). Since the interface and the back face deflect in an identical manner to the same value of 33 mm, it can be concluded that no compression was observed in the second core layer of foam (A800).

For three layers of core gradation, the midpoint deflection for the front face (front skin), interface 1 (between first and second core layer), interface 2 (between second and third core layer), and back face (back skin) were plotted and can be seen in Fig. 10.10c. It is evident from the figure that at $t = 1{,}600$ μs, the front face has deflected to approximately 46 mm, while interface 1 deflects to approximately 37 mm, and interface 2 as well as the back face deflect to approximately 31 mm. The difference between the deflection of the front face and the deflection of interface 1 constitutes the total amount of compression in the first core layer of foam (A300). Therefore, it can be seen that A300 foam compressed approximately 9 mm, which is ~75 % of its original thickness (12.70 mm). Looking at the difference between the deflection of interface 1 and the deflection of interface 2, the amount of compression observed in the second core layer of foam (A500) can be obtained. It can be seen that the A500 foam compressed approximately 6 mm, or ~50 % of its original thickness (12.70 mm). Again, since interface 2 and the back face deflect in an identical manner to the same value of 31 mm, it can be concluded that no compression was observed in the third core layer of foam (A800).

Finally, when using four layers of core gradation, the midpoint deflection of the front face (front skin), interface 1 (between first and second core layer), interface 2 (between second and third core layer), interface 3 (between third and fourth core layer), and back face (back skin) were plotted and are shown in Fig. 10.10d. It can be seen from the figure that at $t = 1{,}600$ μs, the front face has deflected to approximately 46 mm, while interface 1 has deflected to approximately 39 mm, interface 2 has deflected to approximately 33 mm, and interface 3 and the back facesheet deflect to approximately 29 mm. The difference between the deflection of the front face and the deflection of interface 1 signifies the total amount of compression in the first core layer of foam (A300). Therefore, it can be seen that A300 foam compresses approximately 7 mm, which is ~70 % of its original thickness (9.53 mm). Looking at the difference between the deflection of interface 1 and the deflection of interface 2, the amount of compression observed in the second core layer of foam (A400) can be obtained. It can be seen that the A400 foam compresses approximately 6 mm, or ~65 % of its original thickness (9.53 mm). The difference between the deflection of interface 2 and the deflection of interface 3 indicates the total amount of compression observed in the third layer of foam (A500). Therefore, it is evident that the A500 foam compressed approximately 4 mm, which is ~40 % of its original

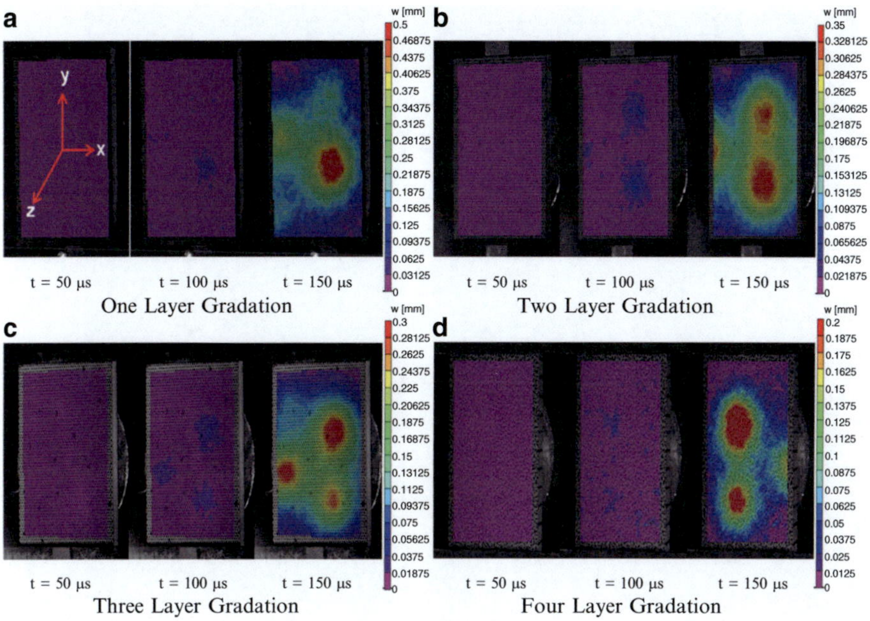

Fig. 10.11 Localized areas of larger deflections (loading) during fluid–structure interaction. (**a**) One-layer gradation. (**b**) Two-layer gradation. (**c**) Three-layer gradation. (**d**) Four-layer gradation

thickness (9.53 mm). Since interface 3 and the back face deflect in an identical manner to the same value of 29 mm, it can be concluded that no compression was observed in the fourth core layer of foam (A800).

It should be noted that for all of the configurations studied, the core layers were graded monotonically by increasing the acoustic wave impedance, and therefore allowing for a stepwise compression of the core. This stepwise compression is more evident in the three- and four-layer core configurations, i.e., Fig. 10.10c, d.

10.4.1.3 DIC Analysis

Utilizing the DIC technique, the full-field deflection, in-plane strain, and particle velocity of the back facesheet of each configuration were generated. Figures 10.11 and 10.12 show the full-field results for the back facesheet of all core gradations. Figure 10.11 shows the full-field out-of-plane deflection (W) during the initial fluid–structure interaction ($t \leq \sim 250$ μs, Wang 2010; Wang et al. 2012), with an emphasis on the shape of the loading, as indicated by the localized areas of larger deflection. Note that the scale used to represent each core gradation is different in order to highlight these areas. For one-layer core gradation, as shown in Fig. 10.11a, by $\sim t = 150$ μs, the loading can be observed as a circular region in the center of the back facesheet. For two-, three- and four-layer core gradation, as shown in Figure 10.11b, c, and d,

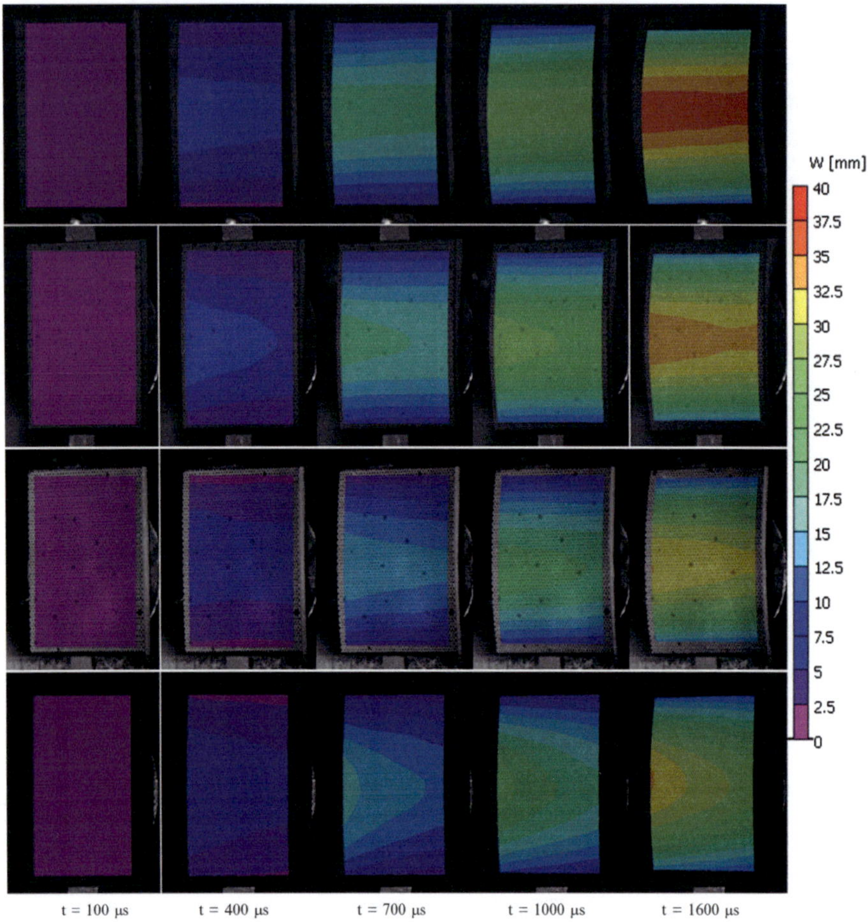

Fig. 10.12 Full-field out-of-plane deflection (*W*) for (**a**) one layer, (**b**) two layers, (**c**) three layers, and (**d**) four layers of core gradation

respectively, by $t = 150$ μs the loading is more dispersed across the back facesheet, resulting in two and even three areas of localized deflection (loading). Therefore, it can be concluded that utilizing multiple layers of core gradation, and thus introducing more material interfaces, aids in dispersing the initial loading on the structure, resulting in up to three areas of localized deflection (loading) on the back facesheets.

Figure 10.12 shows the full-field out-of-plane deflection (*W*) for all core layer gradations. Note the scale ranges from 0 mm (purple) to 40 mm (red). It is evident from the figure that the back face of all core configurations exhibits limited out-of-plane deflection prior to $t = 400$ μs. Between $t = 400$ μs and $t = 1{,}600$ μs, all four of these configurations continue to bend and show deflections. For the sandwich composite with one layer of core gradation, as shown in Fig. 10.12a, it can be observed that at $t = 1{,}600$ μs, the central region of the panel has deflected approximately 36 mm.

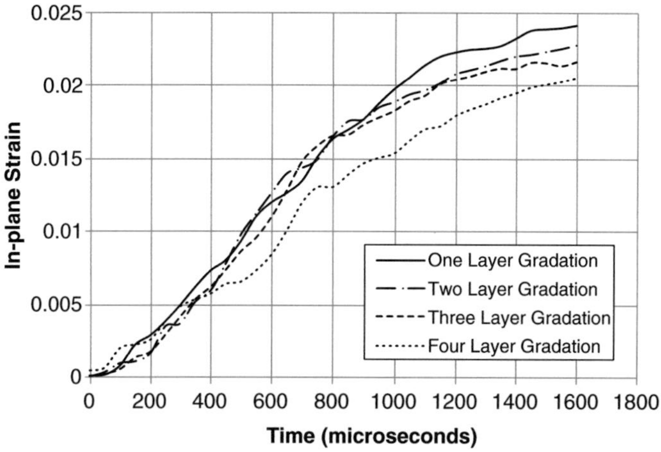

Fig. 10.13 In-plane strain (ε_{yy}) of all core configurations

When using two layers of core gradation (Fig. 10.12b), it can be seen that at $t=1{,}600$ μs, the central region of the panel has deflected approximately 34 mm. For the sandwich composite with three layers of core gradation, Fig. 10.12c, at $t=1{,}600$ μs, the central region has deflected approximately 32 mm. Finally for the sandwich composite with four layers of core gradation (Fig. 10.12d), the central region of the panel has deflected 30 mm at $t=1{,}600$ μs. Therefore, it can be concluded that utilizing more layers of core layer gradation, i.e., two, three, and four layers, the deflection in the central region of the sandwich panel can be reduced 6, 11, and 17 %, respectively, in comparison to one layer of core gradation.

Using the point inspection tool from the DIC software, a point directly in the center of the back face of each specimen was chosen. The out-of-plane deflection (W) showed excellent agreement with the results generated utilizing the high-speed images and therefore only the in-plane strain (ε_{yy}) and out-of-plane velocity ($\mathrm{d}W/\mathrm{d}t$) results are shown. Figures 10.13 and 10.14 show the in-plane strain and out-of-plane velocity values obtained. Looking at the in-plane strain values (Fig. 10.13), it can be seen that at $t=1{,}600$ μs, the maximum in-plane strain value at the central point of the back facesheet for one-layer core gradation is approximately 2.4 %. When using more layers of core gradation, i.e., two, three, and four layers, at $t=1{,}600$ μs the maximum in-plane strain values are reduced to 2.3, 2.2, and 2.1 %, respectively. Therefore, it can be concluded that using more layers of core gradation, the maximum in-plane strain values are reduced by 4, 8, and 12.5 % for two-, three-, and four-layer gradation in comparison to one layer of core gradation.

Figure 10.14 shows the out-of-plane velocity ($\mathrm{d}W/\mathrm{d}t$) for all core gradations. For one layer of core gradation, the maximum velocity is reached at $t=500$ μs and is approximately 32 m/s. Using two layers of core gradation, the maximum velocity is reached at approximately the same time, $t=500$ μs, but the velocity is slightly lower (31 m/s). For the sandwich composite with three layers of core gradation, the

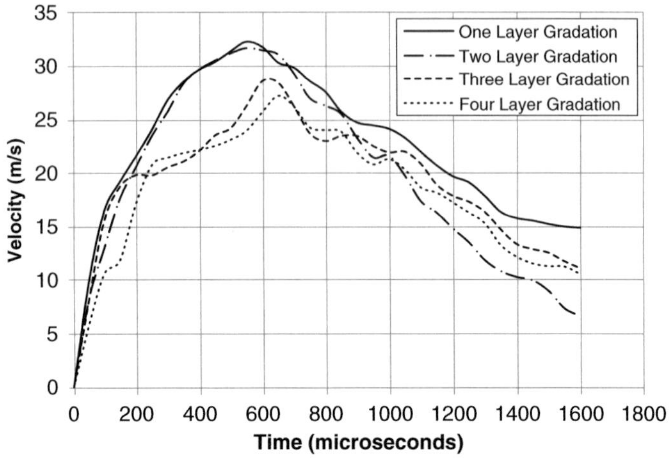

Fig. 10.14 Out-of-plane velocity (dW/dt) of all core configurations

maximum out-of-plane velocity isn't achieved until $t=600$ μs and is approximately 29 m/s. Utilizing four layers of core gradation, the maximum out-of-plane velocity of 27 m/s is reached at approximately $t=700$ μs. Therefore, two things can be concluded when using more layers of core gradation, i.e., two, three, and four layers: (1) the maximum value of the out-of-plane velocity is reduced approximately 15 %, from 32 to 27 m/s, and (2) the time at which the maximum value of the out-of-plane velocity is achieved is delayed up to 200 μs.

10.4.1.4 Postmortem Analysis

After the blast loading event occurred, the damage patterns of the sandwich composites with four different core layer arrangements were visually examined and recorded using a high-resolution digital camera and are shown in Fig. 10.15. When the sandwich composite with one-layer core gradation was subjected to highly transient loading, as shown in Fig. 10.15a, the damage was confined to the areas where the supports were located in the shock tube and core cracking is visible in these two areas. The core cracks propagated completely through the foam core. Core delamination is visible between the two core layers of A500 foam. Also one of the core cracks lead to back skin delamination, where the core separated from the back facesheet. Some core compression is also visible in the first core layer of A500 foam.

For the sandwich composite with two layers of core gradation, the damage patterns after being subjected to the shock loading are shown in Fig. 10.15b. For this core configuration, the damage was again confined to the areas where the supports were located in the shock tube and core cracking is evident. The core cracks propagated completely through the foam core. Skin delamination is obvious between the front facesheet and the foam core, as well as back skin delamination between the back

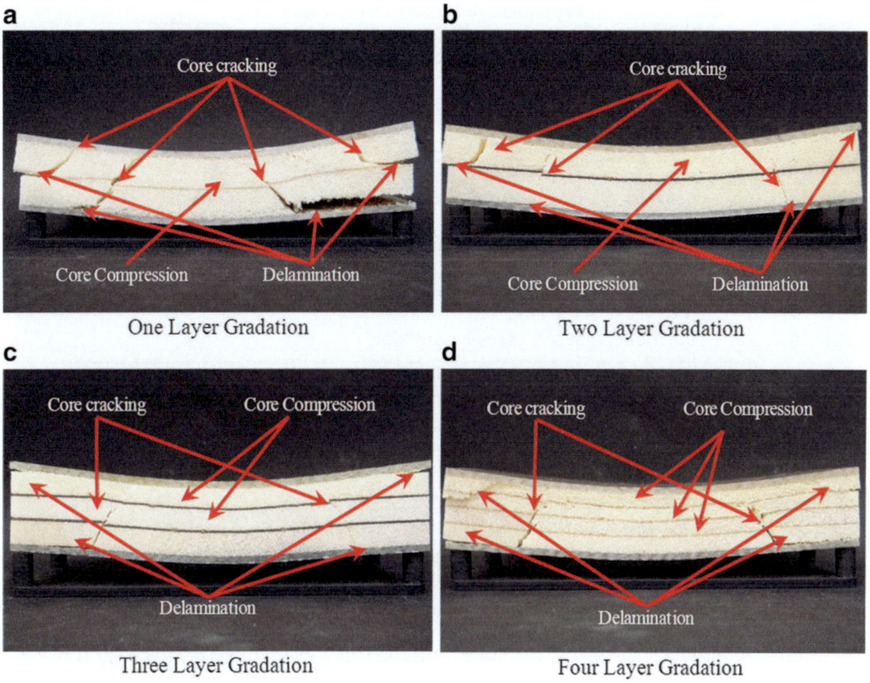

Fig. 10.15 Visual examination of all core configurations after being subjected to high-intensity blast loading. (**a**) One-layer gradation, (**b**) two-layer gradation, (**c**) three-layer gradation, (**d**) four-layer gradation

facesheet and the foam core. Core delamination between the first and second core layers of foam, A300 and A800, respectively, is also evident, along with core compression in the first core layer of foam (A300).

Figure 10.15c shows the damage patterns of the sandwich composite with three layers of core gradation after the blast loading event occurred. Again, the damage to this core configuration was confined to the areas where the supports were located in the shock tube and core cracking is visible in these two areas. These core cracks propagated completely through the foam core. Also skin delamination is visible between the front facesheet and the foam core, as well as back skin delamination located between the back facesheet and the foam core. Core compression is also evident in both the first and second core layers of foam, A300 and A500, respectively.

When the sandwich composite with four layers of core gradation was subjected to the shock loading, as shown in Fig. 10.15d, the damage was again confined to the areas where the supports from the shock tube were located and core cracking is evident in these two areas. Unlike the previous three configurations, the core cracks did not propagate completely through the foam core. Core delamination is obvious between the first and second core layers of foam, A300 and A400, as well as back skin

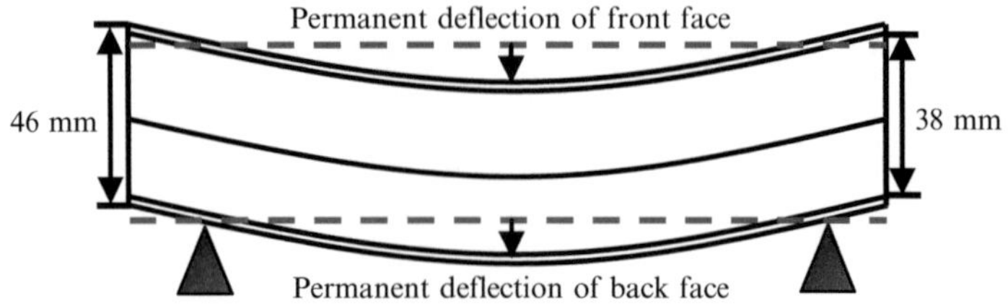

Fig. 10.16 Specimen schematic for permanent deflection measurements (between simple-supports)

Table 10.4 Permanent deflection and core compression

	1 Layer	2 Layer	3 Layer	4 Layer
Permanent front face deflection (mm)	11.1	14.1	12.9	12.4
Permanent back face deflection (mm)	8.5	9.2	7.3	6.1
Final core thickness (mm)	35.4	33.1	32.4	31.7
Permanent core compression (mm/%)	2.6 (7 %)	4.9 (13 %)	5.6 (15 %)	6.3 (17 %)
Real-time core compression (mm/%)	11 (30 %)	13 (35 %)	15 (40 %)	17 (45 %)
Recovered (mm/%)	8.4 (23 %)	8.1 (22 %)	9.4 (25 %)	10.7 (28 %)

delamination between the back facesheet and the foam core. Core compression is very obvious in this configuration. The first, second, and third layers of foam, A300, A400, and A500, respectively, all exhibit various amounts of core compression.

10.4.1.5 Permanent Deformation

The permanent deflection (deformation) for each graded core configuration was measured after the shock loading experiment. A schematic of the specimen and how the measurements were taken can be seen in Fig. 10.16. The distance between the top dotted line (red) and the front surface of the front facesheet is defined as the permanent deflection of the front face. Similarly, the distance between the bottom dotted line and the top surface of the back facesheet is defined as the permanent deflection of the back face. Subtracting the total permanent deflection of the back face from the front face, the final core thickness and thus total core compression (permanent) can be obtained. These values are shown in Table 10.4.

For one-layer core gradation, the front face exhibited a permanent deflection of ~11.1 mm, while the back face exhibited ~8.5 mm. By subtracting the back face deflection from the font face deflection, the total amount of permanent core compression can be calculated. For the sandwich panel with one layer of core gradation, the core showed approximately 2.6 mm (7 %) of permanent core compression. When using two layers of core gradation, the front face showed ~14.1 mm of permanent deflection, while the back face showed approximately ~9.2 mm. Therefore it can be concluded that the foam core itself exhibited approximately 4.9 mm (13 %) of permanent compression. For three layers of core gradation, the front face exhibited ~12.9 mm of permanent deflection, while the back face exhibited ~7.3. This results in a total core compression of approximately 5.6 mm (15 %). When using four layers of core gradation, the front face shows approximately 12.4 mm of permanent deflection, while the back face shows 6.1 mm. Consequently, the total permanent compression of the core is 6.3 mm (17 %). Therefore, increasing the number of monotonically graded core layers results in higher levels of permanent core compression, but lower levels of permanent back face deflections.

The total core compression observed during the experiment, as measured from the high-speed images and shown in Fig. 10.9, is listed in Table 10.4 (real-time core compression). Subtracting the total permanent core compression from the real-time core compression, the total amount of core compression recovered can be measured. When using one and two layers of core gradation, the amount of core compression recovered is almost the same, approximately 8.4 and 8.1 mm, respectively. When using three and four layers of core gradation, the total amount of core compression recovered is 9.4 and 10.7 mm. Thus, higher levels of core gradation allow for larger amounts of real-time compression to be recovered.

The postmortem images in Fig. 10.15, along with the results generated in Table 10.4, indicate the ability of functionally graded sandwich composites with higher levels of core gradation (three and four) to better mitigate blast energy. The sandwich panels with three and four layers of core gradation exhibited the largest amount of total core compression, 5.6 mm (15 %) and 6.3 mm (17 %), but the least amount of permanent back face deflection, 7.3 and 6.1 mm. Also, these sandwich panels showed the largest amount of real-time core compression recovered, 9.4 mm (25 %) and 10.7 mm (28 %).

10.4.2 Effects of Polyurea Location

With its ability to improve structural performance and damage resistance of structures, as well as effectively dissipate blast energy, the application of polyurea to sandwich structures has become a new area of interest. The following sections reveal the importance of polyurea and the effect of its location on the blast response of sandwich composites with a functionally graded SAN foam core and polyurea interlayer.

10.4.2.1 Real-Time Deformation

The real-time observations of the transient behavior for both core configurations subjected to shock wave loading are shown in Figs. 10.17 and 10.18. The shock wave propagates from the right side of the image to the left side and some detailed deformation and failure mechanisms are pointed out in the figures. The time frames used to represent the blast loading event are chosen in a manner so that they can be correlated to the time in which these damage mechanisms were first observed.

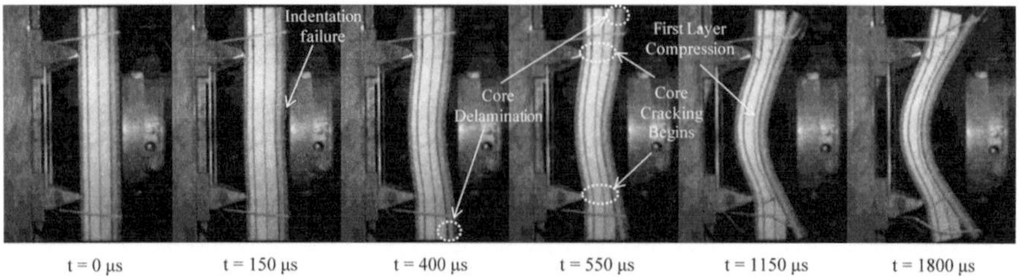

Fig. 10.17 High-speed images for configuration 1 (PU/A300/A500/A800)

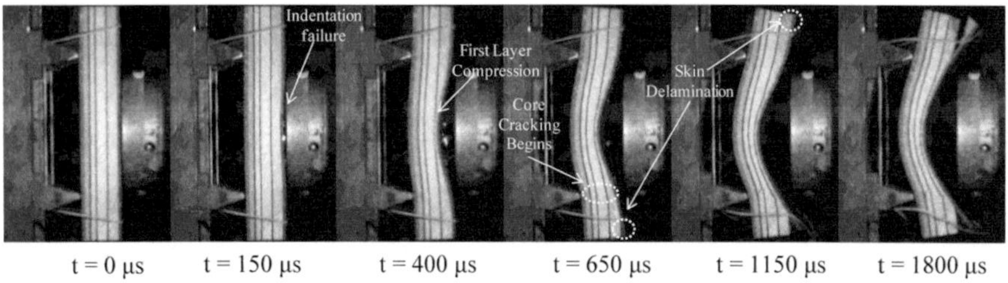

Fig. 10.18 High-speed images for configuration 2 (A300/A500/A800/PU)

The real-time blast loading response of configuration 1 (PU/A300/A500/A800) is shown in Fig. 10.17. It can be observed from the images that indentation failure (initiation of core compression) begins at approximately $t = 150$ μs. Following indentation failure, core delamination is first observed at $t = 400$ μs, and occurs at the bottom of the specimen between the polyurea interlayer and the first layer of foam core. By $t = 550$ μs, more core delamination is observed at the top of the specimen again between the polyurea interlayer and the first layer of foam core. Also at this time, two central core cracks have initiated where the supports are located, and heavy core compression is present in the first layer of the foam core (A300). By $t = 1{,}150$ μs, the first layer of foam core (A300) has reached a maximum level of compression (8 mm), approximately 75 % of its original thickness (12.7 mm). After this time, the response is global bending of the specimen and by $t = 1{,}800$ μs, no new failure mechanisms were observed. Also the core cracks have propagated through the foam core to the polyurea interlayer and there is heavy core delamination between the polyurea interlayer and the foam core.

The real-time blast loading response of configuration 2 (A300/A500/A800/PU) is shown in Fig. 10.18. It is evident from the figure that indentation failure begins at $t = 150$ μs. After indentation failure is observed, heavy core compression is observed in the first core layer (A300 foam). By $t = 650$ μs the first layer of foam (A300) has compressed to a maximum, reaching its densification level, and the shock wave has now propagated into the second foam core layer (A500), initiating compression of this core layer. Also at this time, a core crack has initiated at the bottom of the specimen where the support is located. Skin delamination is evident between the front skin and the foam core, and is located at the bottom of the specimen. At $t=1{,}150$ μs, skin delamination can be observed at the top of the specimen between the front facesheet and the foam core. Also at this time, the compression in the second foam core layer has increased to its maximum and no more compression is observed in the core, resulting in a global bending response. Between $t=1{,}150$ μs and $t=1{,}800$ μs, no new failure mechanisms were observed. The core crack continues to propagate through the third layer of the foam core (A800) and skin delamination at the bottom of the specimen has increased between the front facesheet and the foam core.

Comparing the deformation mechanisms observed in configuration 1 and configuration 2, the location of the polyurea interlayer affects the order and level of different failure mechanisms, such as core compression, core cracking, and interface delamination, as well as the time at which they are first observed. In configuration 1, indentation failure (core compression) is followed by delamination in the core and then core cracking. Unlike configuration 1, the indentation failure of configuration 2 is followed by heavy core compression, core cracking, and then skin delamination. Comparing both configurations, the indentation failure is observed at approximately the same time, while core cracking and delamination initiate in configuration 1 earlier in the deformation, approximately 100 and 250 μs, respectively, than in configuration 2.

The location of the polyurea interlayer also affects the core deformation mode for each configuration. In configuration 1 the initial blast loading is uniformly distributed over the polyurea layer, resulting in a global uniform compression of the first layer of the foam core (A300). On the contrary, in configuration 2 the initial impulse is non-uniformly distributed into the foam core, resulting in a local compression in the central region of the first layer of the foam core (A300) where the shock loading was applied. This indicates that the polyurea interlayer has the ability to disperse the shock loading.

Also the deformation shape for both configurations is much different. For configuration 1, the specimen exhibits a double-winged deformation shape until approximately $t=400$–500 μs, and then the polyurea layer begins to delaminate from the core, exhibiting a shape much like a specimen in pure bending. For configuration 2, the specimen exhibits a double-winged deformation shape throughout the entire blast loading event. Therefore, configuration 2 has the ability to support the shear stresses that are present during the event, while configuration 1 could not.

10.4.2.2 Deflection

The midpoint deflections of the constituents of sandwich composites with different core configuration were obtained from the high-speed side-view images and shown in Fig. 10.19. For configuration 1, the midpoint deflection of the front face (front skin), interface 1 (between first and second core layer), interface 2 (between second and third core layer), interface 3 (between third and fourth core layer), and back face (back skin) were plotted and are shown in Fig. 10.19a. The front face and interface 1 deflect in the same manner to the same value of approximately 43 mm at

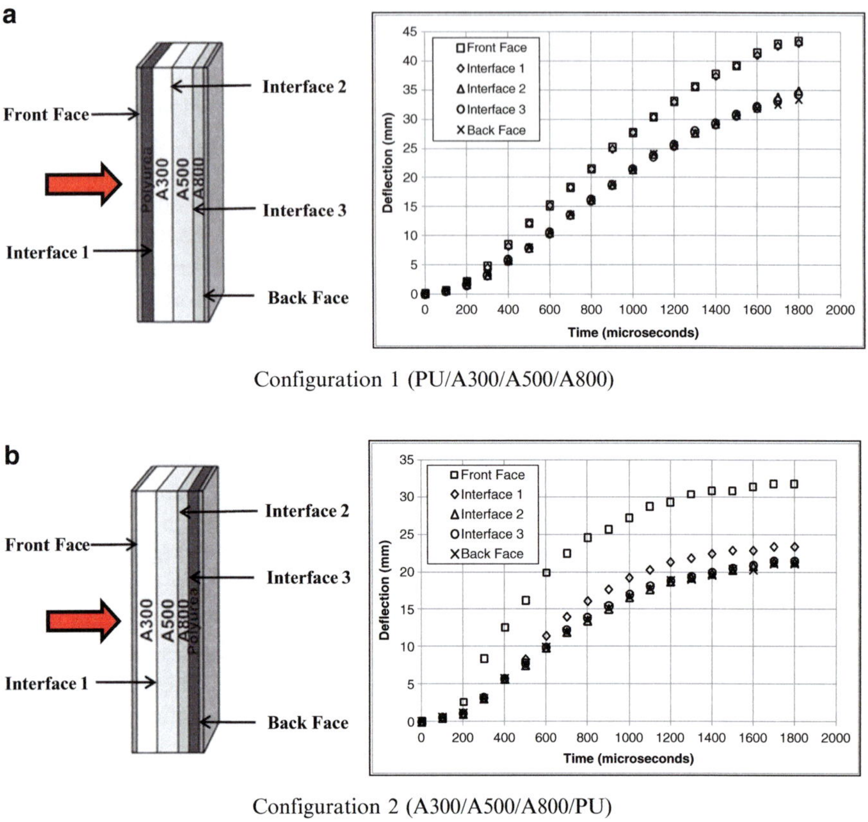

Fig. 10.19 Midpoint deflection of both configurations. (**a**) Configuration 1 (PU/A300/A500/A800). (**b**) Configuration 2 (A300/A500/A800/PU)

$t = 1{,}800$ μs, while interface 2, interface 3, and back face deflect in a similar manner to approximately 34 mm. Since the front face and interface 1 deflect in the same manner to the same value (43 mm), it signifies that the polyurea interlayer, which is located between the front face and interface 1, exhibits no compression. The difference between the deflection of interface 1 and interface 2 indicates the total amount of compression observed in the second core layer, which is the first layer of foam (A300). It can be seen that the A300 layer of foam compressed approximately 9 mm, which is 75 % of its original thickness (12.7 mm). Since interface 2, interface 3, and back face all deflected in a similar manner to the same value of approximately 34 mm, it can be concluded that the A500 foam layer (located between interface 2 and interface 3) and the A800 foam layer (located between interface 3 and back face) showed no compression. Therefore the core layer arrangement of configuration 1 allows for compression only in the A300 layer of foam and has a front face and back face deflection of approximately 43 and 34 mm, respectively.

For configuration 2, the midpoint deflection of the front face (front skin), interface 1 (between first and second core layer), interface 2 (between second and

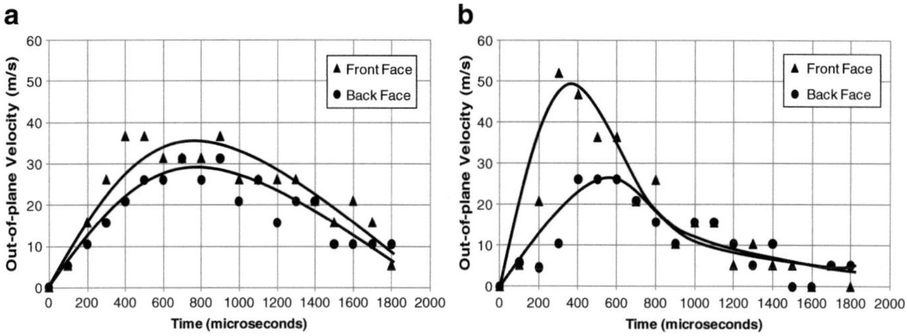

Configuration 1 (PU/A300/A500/A800)
(Hard core-type response)

Configuration 2 (A300/A500/A800/PU)
(Soft core-type response)

Fig. 10.20 Front and back face out-of-plane velocities for both configurations. (**a**) Configuration 1 (PU/A300/A500/A800) (hard core-type response). (**b**) Configuration 2 (PU/A300/A500/A800) (soft core-type response)

third core layer), interface 3 (between third and fourth core layer), and back face (back skin) were plotted and are shown in Fig. 10.19b. The front face deflected to approximately 33 mm $t = 1{,}800$ μs, while interface 1 deflected to approximately 24 mm, and interface 2, interface 3, and back face deflected in the same manner to a value of 21 mm. The difference between the front face deflection and the deflection of interface 1 signifies the amount of compression in the first core layer (A300 foam). Therefore it can be observed that the A300 foam layer compressed approximately 9 mm, or 75 % of its original thickness (12.7 mm). Again, noting the difference between the deflection of interface 1 and interface 2 the amount of compression in the second core layer (A500 foam) can be obtained. By inspection the A500 foam core layer compresses approximately 3 mm, which is approximately 25 % of its original thickness. Finally, since interface 2, interface 3, and back face all deflected in a similar manner to the same value of approximately 21 mm, it can be concluded that there was no compression in the third and fourth core layer (A500 foam layer and the polyurea interlayer). Therefore the core arrangement of configuration 2 allows for a stepwise compression through the core and the front face and back face deflect to approximately 33 and 21 mm, respectively.

From the midpoint deflection data in Fig. 10.19a, b, the average midpoint velocities of the front face and back face for both configurations can be obtained by differentiating the front face and back face deflection with respect to time, and are shown in Fig. 10.20. It can be seen in the figure that in configuration 1, the front face and back face reach a maximum velocity together early in the deformation history, $\sim t = 700$ μs, converging on the same common velocity (~30 m/s) and then decelerating together. The front face and back face of configuration 2 reach maximum velocities at different times, $\sim t = 400$ μs and $\sim t = 600$ μs, respectively, and then share a common velocity of ~20 m/s much later in the event (~800 μs), before decelerating together. This suggests that the back face was beginning to decelerate, while the core was still compressing. Such responses and phenomena have been investigated by Liang et al. (2007) and Tilbrook et al. (2006). When the front and back face velocities equalize early in the

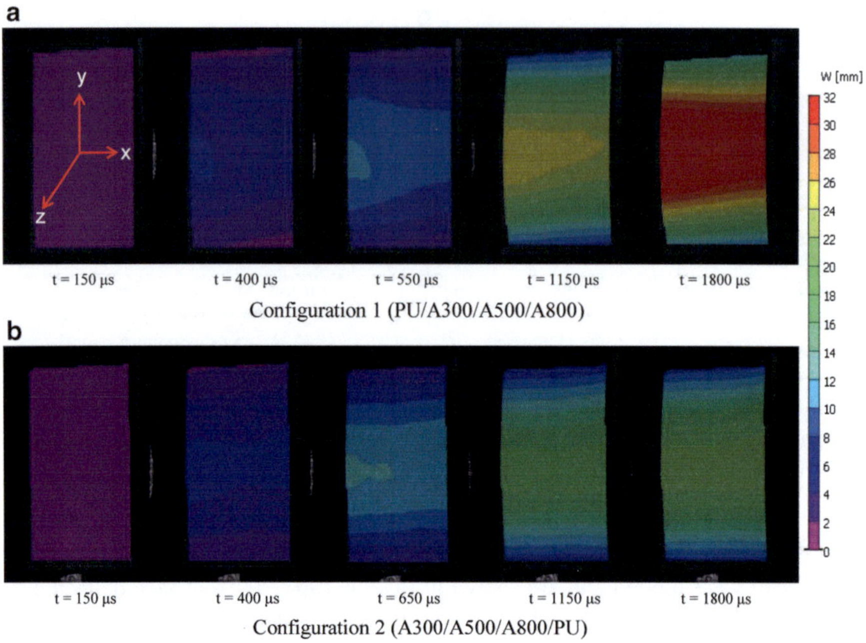

Fig. 10.21 Full-field out-of-plane deflection (*W*) of both configurations. (**a**) Configuration 1 (PU/A300/A500/A800). (**b**) Configuration 2 (A300/A500/A800/PU)

deformation history, this response is labeled as a hard core-type response. In contrast, when the back face begins to decelerate while the core is still compressing, this response is labeled as a soft core-type response. Therefore it can be concluded that configuration 1 exhibits a hard core-type response, while configuration 2 exhibits a soft core-type response. The authors (Liang et al. 2007; Tilbrook et al. 2006) suggested that the optimal performance of sandwich beams is attained for soft core designs, which allows for a reduction in the transmitted impulse during the initial fluid–structure interaction stage.

10.4.2.3 DIC Analysis

Utilizing the DIC technique, the deflection, in-plane strain, and particle velocity contours of the back facesheet for each configuration were generated. Figures 10.21 and 10.22 show the full-field results for the back facesheet of configuration 1 and configuration 2, respectively. Figure 10.21 shows the full-field out-of-plane deflection (*W*) with a scale of 0 mm (purple) to 32 mm (red). It is evident from the figure that for configuration 1, as shown in Fig. 10.14a, the back face exhibits very little out-of-plane deflection until approximately $t=400$ μs. Between $t=400$ μs and $t=1,800$ μs, the panel continues to show deflection. By $t=1,800$ μs, it can be observed that the central region of the panel has deflected approximately 32 mm. For configuration 2, as shown in Fig. 10.14b, the back face shows very little out-of-plane

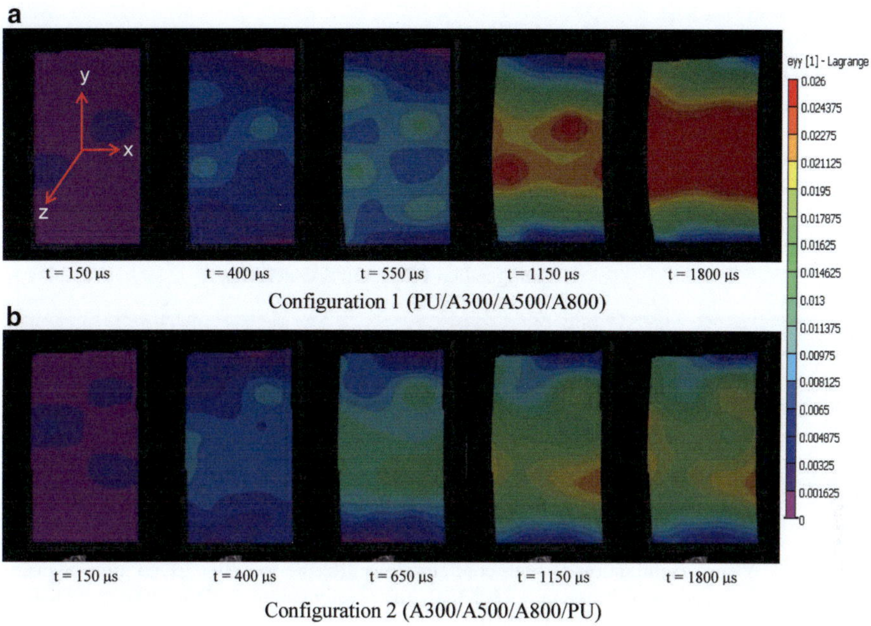

Fig. 10.22 Full-field in-plane strain (ε_{yy}) of both configurations. (**a**) Configuration 1 (PU/A300/A500/A800). (**b**) Configuration 2 (A300/A500/A800/PU)

deflection until $t=400$ μs. Between $t=400$ μs and $t=1,800$ μs, the panel continues to exhibit deflection. By $t=1,800$ μs, the central region of the panel has deflected approximately 22 mm. Therefore, configuration 2 deflects approximately 35 % less than configuration 1.

Figure 10.22 shows the full-field in-plane-strain (ε_{yy}) for both configurations with a scale of 0 (purple) to 0.026 (red), or 0–2.6 %, respectively. For configuration 1, as shown in Fig. 10.15a, the back face exhibits very minimal in-plane strain (ε_{yy}) until approximately $t=150$ μs. Between $t=150$ μs and $t=1800$ μs, the in-plane strain (ε_{yy}) continues to increase. By $t=1,800$ μs, it can be observed that the panel shows a maximum in-plane strain (ε_{yy}) across the central region of the panel of approximately 2.6 %. For configuration 2, as shown in Fig. 10.15b, the back face shows very minimal in-plane strain (ε_{yy}) until approximately $t=150$ μs. Between $t=150$ μs and $t=1,800$ μs, the in-plane strain (ε_{yy}) continues to increase. By $t=1,800$ μs, the panel shows a maximum in-plane strain (ε_{yy}) across the central region of the panel of approximately 1.625 %. As a result, configuration 2 exhibits approximately 35 % in-plane strain than configuration 1.

10.4.2.4 Postmortem Analysis

After the blast loading event occurred, the damage patterns of both configuration 1 and configuration 2 were visually examined and recorded using a high-resolution

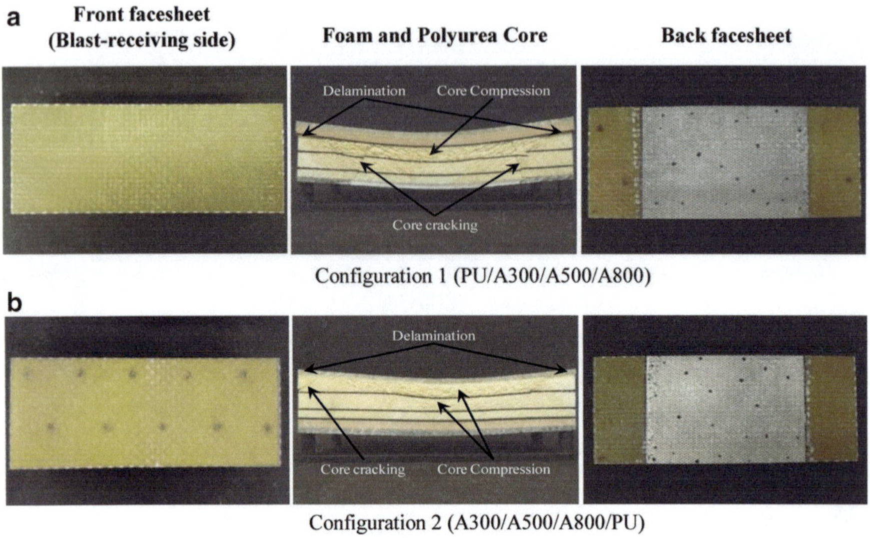

Fig. 10.23 Visual examination of both configurations after being subjected to high-intensity blast load (reflected peak pressure ~5.0 MPa). (**a**) Configuration 1 (PU/A300/A500/A800). (**b**) Configuration 2 (A300/A500/A800/PU)

digital camera and are shown in Fig. 10.23. When configuration 1 was subjected to transient shock wave loading, as shown in Fig. 10.23a, the damage was confined to the areas where the supports were located in the shock tube and core cracking is visible in these two areas. The core cracks propagated completely through the foam core to the polyurea interlayer. Core delamination is visible between the polyurea interlayer, and the first layer of the foam core (A300). Core compression is visible in the first core layer of A300 foam.

When configuration 2 was subjected to transient shock wave loading, the damage patterns can be seen in Fig. 10.23b. For this configuration, very little core damage was observed. Core delamination between the first two layers of the foam core (A300 and A500) led to a crack that propagated through the first foam core layer (A300) to the front facesheet. Skin delamination was evident between the front face and the first foam core layer (A300). Also core compression can be observed in the first two layers of the foam core (A300 and A500).

Figure 10.24 shows the damage patterns of both configuration 1 and configuration 2 after they were subjected to the higher levels of blast loading (incident peak pressure ~1.5 MPa, reflected peak pressure ~7.5 MPa, wave velocity of 1,300 m/s). It can be seen from the figure that when configuration 1, as seen in Fig. 10.24a, was subjected to a higher level of blast loading, the core exhibited heavy core cracking which led to catastrophic failure. The font face showed heavy fiber delamination and cracking across the central region, while the back face delaminated completely from the core.

Configuration 2 on the other hand, as seen in Fig. 10.24b, remained structurally intact after the higher level of blast loading. The front face showed minor fiber

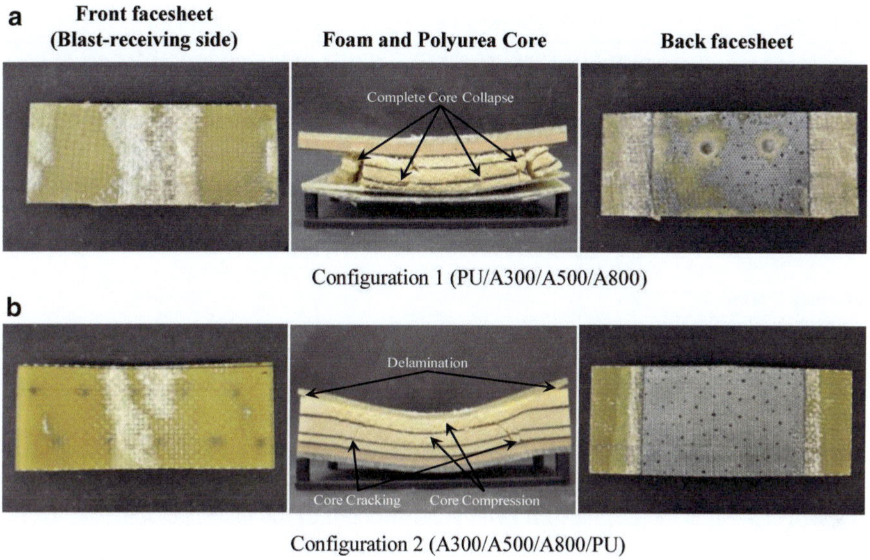

Fig. 10.24 Visual examination of both configurations after being subjected to high-intensity blast load (reflected peak pressure ~7.5 MPa). (**a**) Configuration 1 (PU/A300/A500/A800). (**b**) Configuration 2 (A300/A500/A800/PU)

delamination, while the core exhibited cracking along the central region where the supports were located. Minor front skin delamination was evident between the front face and the first foam core layer (A300). Also core compression can be observed in the first and second core layers of foam, A300 and A500, respectively.

10.4.3 Effects of Pre-loading

Sandwich composites have important applications in the naval and aerospace industry. During normal service, these sandwich structures are sometimes subjected to dynamic loads which can occur when these structures are exposed to pre-loadings. The following sections reveal the pre-loading effects on sandwich composites when subjected to air blasts.

The in-plane buckling behavior of sandwich composites has been studied for years (Fleck and Sridhar 2002; Mamalis et al. 2005). There are three main buckling modes based on the geometries and materials (Fleck and Sridhar 2002): elastic buckling in bending and shear (global buckling), and plastic micro-buckling of the facesheets and face wrinkling. The sketches of the different failure modes of the sandwich composites are shown in Table 10.5, as well as the analytical expression of the critical buckling load and the calculated critical buckling load based on the parameters of the sandwich composites (as shown in Table 10.2). t_f is the facesheet thickness, t_c is the core thickness, b is the total width, and l is the total length. E_f and

Table 10.5 Sketches of the different failure modes of the sandwich composites

Failure mode name	Elastic buckling in bending and in shear (global buckling)	Plastic micro-buckling of the facesheet	Facesheet wrinkling
Failure mode sketch			
Critical buckling load expression	$\dfrac{P_E P_s}{P_E + P_s}$	$2bt\sigma_{fy}$	$0.5bt_f\left(\sqrt[3]{E_f E_c G_c}\right)$
Calculated critical buckling load (kN)	128	76	146

E_c represent the in-plane uniaixal Young's modulus of the facesheet and core materials, respectively. σ_{fy} is the in-plane uniaxial yield strength of the facesheet material. G_c is the shear modulus of the core material. $P_E = \dfrac{4\pi^2(EI)_{eq}}{l^2}$ is the Euler buckling load and $P_S = bcG_c$ is core shear buckling load. $(EI)_{eq}$ is the bending stiffness of the sandwich panel:

$$(EI)_{eq} \approx \frac{btc^2 E_f}{2} \tag{10.3}$$

From Table 10.5, the mode at which the sandwich panel is most likely to fail is plastic micro-buckling of the facesheets. Based on the critical buckling load 76 kN, three different in-plane compressive loading levels were chosen for the present study: 0 kN (no loading), 15 kN (~20 % of critical buckling load), and 25 kN (~33 % of critical buckling load).

10.4.3.1 Real-Time Deformation

The real-time side-view images of sandwich composites subjected to the same amplitude and duration of the incoming blast wave and with different levels of initial compressive loading on them are shown in Fig. 10.25. The shock wave propagates from the right side of the image to the left side. Some deformation details are pointed out on the images.

From the images, it can be clearly seen that the initial deformation modes for the sandwich composite with different levels of compressive loading are very similar. Prior to 400 μs all the specimens show global bending with a typical double-wing shape, which means that the core is under intense shear loading. At almost 350 μs after shock loading, shear cracks in the core begin to emerge in specimens with no in-plane loading (0 kN). The in-plane compressive loading tends to suppress the formation of core cracks. In the specimens with 15 and 25 kN in-plane loading, the shear cracks in the core only begin to appear around 1,200 and 900 μs, respectively.

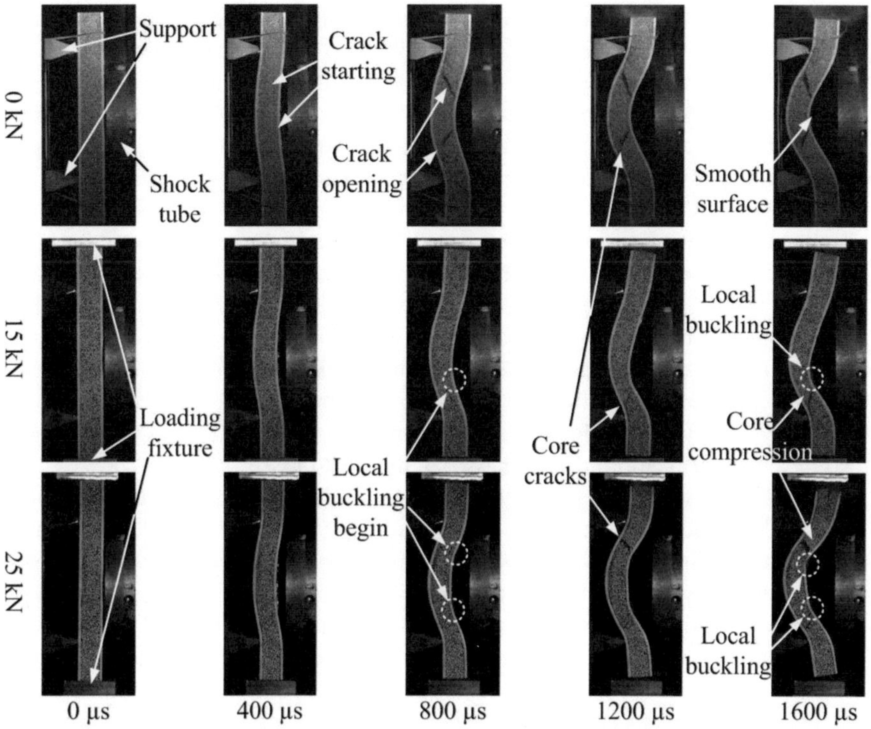

Fig. 10.25 Real-time side-view deformation of sandwich composites with initial compressive loadings

The deformation modes of the specimens with different in-plane loadings also change after 400 μs. The specimen without in-plane loading (0 kN) continues bending symmetrically. The front facesheet shows a profile with a smooth curvature, implying no local buckling in the front face. For the specimen with the 15 kN compressive loading, the initial deformation with a symmetric profile shifts to an asymmetric mode. The section of the front facesheet close to the lower support exhibits more curvature than the section close to the upper support. This asymmetrical phenomenon indicates that there is local buckling at the lower section of the front facesheet. At approximately 1,600 μs, the debonding of the front facesheet from the core gives clear evidence of local buckling (shown by the white circle). Meanwhile, the curved profile in the middle section (right above the local buckling position) of the front facesheet changes to a flat profile, indicating that the bending moment applied on this section has been released. For the specimen with the 25 kN compressive loading, there are two obvious kinks in the front facesheet (shown by the white circle). The middle section has a flat profile between two kinks implying that there are two failure hinges at kink positions and these failure hinges release the moment applied on this section. These two failure hinges are induced by local buckling in the facesheet. From the photographs it can also be seen that kink formation occurs earlier in specimens with larger in-plane compressive loading.

On the other hand, the deformation behaviors of the back facesheets for all sandwich panels are similar. They all show smooth curvatures, indicating that there is no buckling in the back facesheet. This is because the shock loading does not apply on the back face directly. As mentioned in Fleck and Deshpande (2004), the motion of the sandwich structure starts from the front facesheet upon which the shock loading is directly applied. The load is only transmitted to the back facesheet after the stress wave has passed through the core and the core is compressed. This implies two facts. First, the load applied on the back face gets dispersed as it transmits through the core, and as such will not be concentrated. Second, the intensity of load transmitted to the back face depends on the core compressive properties. As the core foam material has a long low-level stress plateau in its constitutive behavior, the load applied on the back face will be much lower than the shock loading applied directly on the front face. Therefore, the combined effect of the in-plane compressive loading and the local transverse shock wave loading (as discussed in depth in Sect. 10.4.3.4) only changes the global and local deformation behavior of the front face.

It needs to be noted that in Fig. 10.25 the ends of the specimen without in-plane loading almost do not rotate in the very beginning (before 400 μs). This indicates that the flat loading head will probably not constrain the end rotation of the specimens when applying the in-plane compressive loading. After 400 μs, the ends of the specimens with in-plane loadings actually disconnect from the loading head. The position of the loading head remains fixed, whereas the vertical dimension of the specimen shrinks due to global bending. Therefore, the constraint of end rotation is minimized in the present study and the simply supported boundary as mentioned in Sect. 10.3.1.1 is approximately achieved.

10.4.3.2 Deflection

The deflections of the middle edge point located on the front and back faces obtained from the side-view high-speed images are shown in Fig. 10.26. Prior to 400 μs, the deflections of all panels are almost identical. After 400 μs, the specimens with the 0 and 15 kN compressive loading have similar deflections while the deflection of the specimen with the 25 kN compressive loading is higher. This indicates that only at high in-plane compressive loading level (25 kN, 33 % of the critical load of the structure buckling) will the deflections along the middle line of the facesheets be significantly affected. Moreover, the middle edge point deflections of the front and back faces for each panel are almost overlapping. This means that there is negligible core compression in the core thickness along the middle line (symmetry line), though there were obvious compressive regions in the core around the buckling area of the core, near to the front face.

10.4.3.3 DIC Analysis

Figure 10.27 shows the back face out-of-plane (z-direction) deflection contours of sandwich composites subjected to blast loading with different levels of initial

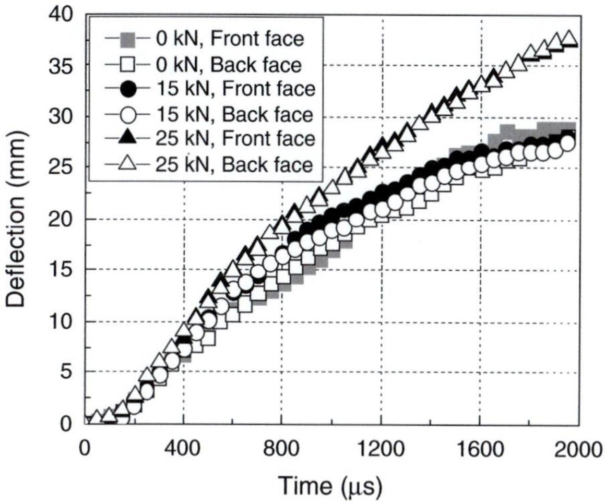

Fig. 10.26 Midpoint deflections of front and back face for sandwich composites compressive with pre-loading

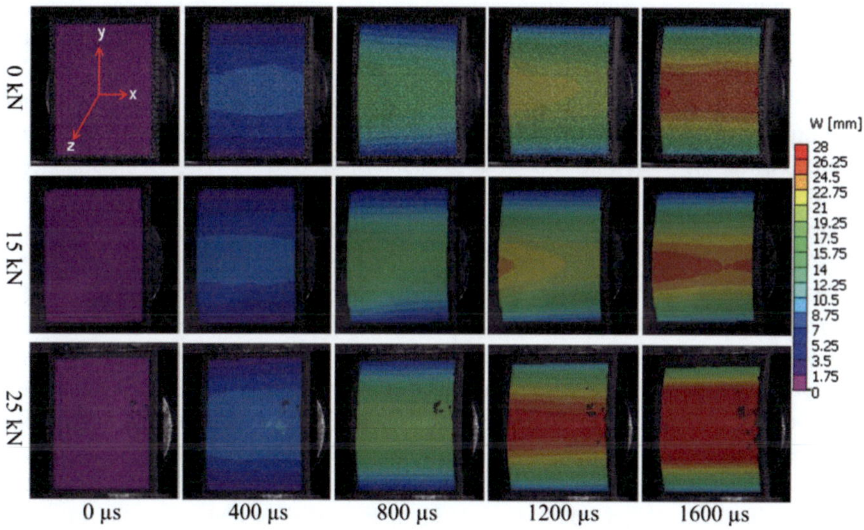

Fig. 10.27 Full-field out-of-plane deflection (*W*) of sandwich composites with compressive pre-loading

compressive loading from the DIC analysis. The support knife edges are located at the top and bottom of the images and parallel to the *x*-direction. The color stripes, which are parallel to the support line, show that the deflections of the points through the width of the panels are almost identical. This indicates that the specimens acted like a beam under the current support conditions. It can also be seen that the deflections of the specimens without compressive loading (0 kN) are comparable with the

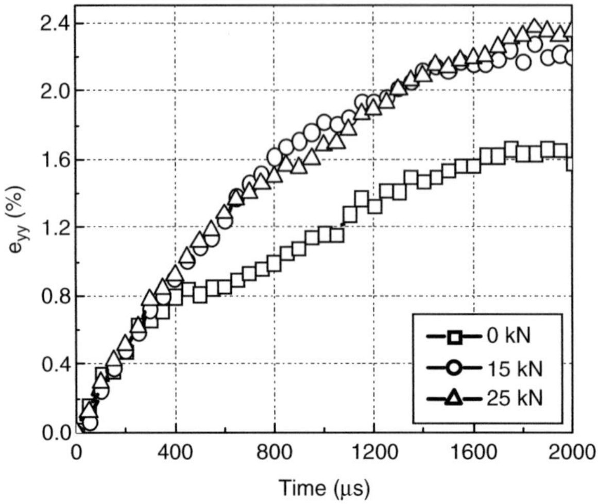

Fig. 10.28 In-plane strain on the back face at the center point

15 kN compressive loading specimens, with the compressed specimens showing larger deflection after 800 μs. The specimen with the 25 kN compressive loading has higher deflections.

The in-plane strains e_{yy} of the center point on the back face from the DIC technique are shown in Fig. 10.28. It can be clearly seen that the trend of the in-plane strains is much different from that of the out-of-plane deflections. Though the deflections are almost identical, the back face in-plane strain of the specimen with the 15 kN compressive loading is much higher than the one without compressive loading. This is due to the different failure modes of the specimens under different in-plane compressive loading. The in-plane strain e_{yy} on the back facesheet depends on the loads transmitted from two physical mechanisms: the deformation of the core itself and the global bending of the sandwich structure. Due to the constitutive behavior of the core foam material (the stress plateau after the yield point), the load transmitted to the back facesheet by the core has a limit value. However, the load transmitted by the interfaces is most likely based on the global deflection of the sandwich structure, which means that this load will keep increasing as long as the interfaces do not totally delaminate. After delamination this load will drop and affect the in-plane strain on the back facesheets. Therefore, in the beginning with small global deflection, i.e., prior to 400 μs, the effects of the core and the interfaces are comparable. When the core cracks appear in the specimen without in-plane load around this time, there is an obvious drop in the e_{yy} verses time curve. On the other hand, the global bending of the sandwich structure tends to increase the in-plane strain. The combined effect causes the in-plane strain e_{yy} to keep increasing with a smaller slope after initial drop. For the specimens with 15 and 25 kN loadings, the core cracks appeared much later (1,200 and 900 μs, respectively) with a relatively large global deflection. This implies that when the core cracks appear, their effect will be small

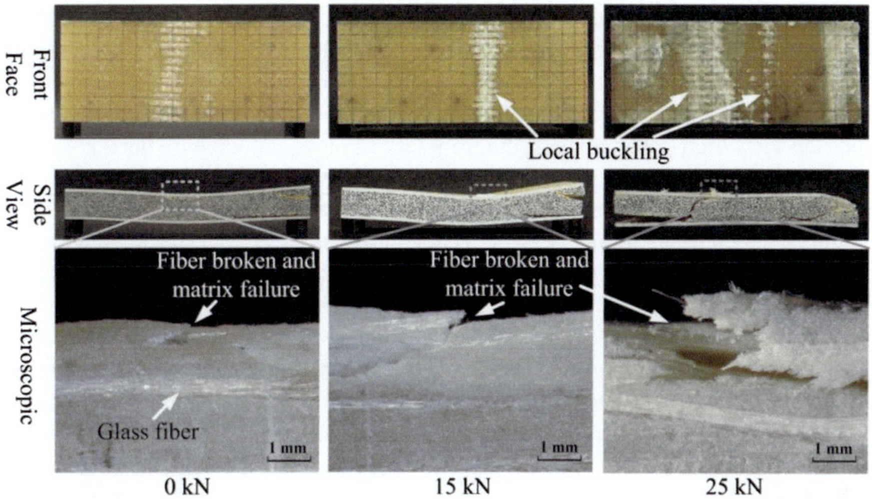

Fig. 10.29 Postmortem images of sandwich composites with compressive pre-loading

and not that obvious. It can be clearly seen in Fig. 10.28 that the in-plane strain curves have small kinks around 1,200 and 900 μs for the specimen with 15 and 25 kN in-plane loadings, respectively. These kinks mean that the core cracks do affect the in-plane strain on the back facesheet, but the effect is not as strong as when the core cracks appeared at 400 μs. Two other features are significant in the strain plots. First, prior to 400 μs the strain e_{yy} of the back facesheet is almost identical for the specimens with and without in-plane loadings. The reason has been mentioned in the discussion of Fig. 10.25 which states that the specimens with different levels of compressive loading have similar deformation modes during this time period. Second, the in-plane strains of the specimens with 15 and 25 kN in-plane loading are comparable during the whole time period until 2,000 μs while they have different back face deflections (see Fig. 10.26). This is because the specimen with 15 kN in-plane loading has one crack in the core while the specimen with 25 kN in-plane loading has two cracks which means that the latter should have less effect on the in-plane strain of the back facesheet. Therefore, though the deflection of the specimen with 25 kN in-plane loading is larger, the combined effect of the damage in the core and the global bending makes the in-plane strain of the back face comparable with that of the specimen with 15 kN in-plane loading.

10.4.3.4 Postmortem Analysis

The damage patterns of the sandwich composites after the shock loading were visually examined and recorded using a high-resolution digital camera and are shown in Fig. 10.29. Since the back facesheets do not show any change after the experiments, they are not shown here.

From the front facesheet images, the local buckling positions are apparent. Note that the yellow color is the original color of the specimen and the white color signifies fiber delamination and facesheet buckling. For the specimen with the 0 kN compressive loading, there is only slight fiber delamination on the front face. From the side-view high-speed images, the front face of the specimen with the 0 kN compressive loading shows a profile with a smooth curvature during the whole shock loading process. Therefore, the delamination is due to the large deflection bending of the sandwich panel during the dynamic loading process. For the specimen with the 15 kN compressive loading, there is a region of heavy fiber delamination on the front face that correlates to the buckling section with large curvature in the side-view high-speed images. For the panel with the 25 kN compressive loading, heavy fiber delamination and buckling occurred at two positions on both sides of the center of the specimen that correlates to the two kinks in the side-view high-speed images. The white area at the end of the specimen is not due to local buckling induced by the compressive loading but is due to a collision between the specimen and the shock tube during the large deflection bending process of the specimen that occurs much later in the event.

From the side-view postmortem images, the permanent back facesheet deflection of the specimen with the 0 kN compressive loading (3.4 mm) is less than that of the specimen with the 15 kN compressive loading (4.8 mm). The specimen with the 25 kN compressive loading showed heavy delamination between the core and back facesheet, which released the stress in the back facesheet. Thus, it showed the least permanent back facesheet deflection. It should be noted that the opened cracks in the core of the specimen without in-plane loading (Fig. 10.25) almost closed after the experiment as shown in Fig. 10.29 (side view). This shows that the core has an ability to stand large shear deformation.

Microscopic analysis of the buckling region observed in the sandwich panels was done using a Nikon SMZ microscope. These microimages are shown in Fig. 10.29. For the specimen with the 0 kN compressive loading only a small crack appears in the front face when delamination occurred. For the specimen with 15 kN compressive loading, the crack crossed the first two fiber layers of the front facesheet. For the specimen with the 25 kN compressive loading, the crack had completely opened and propagated into the facesheet by more than half of its thickness. The edge of the crack shows evidence of tearing.

10.4.3.5 Failure Mechanisms

The experimental results show that compressive loading significantly affects the blast performance of sandwich composites. Failure and buckling occurred at specific positions on the front facesheet. In this section, a qualitative model is described to explain the dynamic failure process of sandwich composites with compressive loadings.

A sketch of the response and failure process of sandwich composites under in-plane compressive loadings when they are subjected to transverse shock wave loading is shown in Fig. 10.30. When the shock wave first impinges on the sandwich

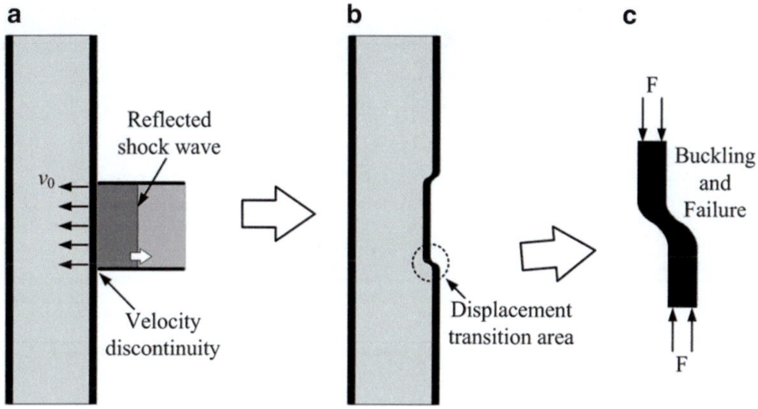

Fig. 10.30 A sketch of the response and failure process of sandwich composites under in-plane compressive loading and transverse shock wave loading

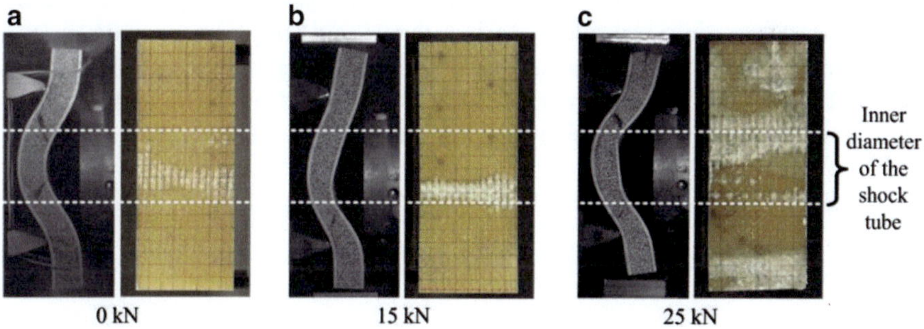

Fig. 10.31 Failure positions for the sandwich panel with different compressive loading

panel, it gives an initial velocity field to the front face (Fleck and Deshpande 2004). Since the loading area is smaller than the front face area, the front face will have different velocities at different locations, as shown in Fig. 10.30a. This velocity difference will create some transition areas around thc loading area due to the displacement difference after a very short time period, as shown in Fig. 10.30b. Then, the flat shape of the front face is destroyed. The in-plane compressive loading on the front face generates shear loading at the transition areas as shown in Fig. 10.30c. This shear loading significantly reduces the critical buckling load of the front face and reduces the blast resistance of the sandwich composite.

Correlation between the high-speed side-view images with postmortem images for the three compressive loading states is shown in Fig. 10.31. The lines show the boundaries of the shock loading area. It can be clearly seen that the failure positions on the front face of the sandwich panels are located around the boundaries of the shock loading area. These areas are the transition areas seen in Fig. 10.30b.

In the experiments, the failure areas are not symmetric to the loading area, though the transition areas are. This is because the loading, support, and composite material

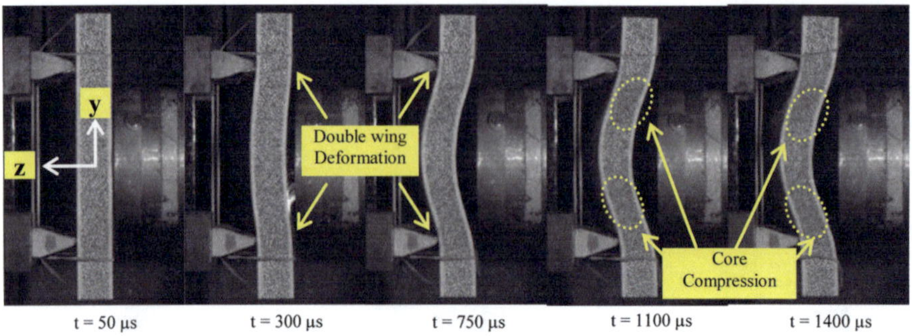

Fig. 10.32 Real-time deformation of the sandwich composite at room temperature (22 °C at core)

properties may not be perfect. The failure always occurred at one transition position first. Subsequent to this, if the in-plane compressive loading was not high enough, the second buckling site did not develop and the loading applied on the symmetric position of the front face was released. Thus, there is only one failure position on the front face (Fig. 10.31b). If the in-plane compressive loading is high enough, as is the case with the 25 kN loading, another failure site develops before loading is totally released as seen on the front face of Fig. 10.31c.

10.4.4 Environmental Effects

Aside from pre-loadings, during normal service (naval and aerospace industry), dynamic loads can occur when these structures are exposed to varying environmental conditions (high and low temperatures). The following sections reveal the effects of the environment on sandwich composites when subjected to dynamic loadings such as air blast.

10.4.4.1 Real-Time Deformation

The real-time observation of the transient behavior of sandwich composites under shock wave loading at room temperature (22 °C), high temperature (80 °C), and low temperature (–40 °C) is shown in Figs. 10.32, 10.33, and 10.34, respectively.

At room temperature, the sandwich composite exhibited a typical double-wing formation but no visible failure was seen during the blast event as shown in Fig. 10.32. At $t=0$ μs, the shock wave impinged on the specimen and the fluid structure interaction occurred between the incident gas and the specimen. It takes approximately 50 μs for the stress wave to propagate in the specimen. Both front and back faces of the specimen begin to show deformation around this time and by the end of the fluid–structure interaction time of approximately 150 μs (Wang 2010;

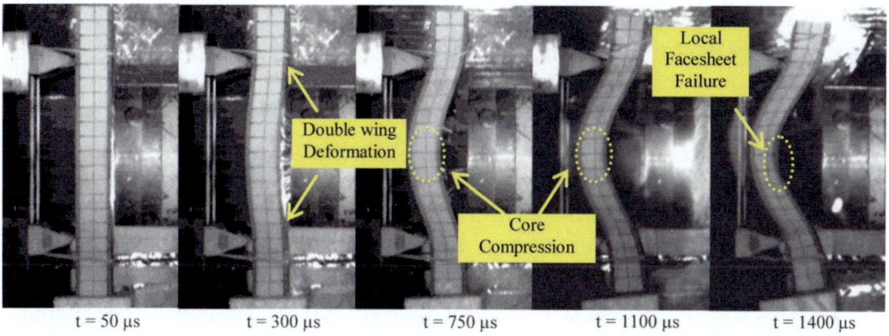

Fig. 10.33 Real-time deformation of the sandwich composite at high temperature (80 °C at core)

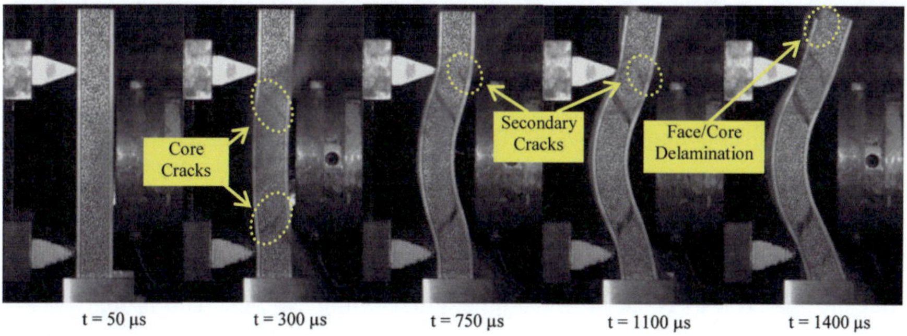

Fig. 10.34 Real-time deformation of the sandwich composite at low temperature (–40 °C at core)

Wang et al. 2012), the specimen deformed by approximately 2.5 mm on both the front and back face. The sandwich specimen was observed to be deforming in a double-wing manner (as seen from the image at t=300 μs), which corresponds to high shear stresses in the core. By t=750 μs, the specimen had deflected approximately 15 mm on the back face. The compression in the core initiated at t=1,100 μs. At t=1,400 μs, a compressive strain of 10 % at a strain rate of 150/s was observed in the core. It can also be observed that the core compression only occurred in the region of high shear loads.

The blast response of the sandwich composite at a core temperature of 80 °C is shown in Fig. 10.33. At t=0 μs, the shock wave impinged on the sandwich specimen. After the fluid–structure interaction between the specimen and gas, the specimen began to deform in a double-wing manner causing core compression after t=300 μs. A decrease in the flow stress value of core material at high temperature (due to thermal softening of Corecell M100 foam) allowed for high core compression during blast event. At 80 °C, the center of the core was compressed by 6 % by t=750 μs and later, a nonuniform compression at the center line in the core was observed. A local core compression in the proximity of the front face developed due to the failure of the facesheet along the center line at approximately t=1,100 μs. Compressive stresses in the front face and tensile stresses in the back face of the specimen were generated during global bending deformation under blast loading. At room temperature, these compressive and tensile stresses were not high enough to reach the compressive or tensile failure strength of the facesheets. However, at high temperatures, the softening of the resin decreased the compressive and tensile strength of the facesheet, which resulted in the failure of the back and front facesheet along the line of maximum in-plane stresses (Shenghu et al. 2009). Therefore, the core compression increased monotonically with time and at $t = 1{,}400$ μs, the core was compressed approximately 30 % along the center line.

The blast loading response of the sandwich composite at a core temperature of −40 °C is shown in Fig. 10.34. The shock wave impacted the sandwich specimen at $t = 0$ μs. At $t = 300$ μs, primary core cracks were formed in the high-shear-stress regions of the core and propagated through the core thickness within 50 μs. This phenomenon of shear crack formation has been demonstrated previously by Wang (2010) and Wang et al. (2012). It can also be explained by the increase in the shear modulus of foam materials as temperature is decreased (Shariatmadari et al. 2012). Also, there is no significant change in the shear strength of the polymeric foams at low temperatures (Benderly and Putter 2004). Therefore, the change in shear modulus at −40 °C resulted in the brittle response of the core material and the critical shear strains for the failure of the core material significantly decreased with the decrease in temperature. As the shock wave loading induced a large amount of shear deformation inside the core of sandwich composite, this shear deformation caused the core material to crack at low temperatures within 300 μs. The primary shear crack formation was followed by secondary crack initiations at $t = 750$ μs. After $t = 750$ μs, the significant failure of the core inhibited the sandwich specimen to deform under shear loading, and the deformation behavior changed from double wing to global bending. This allowed for slipping between the core and facesheets, and as a result, face/core delamination began at $t = 1{,}100$ μs. At $t = 1{,}400$ μs, the core failure occurred from significant core cracking and face/core delamination, which resulted in higher front and back face deflection as compared to 22 °C.

10.4.4.2 Deflection and Core Compression

Split Hopkinson pressure bar (SHPB) experiments on the core and facesheet had shown the thermal softening behavior at high temperatures (80 °C). This softening behavior caused a local failure in sandwich composites along the center line during the global bending response. As a result, the sandwich specimen at 80 °C demonstrated 65 % more back face deflection than at room temperature. When subjected to a low-temperature environment, brittle failure of the core occurred from core cracking in the early stages of the blast loading (~300 μs). This allowed the skin to delaminate from the core which caused the sandwich specimen to begin deforming in global bending. This resulted in 35 % more deflection at −40 °C compared to room-temperature experiments.

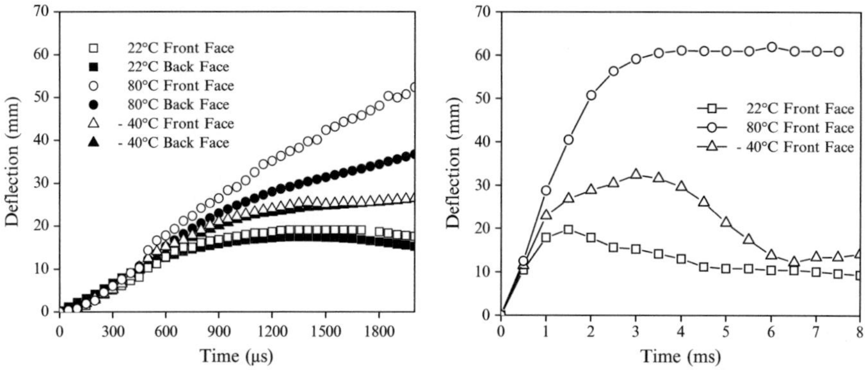

Fig. 10.35 Initial front/back face deflection and front face deflection history of the entire event for the sandwich composite at room, high, and low temperature

Table 10.6 Core compression and compressive strain at time $t = 1{,}200\ \mu s$

Temperature	Core compression (mm)	Compressive strain (ε_{zz}) at center (%)
22 °C	1.9	8
80 °C	8.0	27
–40 °C	0.8	3

The front face deflections are also plotted in Fig. 10.35 to compare the amount of core compression observed during the blast experiments. The core compression and compressive strain (ε_{zz}) at the center of the specimen at $t = 1{,}200\ \mu s$ are listed in Table 10.6. The high temperature softens the core which allows for more local deformation of the front facesheet, leading to local bending failure after significant core compression has been developed. The failure of the core at –40 °C, due to extensive core cracking and face/core delamination, changes the deformation behavior from a double-wing deformation to global bending and inhibits core compression during shock wave loading. This explains the lower values of core compression in the experiments at –40 °C.

10.4.4.3 DIC Analysis

The deflection, in-plane strain (ε_{yy}), and the velocity of the back face for experiment at each temperature environment were generated using 3-D DIC technique. Figure 10.36 shows the full-field out-of-plane displacement (w) with a scale of 0–40 mm. It is evident from thc figure that for room temperature, as shown in Fig. 10.36a, the back face reached a maximum deflection of 20 mm at $t = 1{,}400\ \mu s$ and started reverberating after that. For the high-temperature experiment, as shown in Fig. 10.36b, at $t = 400\ \mu s$, the specimen reached a deflection of 10 mm. The

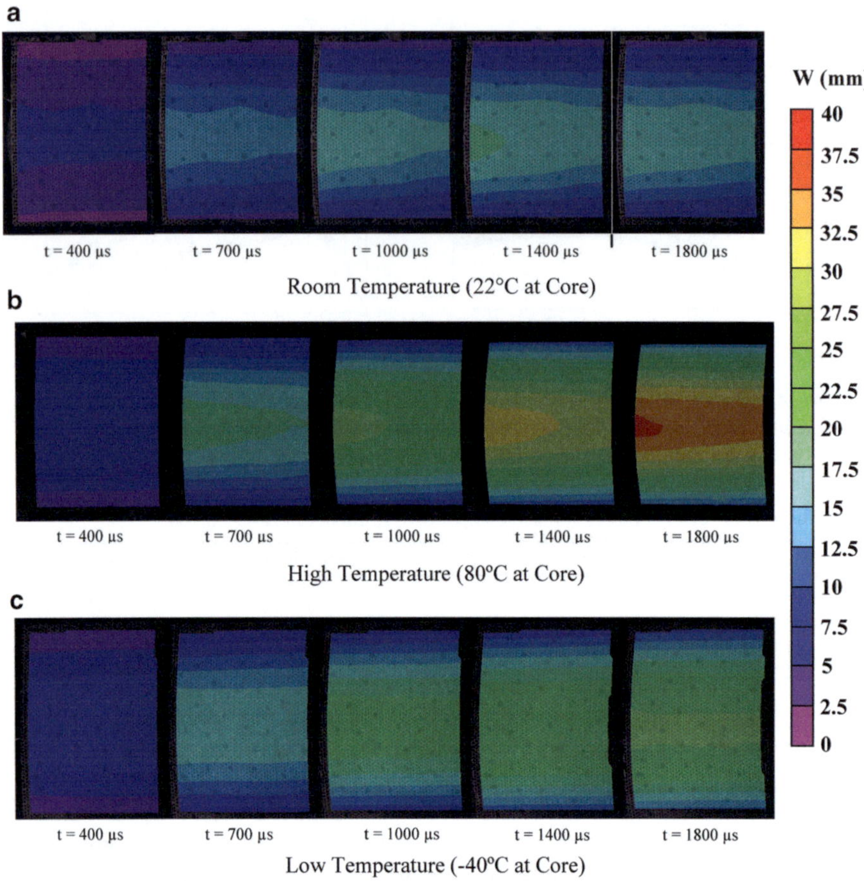

Fig. 10.36 Full-field out-of-plane deflection of the back face for room, high, and low temperature. (**a**) Room temperature (22 °C at core). (**b**) High temperature (80 °C at core). (**c**) Low temperature (–40 °C at core)

specimen continued to deform from $t=400$ to 1,800 μs and showed significant deflection of around 40 mm. For the low-temperature experiment, as shown in Fig. 10.36c, the specimen showed similar deflection as the room-temperature experiment until $t=400$ μs. Later, the low-temperature specimen showed higher deflections than the room-temperature specimen and reached a maximum value of 28 mm. Therefore, the high-temperature experiment and low-temperature experiment showed approximately 35 and 65 % more out-of-plane deflection, respectively, on the back face as compared to the room-temperature experiment at $t=1{,}400$ μs.

The in-plane strain (ε_{yy}) on the back face of the sandwich specimen is plotted in Fig. 10.37a. At room temperature (22 °C), a maximum in-plane strain of 1 % was observed at $t=900$ μs. At high temperature (80 °C), the sandwich specimen deformed under significant core compression and showed a maximum in-plane strain of 2.2 % at $t=1{,}800$ μs. At low temperatures (–40 °C), the sandwich specimen

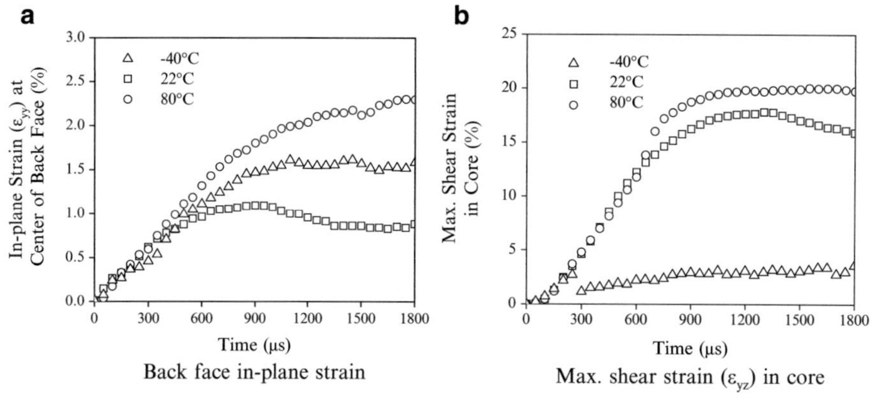

Fig. 10.37 Back face in-plane strain and max shear strain in the core as calculated from DIC. (**a**) Back face in-plane strain. (**b**) Max shear strain (ε_{yz}) in core

exhibited high deflections causing higher in-plane strain values as compared to room temperature, resulting in a maximum in-plane strain of 1.5 %. Figure 10.37a shows that in the initial part of the blast loading (~300–400 μs), the sandwich specimens at all temperatures exhibited similar in-plane strain profiles. After $t=400$ μs, significant core cracking and high core compression at −40 and 80 °C caused the increased values of in-plane strain on the back face, respectively.

During shock wave loading, sandwich composites are subjected to extreme shear loadings. In order to investigate the shear strains present within the sandwich composite, the maximum shear strains (ε_{yz}) in the core were calculated from the side-view images using 2-D DIC technique and are shown in Fig. 10.37b. At 22 °C, the core showed a maximum shear strain of 17 % at $t=1{,}300$ μs, while at high temperature (80 °C), the core reached a maximum shear strain of 20 % at $t=1{,}400$ μs. Similar shear strain profiles were observed up to approximately $t=650$ μs for both 22 and 80 °C. Later in the event, a low flow stress value of the core material at high temperature allowed for higher back face deflections, resulting in higher shear strains inside the core. For low-temperature experiments, the core showed a maximum shear strain of only 3 % at $t=250$ μs. Following maximum shear, the brittle nature of the core at low temperature caused the formation of core cracks in the high shear stress regions. Once the shear cracks propagated through the thickness, the core was not able to absorb any additional shear stresses transferred from the face, causing face/core delamination in the sandwich composite.

10.4.4.4 Postmortem Analysis

The damage patterns in the sandwich specimens were visually examined and recorded using a high-resolution digital camera and are shown in Fig. 10.38. The shock loading at room temperature shows a permanent deflection of 2.9 mm at the

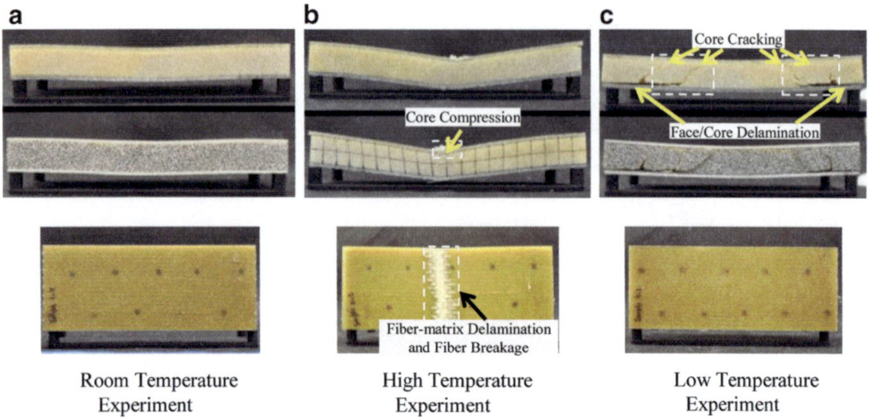

Fig. 10.38 Visual inspection of sandwich specimens after being subjected to blast loading. (**a**) Room-temperature experiment. (**b**) High-temperature experiment. (**c**) Low-temperature experiment

center of front face and there is no evidence of any other damage inside the core and in facesheets.

Excessive fiber-matrix delamination and fiber breakage was observed for high-temperature experiments. This local damage in the facesheet can be attributed to the low heat distortion temperature (82 °C) of the resin used in the preparation of sandwich composites (http://www.derakane.com). Due to the fact that the high-temperature experiments were conducted around the heat distortion temperature of the resin, relatively high center point permanent deflection (~18.5 mm at the center of back face) was observed in the specimen.

When the sandwich specimens were subjected to blast loading under low temperatures, the damage was confined to high shear stress regions and a significant amount of core cracking was observed. The core cracks were generated uniformly through the width of the specimen and propagated completely in the thickness direction. A large amount of skin delamination between the core and the back facesheet was observed. No specific damage in the facesheets and core compression was visible, and the specimen showed a small deflection of 4.9 mm at the center of the back face.

10.5 Summary

It was shown that the number of core layers has an influence on the dynamic response of structures under blast loading. More specifically, by increasing the number of monotonically graded layers, the acoustic wave impedance mismatch between successive layers is reduced. Therefore, the strength of the initial shock wave (stress wave) can be weakened by the time it reaches the back facesheet, resulting in lower back face deflection, in-plane strain, and velocity. More importantly, the overall damage imparted on the structure can be reduced and structural integrity can be maintained. Also, increasing the number of monotonically graded foam core layers, thus introducing more material interfaces, allows for blast wave (stress wave) attenuation through the following mechanisms: (1) stepwise compression of the core (energy dissipation mechanism) and (2) scattering/dispersion of the wave through interface variations. Combining these mechanisms results in lengthened timescales for pressure rises across the samples, allowing for a time-delay of the peak stress arrival, and thus delaying the time of damage initiation.

When the polyurea interlayer is located behind the graded foam core, and in front of the back face (i.e., configuration 2), the core layer arrangement allows for a stepwise compression of the core. Larger compression was visible in the A300 and A500 foam core layers of configuration 2 than configuration 1. This compression lowers the strength of the initial shock wave by the time it reaches the back facesheet and thus the overall deflection, in-plane strain, and velocity were reduced in comparison to the sandwich composite with the polyurea interlayer located behind the front facesheet and in front of the foam core (i.e., configuration 1). Therefore it can be concluded that placing the polyurea interlayer behind the foam core and in front of the back facesheet (configuration 2) improves the blast resistance of the sandwich composite and better maintains structural integrity. The methods used to evaluate the energy as described by Wang et al. (http://www.derakane.com) were implemented and the results analyzed. It was observed that the location of the polyurea layer has a significant positive effect on the response of composite sandwich panels to shock wave loading, both in terms of failure mitigation and energy absorption, if it is placed opposite the blast-receiving side (configuration 2). On the contrary, the presence of polyurea on the blast-receiving side (configuration 1) amplifies the destructive effect of the blast, promoting (rather than mitigating) the failure of the composite sandwich panels.

In-plane compressive loading coupled with the transverse blast loading induced local buckling in the front facesheet of the sandwich composites. Higher levels of compressive loading caused more damage in the front facesheet, larger out-of-plane deflection, and higher in-plane strain on the back facesheet. Consequently, the overall blast resistance of sandwich composites was significantly reduced.

In regard to the environmental effects, the blast performance of sandwich composites at room temperature (22 °C) was superior to its performance at a high temperature (80 °C) and at a low temperature (−40 °C). Specimens did not exhibit shear cracking and face/core delamination. The maximum core compression observed in shock wave experiments was 13 % in the specimens. At elevated temperatures (80 °C) a significant amount of fiber breakage and fiber-matrix delamination occurred in the facesheets due to the reduced stiffness of the facesheet polymer resin at these high temperatures. Specimens also showed the largest front/back face deflection, core compression, in-plane strain, and back face out-of-plane velocity when compared to blast loading experiments at 22 and −40 °C. Specimens exhibited a maximum core compression of 45 %, an in-plane strain of 2.2 %, and a back face velocity reaching 33 m/s at 80 °C. At lower temperatures (−40 °C) the main failure mechanisms observed in the sandwich composites at −40 °C were core cracking and skin delamination. Specimens exhibited larger deflection and in-plane strain as compared to 22 °C. Moreover, specimens demonstrated lower shear strains due to shear crack initiation in the core while the sandwich composites at 22 and 80 °C withstood higher shear strains under blast loading.

Acknowledgments The authors kindly acknowledge the financial support provided by Dr. Yapa D. S. Rajapakse, under the Office of Naval Research (ONR) Grant No. N00014-04-1-0268 and N00014-10-1-0662. The authors acknowledge the support provided by the Department of Homeland Security (DHS) under Cooperative Agreement No. 2008-ST-061-ED0002. Authors also thank Gurit SP Technology and Specialty Products Incorporated (SPI) for providing the material, as well as Dr. Stephen Nolet and TPI Composites for providing the facility for creating the composites used in this study.

References

Amini MR, Isaacs JB, Nemat-Nasser S (2010a) Experimental investigation of response of monolithic and bilayer plates to impulsive loading. Int J Imp Eng 37:82–89

Amini MR, Amirkhizi AV, Nemat-Nasser S (2010b) Numerical modeling of response of monolithic and bilayer plates to impulsive loading. Int J Imp Eng 37:90–102

Amini MR, Isaacs JB, Nemat-Nasser S (2010c) Numerical modeling of effect of polyurea on response of steel plates to impulsive loads in direct pressure-pulse experiments. Mech Mater 42:615–627

Amini MR, Isaacs JB, Nemat-Nasser S (2010d) Investigation of effect of polyurea on response of steel plates to impulsive loads in direct pressure-pulse experiments. Mech Mater 42:628–639

Amirkhizi AV, Isaacs J, McGee J, Nemat-Nasser S (2006) An experimentally-based constitutive model for polyurea, including pressure and temperature effects. Phil Mag 86(36):5847–5866

Apetre NA, Sankar BV, Ambur DR (2006) Low-velocity impact response of sandwich beams with functionally graded core. Int J Solid Struct 43(9):2479–2496

Atas C, Sevim C (2010) On the impact response of sandwich composites with cores of balsa wood and PVC foam. Compos Struct 93:40–48

Bahei-El-Din YA, Dvorak GJ, Fredricksen OJ (2006) A blast-tolerant sandwich plate design with a polyurea interlayer. Int J Solid Struct 43:7644–7658

Benderly D, Putter S (2004) Characterization of the shear/compression failure envelope of Rohacell foam. Polym Test 23(1):51–57

Bibo G, Leicy D, Hogg PJ, Kemp M (1994) High-temperature damage tolerance of carbon fibre-reinforced plastics: part 1: impact characteristics. Composites 25(6):414–424

Choi IH (2008) Low-velocity impact analysis of composite laminates under initial in-plane load. Compos Struct 86:251–257

Dharmasena KP, Wadley HNG, Xue Z, Hutchinson JW (2008) Mechanical response of metallic honeycomb sandwich panel structures to high-intensity dynamic loading. Int J Imp Eng 35(9):1063–1074

Dutta PK (1994) Low-temperature compressive strength of glass-fiber-reinforced polymer composites. J Offshore Mech Arct Eng 116(3):167–172

Erickson MD, Kallmeyer AR, Kellogg KG (2005) Effect of temperature on the low-velocity impact behavior of composite sandwich panels. J Sandwich Struct Mater 7(3):245–264

Fleck NA, Deshpande VS (2004) The resistance of clamped sandwich beams to shock loading. J Appl Mech 71:386–401

Fleck NA, Sridhar I (2002) End compression of sandwich columns. Compos Part A 33:353–359

Gardner N (2012) Novel composite materials and sandwich structures for blast mitigation. PhD Dissertation, University of Rhode Island

Gardner N, Wang E, Shukla A (2012a) Performance of functionally graded sandwich composite beams under shock wave loading. Compos Struct 94(5):1755–1770
Gardner N, Wang E, Kumar P, Shukla A (2012b) Blast mitigation in a sandwich composite using graded core and polyurea interlayer. Exp Mech 52(2):119–133
Gupta S, Shukla A (2012) Blast performance of marines foam core sandwich composites at extreme temperatures. Exp Mech 52(9):1521–1534
Heimbs S, Heller S, Middendorf P, Hahnel F, Weiße J (2009) Low velocity impact on CFRP plates with compressive preload: test and modeling. Int J Imp Eng 36:1182–1193
Hoo Fatt MS, Ouyang X, Dinan RJ (2004) Blast response of walls retrofitted with elastomer coatings. Struct Mater 15:129–138
Hossain MK, Liu QL, O'Toole BJ (2007) Functionally graded foam material system for energy absorption. SAMPE 39th ISTC, Cincinnati, OH, 29 Oct–1 Nov
Jang BZ, Lieu YK, Chang YS, Hwang LR (1987) Cryogenic failure mechanisms of fiber-epoxy composites for energy applications. Polym Compos 8(3):188–198
Kiernan S, Cui L, Gilchrist MD (2009) Propagation of a stress wave through a virtual functionally graded foam. Int J Non-Linear Mech 44:456–468
Kumar P, LeBlanc J, Stargel D, Shukla A (2012) Effect of plate curvature on blast response of aluminum panels. Imp Eng 46:74–85
LeBlanc J, Shukla A, Rousseau C, Bogdanovich A (2007) Shock loading of three-dimensional woven composite materials. Compos Struct 79:344–355
Li Y, Ramesh KT, Chin ESC (2001) Dynamic characterization of layered and graded structures under impulsive loading. Int J Solid Struct 38(34–35):6045–6061
Liang Y, Spuskanyuk AV, Flores SE, Hayhurst DR, Hutchinson JW, McMeeking RM, Evans AG (2007) The response of metallic sandwich panels to water blast. J Appl Mech 74:81–99
Mamalis AG, Manolakos DE, Ioannidis MB, Papapostolou DP (2005) On the crushing response of composite sandwich panels subjected to edgewise compression: experimental. Compos Struct 71:246–257
McShane GJ, Deshpande VS, Fleck NA (2007) The underwater blast resistance of metallic sandwich beams with prismatic lattice cores. J Appl Mech 74:352–364
McShane GJ, Radford DD, Deshpande VS, Fleck NA (2006) The response of clamped sandwich plates with lattice cores subjected to shock loading. Eur J Mech A Solids 25:215–229
Nurick GN, Langdon GS, Chi Y, Jacob N (2009) Behavior of sandwich panels subjected to intense air blast: part 1-experiments. Compos Struct 91(4):433–441
Perl R, O'Rourke R (2001) Terrorist attack on USS cole: background and issues for congress. Emerging technologies: recommendations for counter-terrorism. Institute for Security Technology Studies, Dartmouth College
Perl R (1998) Terrorism: US responses to bombings in Kenya and Tanzania: a new policy direction? Congressional Report. Congressional Research Service, The Library of Congress
Radford DD, Fleck NA, Deshpande VS (2004) The response of clamped sandwich beams subjected to shock loading. Int J Impact Eng 32(6):968–987
Roland CM, Twigg JN, Vu Y, Mott PH (2006) High strain rate mechanical behavior of polyurea. Poly 48(2):574–578
Robb MD, Arnold WS, Marshall IH (1995) The damage tolerance of GRP laminates under biaxial prestress. Compos Struct 32:141–149
Salehi-Khojin A, Mahinfalah M, Bashirzadeh R, Freeman B (2007) Temperature effects on kevlar/hybrid and carbon fiber composite sandwiches under impact loading. Compos Struct 78(2):197–206
Shenghu C, Zhis WU, Xin W (2009) Tensile properties of CFRP and hybrid FRP composites at elevated temperatures. J Compos Mater 43(4):315–330
Shariatmadari MR, English R, Rothwell G (2012) Effects of temperature on the material characteristics of midsole and insole footwear foams subject to quasi-static compressive and shear force loading. Mater Des 37:543–559
http://www.derakane.com.

Sun CT, Chen JK (1985) On the impact of initially stressed composite laminates. J Compos Mater 19:490–504

Sutton M, Orteau JJ, Schreir H (2009) Image correlation for shape, motion and deformation measurements: basic concepts, theory and applications, 1st edn. Springer. doi:10.1007/978-0-387-78747-3, Print ISBN 978-0-387-78746-6, Online ISBN 978-0-387-78747-3

Tagarielli VL, Deshpande VS, Fleck NA (2008) The high strain rate response of PVC foams and end-grain balsa wood. Compos Part B 39:83–91

Tekalur SA, Shukla A, Shivakumar K (2008) Blast resistance of polyurea based layered composite materials. Compos Struct 84:271–281

Tiwari V, Sutton MA, McNeill SR, Xu S, Deng X, Fourney WL, Bretall D (2009) Application of 3D image correlation for full-field transient plate deformation measurements during blast loading. Int J Imp Eng 36:862–874

Tilbrook MT, Deshpande VS, Fleck NA (2006) The impulsive response of sandwich beams: analytical and numerical investigation of regimes of behavior. J Mech Phys Sol 54:2242–2280

Wang Z, Jing L, Ning J, Zhao L (2011) The structural response of clamped sandwich beams subjected to impact loading. Compos Struct 93:1300–1308

Wang E, Gardner N, Shukla A (2009) The blast resistance of sandwich composites with stepwise graded cores. Int J Solid Struct 46:3492–3502

Whittingham B, Marshall IM, Mitrevski T, Jones R (2004) The response of composite structures with pre-stress subject to low velocity impact damage. Compos Struct 66:685–698

Wang E, Shukla A (2012) Blast performance of sandwich composites with in-plane compressive loading. Exp Mech 52(1):49–58

http://www.gurit.com

http://www.specialty-products.com

Wright J (1961) Shock tubes. Wiley, New York

http://www.correlatedsolutions.com

Wang E, Gardner N, Gupta S, Shukla A (2012) Fluid-structure-interaction and its effect on the performance of composite structures under air-blast loading. Int J Multi-Phys 6(3):219–239

Wang E (2010) Blast resistance and energy mitigation behavior of sandwich composite materials. PhD Dissertation, University of Rhode Island, KingstonXue Z, Hutchinson JW (2003) Preliminary assessment of sandwich plates subject to blast loads. Int J Mech Sci 45:687–705

Yi J, Boyce MC, Lee GF, Balizer E (2005) Large deformation rate-dependent stress–strain behavior of polyurea and polyurethanes. Poly 47(1):319–329

Zhu F, Zhao L, Lu G, Wang Z (2008) Deformation and failure of blast loaded metallic sandwich panels—experimental investigations. Int J Imp Eng 35(8):937–951

Chapter 11
Progressive Collapse Resistance of Reinforced Concrete Structures

M. Sasani

Abstract Experimental and analytical studies on a two-span fixed-end RC beam, designed satisfying the ACI structural integrity requirements and subjected to increasing vertical displacement at the middle, demonstrate the development of compressive and tensile membrane actions. Progressive collapse resistance of an actual 11-story reinforced concrete structure following a severe initial damage is studied experimentally and analytically. The initial damage was caused by simultaneous explosion (removal) of four first floor neighboring columns and two second floor perimeter deep beam segments. The structure resisted progressive collapse with a maximum permanent vertical displacement at the top of the exploded columns of only about 56 mm (2.2 in.). Beam growth and in turn the development of the beam axial compressive force are modeled and discussed. It is demonstrated that such axial compressive force can significantly affect progressive collapse resistance of the structure. In the actual 6-story structure discussed in this chapter, the development of bidirectional Vierendeel (frame) action is identified as the dominant mechanism in redistribution of loads. This is because of the existence of moment connections and the interaction between beams with columns and infill walls; critical beams deform in double curvature and provide the shear resistance needed to redistribute the loads.

11.1 Introduction

Manmade or natural hazards may subject structures to conditions that lead to progressive collapse. Progressive collapse is defined as the spread of an initial local failure from element to element eventually resulting in collapse of an entire structure or a disproportionately large part of it (ASCE/SEI 7 2010). Following the

M. Sasani (✉)
Department of Civil and Environmental Engineering, 421 Snell Engineering Center, Northeastern University, 360 Huntington Avenue, Boston, MA 02115, USA
e-mail: sasani@neu.edu

A. Shukla et al. (eds.), *Blast Mitigation: Experimental and Numerical Studies*, DOI 10.1007/978-1-4614-7267-4_11, approaches proposed by Ellingwood and Leyendecker (1978), ASCE/SEI 7 (2010) defines two general methods for structural design of buildings to mitigate damage due to progressive collapse: indirect and direct design methods.

Indirect Design: Incorporates implicit consideration of resistance to progressive collapse through the provision of minimum levels of strength, continuity, and ductility.

Direct Design: Incorporates explicit consideration of resistance to progressive collapse through two methods. One is the Alternative Path Method in which local failure is allowed to occur, but seeks to provide alternative load paths so that the damage is absorbed and major collapse is averted. The other method is the Specific Local Resistance Method that seeks to provide strength to resist failure.

In order to mitigate progressive collapse of structures in the event of damage to a supporting element or an abnormal loading, general building codes and standards have provisions to enhance structural integrity (ACI 318 2011; ASCE/SEI 7 2010). Such provisions can be categorized as an indirect design approach in which implicit consideration is given to improve progressive collapse resistance of structures. Breen (1975) has discussed that improved structural integrity is obtained by provision of integral ties throughout the structure (indirect design) and that the amount of ties can be determined from considerations on debris loading and the amount of damage to be tolerated without determination of the magnitude of the explosive or other abnormal load. Although the indirect design method can reduce the risk of progressive collapse (FEMA 277 1996; Corley 2004; Sozen et al. 1998; Corley et al. 1998), estimation of post-failure performance of structures designed based on such a method is not readily possible.

Direct design is utilized in the design provisions specifically developed for progressive collapse analysis of structures (GSA 2003; DOD 2010). Following the Alfred P. Murrah Federal Building attack in 1995 and the establishment of the Interagency Security Committee (ISC) for development of construction standards for federal buildings subject to terrorist attack, the General Services Administration (GSA 2003) published guidelines in 2000 and 2003 for progressive collapse analysis and design of structures. The GSA (2003) guidelines are primarily based on the Alternative Path Method and mandates instantaneous removal of one load-bearing element with different scenarios, at the initiation of damage. DOD (2010) guidelines for progressive collapse analysis provide two design methods: one employs the Tie Force Method (indirect design), and the other employs the APM (direct design). GSA (2003) and DOD (2010) provide threat-independent methods and use different scenarios for initiation of local failure to examine potential progressive collapse of structures. One of these scenarios is the instantaneous removal of a ground floor column located near the middle of exterior frames. Note that because it is impossible to account for every possible threat, the design philosophy in the above-mentioned guidelines is to allow local damage, but to mitigate the occurrence of progressive collapse.

Based on GSA (2003), in a linear static analysis the following load combination is considered:

$$\text{Load} = 2\left(\text{DL} + 0.25\text{LL}\right) \quad (11.1)$$

where DL and LL are dead load and live load, respectively. For an elastic dynamic or nonlinear analysis, the coefficient 2 in (11.1) is removed. The acceptance criteria for a linear analysis are similar to the criteria in FEMA 356 (2000), which are based on internal force demand–capacity ratios (DCR). GSA (2003) also recommends application of a nonlinear analysis, particularly for buildings having more than ten stories above the grade. For such analysis, acceptance criteria based on rotation or rotation ductility as given in DOD (2001) are provided.

Potential progressive collapse of structures due to terrorist attacks (or accidental loads) and the dynamic redistribution of loads that follows need to be evaluated at the system level.

From an analytical point of view, progressive collapse occurs when a structure has its load pattern or boundary conditions changed due to failure of some structural elements or supports such that other elements are loaded beyond their capacity and fail (Krauthammer et al. 2003). There are a few numerical examples of the application of computational schemes such as the Distinct Element Method (Cundall 1971) and the Discontinuous Deformation Analysis (Shi and Goodman 1984) in collapse analysis of structures (Tosaka et al. 1988; Yarimer 1989; Itoh et al. 1994). Another method for collapse analysis of structures is the Modified Distinct Element Method (Meguro and Hakuno 1994), which, compared to the Finite Element Method, is less accurate for small displacements (Meguro and Tagel-Din 2002). Other methods such as Adaptive Shifted Integration (Isobe and Toi 2000) and External Criteria Screening (Choi and Krauthammer 2003) have been proposed. Applied Element Method (Meguro and Tagel-Din 2002) is another method for collapse analysis of structures which is also reliable for small deformations.

11.2 Progressive Collapse Resistance of Beams in a Reinforced Concrete Frame Structure

In order to study progressive collapse resistance of reinforced concrete (RC) structures at the system level, it is necessary to first understand the behavior of RC elements subjected to large deformations at the element level. In particular, evaluating compressive and tensile (catenary) membrane actions in RC beams requires first an understanding of the behavior at the element level, before studying membrane action in RC beams as part of frame systems. Experimental and analytical studies were carried out to study the behavior of a 3/8th scaled model of a continuous two-span RC beam (as part of the exterior frame of an RC building) subjected to an imposed downward displacement at its center (Bazan 2008).

A seven-story building with ordinary RC frames is designed. The building is assumed to be located on a site class C, very dense soil, and soft rock (IBC 2003) in Atlanta, GA. The plan of the building is shown in Fig. 11.1. The floor is a one-way joist system in the transverse direction. The span in the transverse direction is set equal to 30 ft (9.14 m) in order to have an economical joist floor system (Alsamsam and Kamara 2004). The total depth of the floor system is 20 in. with a

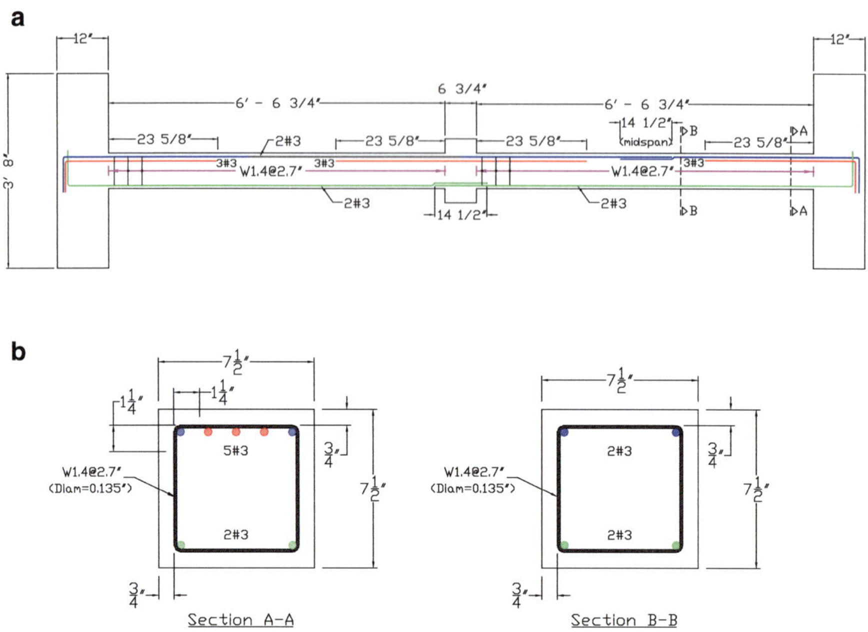

Fig. 11.1 (**a**) Reinforcement detail of 3/8th scaled RC beam specimen. (**b**) Reinforcement detail of RC beam specimen cross section (1 in = 25.4 mm) (Bazan 2008)

solid slab of 4.5 in. The depth of all beams is equal to 20 in. (508 mm; equal to the depth of joist floor system) to minimize the formwork cost. Reinforcement of grade 60 ksi (413 MPa) is used along with concrete compressive strength of 4 ksi (27 MPa). In the design of the building the integrity requirements are satisfied (ACI 318 2011). In order to examine the effects of splices on the development of catenary action in beams and progressive collapse of structures, the longitudinal reinforcements are spliced.

The RC beam shown in Fig. 11.1 is assumed to be part of the RC building described above where an exterior column, supporting the beam at the center, is instantaneously removed by explosion or impact. The removal of such column leads to a dynamic redistribution of loads to the columns neighboring the removed column. This load redistribution is carried out through the floor system and beams bridging over the removed column at all floors. The beams bridging over the removed column can experience large displacements which can lead to distribution of loads through a mechanism of tensile membrane or catenary action.

The behavior of the bridging beam will depend not only on the beam mechanical characteristics, such as dimensions, amount of reinforcement, and mechanical properties of steel and concrete, but also on the rotational and translational stiffness of the beam boundary conditions which are provided by the slab and the beams and columns of the adjacent spans. In order to experimentally study first the compressive and tensile membrane actions of RC beams, without the effects of the variations

Fig. 11.2 RC beam experimental force–displacement response (Bazan 2008)

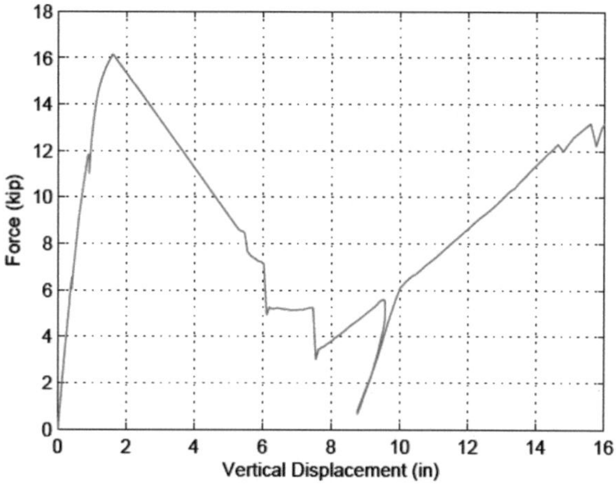

in the rotational and translational stiffness of the beam boundary conditions, the RC beam model is tested under fixed end conditions.

A vertical actuator attached to the center stub of the RC beam imposed a downward vertical displacement while the applied force, the corresponding vertical displacement, and the developed strains and deformations were monitored. Figure 11.2 shows the resisting force versus the applied vertical displacement at the center of the RC beam. The resisting force reaches a maximum value of about 16.2 kip (72 kN) at a vertical displacement of about 1.62 in. (41 mm). After the peak and subsequent drop in the force–displacement relationship, two sudden drops of the resisting force are observed. These sudden drops correspond to the fracture of the two bottom reinforcing bars on the right side of the RC beam center stub.

Strain measurements indicate that yielding of the bottom bars at the center of the beam initiated first on one side of the center stub concentrating subsequent damage and rotation at this location. The two bottom bars to the right side of the center stub reached the yield strain at 0.4 and 0.6 in. (10.2 and 15.2 mm) of vertical displacement, respectively. After the first bar fracture at 6 in. (152 mm) of vertical displacement, the resisting force is nearly constant up to the second bar fracture at 7.5 in. (190 mm). However, after the second bar fracture, the resisting force provided by catenary or tensile membrane action (primarily developed by the top continuous bars) increases at an almost constant rate of 1.3 kip/in. (0.23 kN/mm). After the fracture of the second bottom bar, the resisting force drops to 3 kip (18 % of the peak force) and increases due to catenary action up to about 13 kip (72 kN; 80 % of the peak force) at a vertical displacement of 16 in. (406 mm). One important observation is the fact that the transverse reinforcement, provided to satisfy ACI Integrity Requirements, allowed the top continuous reinforcement to develop catenary or cable-like action without tearing up the concrete cover on top of the reinforcement.

The test was stopped at 9.5 in. (241 mm) of vertical displacement due to limitation of the actuator maximum stroke. The beam was unloaded by disconnecting the actuator from the beam center stub and retracting the actuator. A steel block was attached,

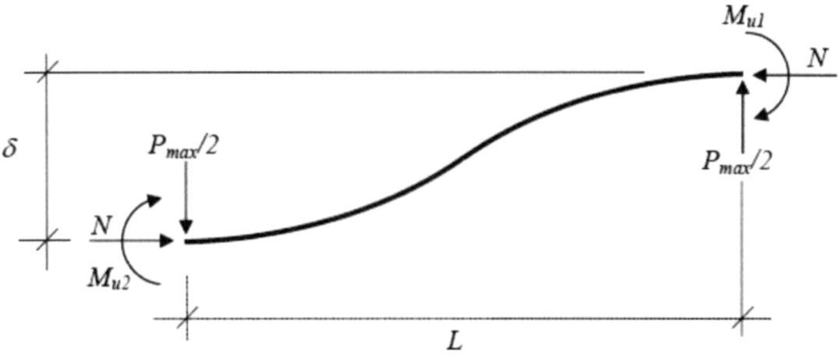

Fig. 11.3 Free body diagram of right half of beam model

as an extension, between the actuator and the beam center stub in order to continue imposing vertical displacement on the RC beam. The RC beam was reloaded and the vertical displacement was increased.

In order to study the behavior of the RC beam under the applied vertical displacement during the test, we will evaluate the force–displacement response shown in Fig. 11.2. Let us examine the behavior of the beam up to the first peak. As the flexural cracks form and tensile rebars yield, the mid-height of the beam section will be under tensile strain. As a result, the beam has a tendency to grow in length. This phenomenon is known as beam growth (Fenwick and Megget 1993; Kim et al. 2004). Because the beam ends were constrained, axial compressive force developed in the beam.

In order to examine the effects of the axial compressive forces on the behavior of the RC beam specimen, we can analytically estimate the beam section flexural strengths and in turn the maximum load carrying capacity of the RC beam considering no axial force and compare it with the maximum applied load during the experiment. Section analyses are carried out for the beam end section (section AA in Fig. 11.1) subjected to negative and positive bending moments. The stress–strain relationships of steel and concrete for section analysis are obtained from material tests. Considering the maximum negative moment at the ends of the two-span beam shown at the ends and the maximum positive moment of the beam section at the face of the center stub, the maximum vertical force than can be applied to the center stub is found to be only about 10 kip (44 kN). This value is by far less than the force recorded during the experiment.

One can account for the axial compressive force developed in the beam as a result of the beam tendency to grow in estimating the vertical load carrying capacity of the beam. This is shown in Fig. 11.3. The axial compressive force has counteracting effects on the beam's vertical load carrying capacity. On the one hand, as the axial compressive force increases, the beam moment capacity increases (and later decreases) as expected from the axial force–moment interaction diagrams of the beam section for positive and negative moments. On the other hand, the so-called P-Δ effect reduces the beam vertical load carrying capacity. Figure 11.4 shows the load carrying capacity of the beam as a function of the axial compressive force in

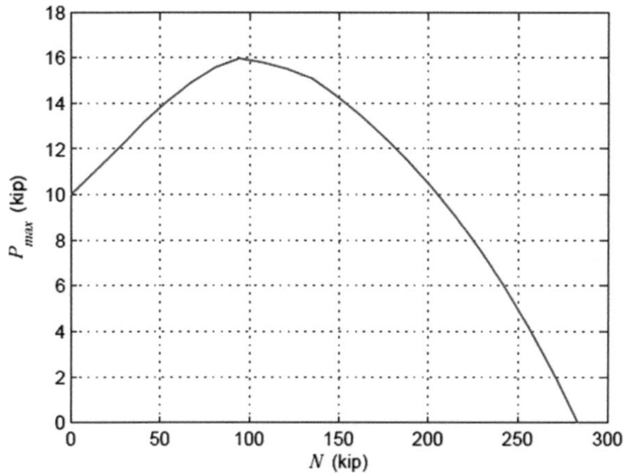

Fig. 11.4 Resulting beam flexural capacity P_{max} for different levels of axial compressive force N(1 kip = 4.44 kN)

the beam. As can be seen, considering the beam axial compressive force, the beam can carry up to about 16 kip (72 kN). The slightly larger load of 16.2 kip (71 kN) recorded during the experiment can be in part due to the confining effects of the transverse reinforcement, which are neglected in this analysis.

In actual structures, the beam end constraints are not complete, still axial compressive forces will develop as a result of the beam's tendency to grow, which will be discussed later in this chapter. Ignoring this effect can significantly underestimate progressive collapse resistance of RC structures.

11.3 Progressive Collapse Resistance of Actual RC Structures

In this section, progressive collapse resistance of actual structures is evaluated.

11.3.1 An 11-Story Structure in Houston, TX

Analytical and experimental behavior of reinforced concrete structures experiencing removal of some load-bearing elements has previously been evaluated (Sasani et al. 2007; Sasani and Sagiroglu 2010). In this section, progressive collapse resistance of the Crowne Plaza hotel, an 11-story structure above the ground located in Houston, TX, is studied (Sasani et al. 2011). The hotel was demolished by implosion after the experiment was concluded. Figure 11.5 shows a view of the hotel, which had an ordinary RC frame structure. The height of the first story was about 30 ft (9.14 m).

Fig. 11.5 Pictures of 11-story building in Houston, TX (aerial image of Houston, TX, 2006, reprinted with permission from the Sanborn Library, LLC, © 2007 Google)

The height of the stories above was 10 ft (3.05 m). Figure 11.6 shows the plan of the second floor of the structure. The second floor was a 6 in. (0.15 m) thick two-way slab system supported by 81 in. (2.06 m) deep beams running in both longitudinal and transverse directions. The width of all the longitudinal beams was 36 in. (0.914 m). The width of transverse beams between axes 3 and 8 was 54 in. (1.372 m). The cross section of the columns in the first floor between axes 4 and 7 was 36 in. × 36 in. (0.914 m × 0.914 m). All columns on floors above were sitting on the longitudinal beams in the second floor. The stories above were made of 6 in. (0.15 m) thick two-way flat slabs supported by columns, which were 11 in. (0.279 m) in the longitudinal and 13 in. (0.330 m) to 16 in. (0.406 m) in the transverse direction of the structure. The structure had an elevator case at the central part of the building and two staircases at two corners.

In order to analytically evaluate the potential progressive collapse of the structure, the mechanical properties of the steel and concrete are necessary. Compression and tension tests on concrete samples and tension tests on steel samples from the building were conducted. Through compressive tests on two concrete samples obtained from the building, the concrete compressive strength measured was $f'_c = 3.9$ ksi (26.9 MPa) for a standard concrete cylinder. From flexural tests based on ASTM C 293, the modulus of rupture was determined to be approximately $f_r = 500$ psi (3.4 MPa). The modulus of elasticity of the concrete was estimated to be 3,600 ksi (24,800 MPa). Based on tension tests of two steel samples, the yield and ultimate stresses for the beam and slab flexural reinforcement were found to be $f_y = 75$ ksi (517 MPa) and $f_u = 110$ ksi (758 MPa), respectively. The steel ultimate tensile strain was measured at 0.1. The modulus of elasticity of steel was set equal to 29,000 ksi (200,000 MPa).

In order to evaluate the effects of axial compressive force developed in beams in actual structures, the recorded response of the Crowne Plaza hotel following the

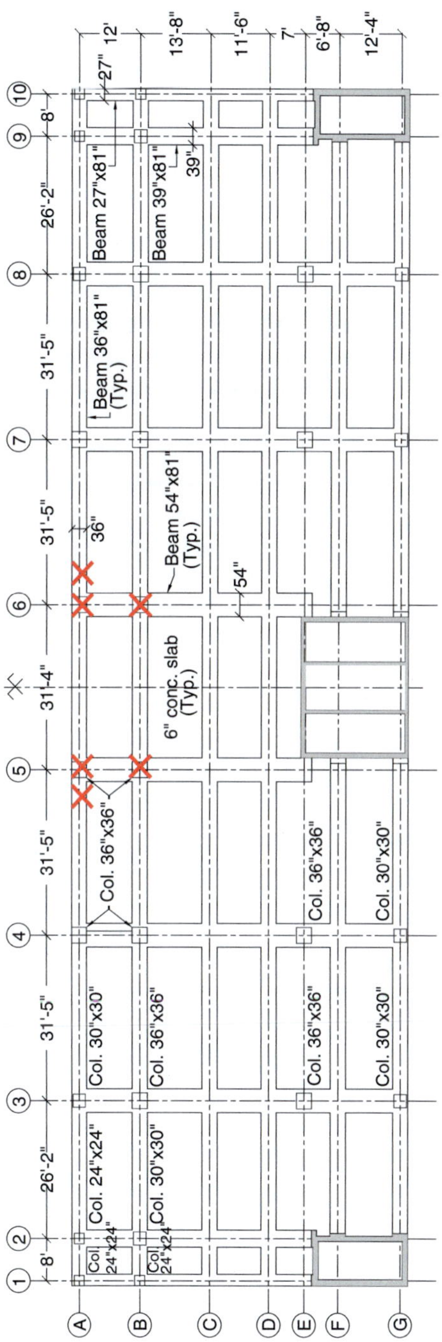

Fig. 11.6 Floor plan with locations of explosions modeling initial damage shown by red crosses (1 in. = 25.4 mm)

Fig. 11.7 Localized plastic hinges of length *Lp*, modeling (*left*) flexural deformation, and (*right*) flexural and axial deformations

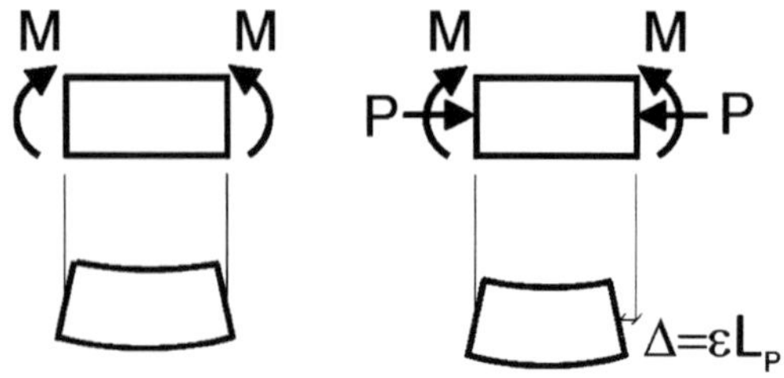

initial damage shown in Fig. 11.6 is compared with analytical results obtained from two different models of the structure. The initial damage, as shown in Fig. 11.6, was caused in part by simultaneous explosion (removal) of four adjacent columns, two of which were exterior columns. In addition to the removal of these columns, two beam sections were also exploded.

A three-dimensional finite element model of the structure is developed. Geometric nonlinearity is accounted for. Beam–column elements with localized plastic hinges are used to model material nonlinearity in beams and columns. In order to account for the material nonlinearity of the slab elements between axes 4 and 7 and axes *A* and *D* in longitudinal and transverse directions, respectively (see Fig. 11.6), the slab is modeled by a two-directional grid made of beam elements with rectangular cross sections having localized plastic hinges. The other portions of the slab in the floor, where the behavior is expected to be linear, are modeled using elastic shell elements.

The beams in the second floor are modeled with T- and L-sections and the slab elements are modeled with rectangular sections. The effective flange width on each side of beams was set equal to four times the slab thickness, 24 in. (0.610 m) (ACI 318 2011). For rectangular sections modeling the slab, because of the difference between shear flow in the slab and that in the beam elements modeling the slab, the torsional stiffness was set to ½ that of gross sections (MacLeod 1990). To find a reasonable value for the torsional stiffness of the beam elements, a run with the full torsional stiffness was implemented and for all the beam elements with torque moments larger than the cracking moment, the cracked torsional stiffness is used. A mass proportional damping ratio of $\xi = 0.05$ is used. Two variations of the building model are considered here. In these models two types of plastic hinges are used. These plastic hinges are described below.

Figure 11.7a shows a localized plastic hinge modeling flexural plastic deformation of a beam element over a length LP. Such a plastic hinge is called moment plastic hinge (MPH) in this chapter and is sometimes used in progressive collapse analysis of structures. The beam axial plastic deformation in this case is not accounted for. Furthermore, such a plastic hinge is used in conjunction with a linear beam element; therefore the effect of cracking on axial deformation of the beam is not accounted for either. The behavior of an MPH up to yield point is usually rigid, which means that the hinge has no effect on the response of the element up to the yield point

(SAP2000 2009). After the yield point however, the hinge provides plastic deformation to the element (in the vicinity of the hinge). So, the behavior of moment plastic hinges is called rigid-plastic.

The moment–rotation relationship for an MPH is obtained under a constant axial force (usually zero) and the interaction of the axial force and the moment (P–M interaction) is not considered. To account for the effect of the axial force and moment interaction, the PMM (the axial force and the bending moment in two directions) plastic hinges are also available in SAP2000. In the PMM plastic hinges, the P–M interaction is partially considered at the force level, that is, the deformation interaction is not accounted for.

RC beams experiencing cracking and rebar yielding, however, tend to elongate. The floor system and the structure as a whole constrain this tendency, which leads to the development of axial compressive force in the beam. Effects of axial deformation and in turn the axial compressive force in beams and slabs on progressive collapse resistance of structures will be neglected if MPHs or PMMs are used. Figure 11.7b shows another type of plastic hinge, called moment-axial plastic hinge, MAPH, which accounts for the interaction between the flexural and axial deformations. If an MAPH is used to define the hinge section, the section cracking can be accounted for. In SAP2000 a fiber plastic hinge is available, which is similar to an MAPH. In fiber plastic hinges, the geometry and the material properties (stress–strain curves) of the materials constituting the section are defined in the program. At each step of nonlinear dynamic analysis, the program uses the material constitutive laws (stress–strain relationships) to obtain the section flexural and axial forces and stiffness and in turn those of the elements to carry out the analysis using equilibrium and compatibility equations. Therefore, the interaction between the axial force and the moments is accounted for in terms of both force and deformation.

In MAPHs, because the program uses stress–strain relationships of the materials to carry out section analysis and in turn obtain the hinge behavior, the hinge is not rigid-plastic. That is, the so-called elastic response is included in the hinge deformations (axial and rotational). The program uses not only the elastic beam formulation but also the hinge characteristics. Therefore, the flexibility of the region represented by the hinge (plastic hinge length) is double counted. In order to resolve this issue, the element axial and flexural stiffness is practically set to infinity (large numbers), which in effect diminishes the effect of the elastic element flexibility. Plastic hinges are used in the elements wherever was necessary. The lengths of the hinges in beams are set equal to the distance between the hinges (less than half of the beam depth). For the slab elements the length of the hinges is set equal to 3 in. (0.076 m) (half of the thickness of the slab).

Figure 11.8 compares the vertical displacements at the top of the first floor removed column B-6 obtained from the two models, one with MPH and the other with MAPH. The permanent vertical displacement (at the end of vibration) of the second floor at the locations of removed column B-6 using MAPHs is about 1.8 in. (46 mm). The calculated displacement in the MPH model is approximately 3.2 in. (81 mm). The estimated vertical displacements using MPHs are approximately

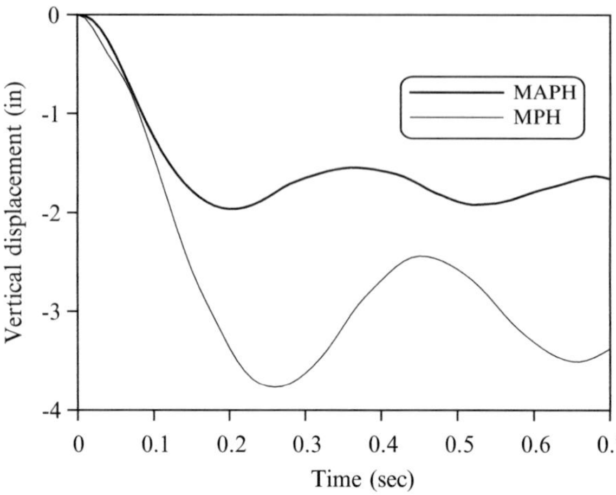

Fig. 11.8 Vertical displacement versus time of joint B-6 on second floor (1 in. = 25.4 mm)

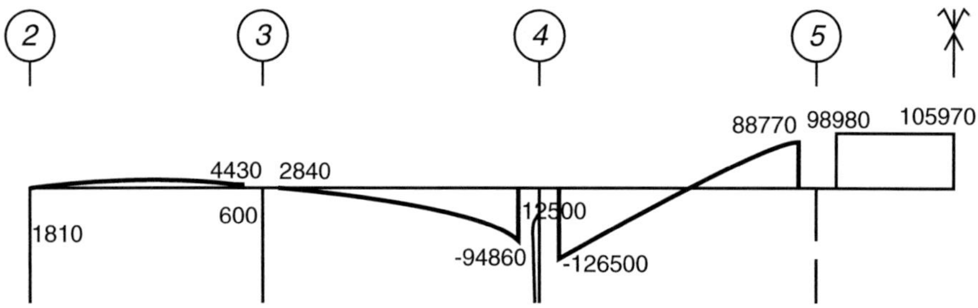

Fig. 11.9 Bending moment diagram in beam of axis *B* (kip in.) after column removal at peak vertical displacement (1 kip in. = 0.113 kN m)

80 % more than that obtained from the model with MAPHs. The vertical displacements at the top of the removed columns were measured at the site using several displacement sensors. The measured permanent vertical displacement in the second floor at the locations of removed column B-6 was about 1.7 in. (43 mm), which is in reasonable agreement with the analytically obtained displacements using MAPHs. As can be seen, the model with MPHs significantly overestimates the permanent vertical displacements of the structure in relation to experimentally obtained data. Therefore it is concluded that the structural model with MPHs, which ignores the axial deformations of beams, does not properly represent the structural behavior.

Figure 11.9 demonstrates the bending moment diagram of the beam in axis *B* after column removal (at the peak displacement of the midpoint of the beam). After column removal, the central three-span beams in axis *B* act like a single-span beam. The maximum negative moment at the face of axis 4 changes from 11,940 kip in. (1,349 kN m) under gravity loads before column removal to 126,500 kip in.

(14,292 kN m) after column removal. The maximum positive moment at the midspan changed from 5,340 kip in. (603 kN m) to 105,970 kip in. (11,972 kN m) after column removal. These values represent about 11 and 20 times increase in the negative (face of axis 4) and positive (midpoint) moments of the beam, respectively. The moment-curvature of the section under negative moment and subjected to zero axial load shows all the tensile rebars of the beam section yield under the action of My = 95,000 kip in. (10,733 kN m). However, the analysis results show that the tensile rebar yielding occurs under a moment of about 131,000 kip in. (14,800 kN m), which is 38 % more. This large increase in the yielding moment of the section is due to the effect of an axial compressive force equal to 730 kip (3,244 kN) on the section.

11.3.2 A 6-Story Structure in San Diego, CA

In this section progressive collapse resistance of Hotel San Diego, which was a 6-story structure above the ground located in San Diego, CA, is studied (Sasani 2008; Sasani and Sagiroglu 2008). The hotel was demolished by implosion after the experiment was concluded. Figure 11.10 shows the first floor of the hotel.

An Annex, including the part of the structure between axes *A* and *H*, and between axes 7 and 13, including east and west wings (see Fig. 11.10), was added to the original hotel. In the experimental evaluation of progressive collapse resistance of the hotel, the east wing between axes *A* and *G* and 7 and 9 was studied. Columns A7 and A8 in the first (ground) floor were simultaneously exploded (suddenly removed) prior to the building demolition by implosion. Figure 11.11 shows the south elevation of the hotel and the exploded columns, in which the two exploded columns are also identified.

In case of an explosion caused by manmade hazards, the air-blast shock wave is the primary damage mechanism. Damage due to the air-blast may be divided into direct air-blast effects and progressive collapse. Direct air-blast effects are damage caused by the high-intensity pressure that induces localized failure to nearby buildings. Such localized failure depends on the size of the explosion, its distance to the explosion, and the building characteristics. Damaged structures may confine the damage to the initially affected zone; otherwise the collapse may propagate. The aim of this chapter is to study potential progressive collapse of the structure following initial local failure without the direct air-blast effects.

In order to estimate the 3-dimensional displacement components of beam–column joint A7 on the second floor, three potentiometers were used to measure the changes in distances R_1, R_2, and R_3 (see Fig. 11.12). Each of these three new distances determines (independently from each other) a sphere with radius R_i ($i = 1, 2, 3$) in which joint A7 is located after column removals.

Figure 11.12 schematically shows the location of the potentiometers. The three-dimensional displacement of joint A7 was estimated using the potentiometer recordings. Figure 11.13 shows the vertical displacement history of the joint. Negative

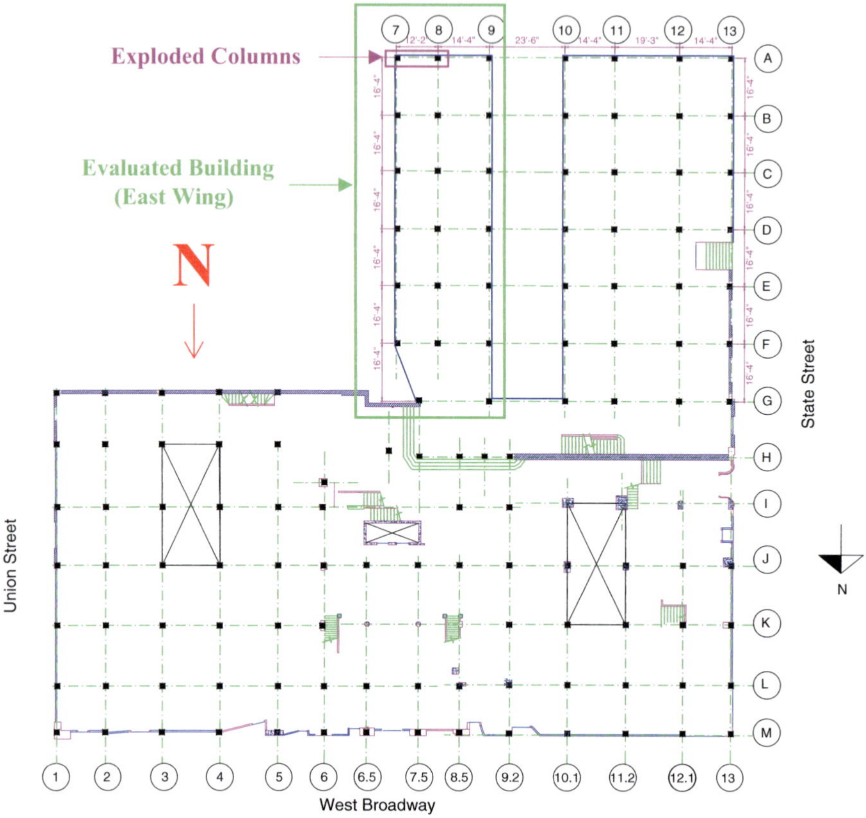

Fig. 11.10 Plan of first floor of Hotel San Diego

values show downward displacements. The columns were removed at time 0.0 s. The figure displays a maximum downward vertical displacement of 0.242 in. (6.1 mm) at 0.069 s. A permanent vertical displacement of about 0.242 in. (6.1 mm) at the end of vibration is recorded, which is equal to the maximum displacement. The horizontal displacements in the transverse (east–west) and longitudinal (north–south) directions at the time of the maximum displacement are estimated at 0.036 in. (0.91 mm) and 0.015 in. (0.38 mm), respectively, which are as expected considerably smaller than the maximum vertical displacement.

The analytically estimated deformed shape of the structure at the maximum vertical displacement is shown in Fig. 11.14 with a magnification factor of 200. The experimentally measured deformed shape over the end regions of beams A7–A8 is presented in the figure by solid curves. A total of seven potentiometers were located at the top and bottom of the end regions of the second floor beams A7–A8, which was one of the most critical elements in the redistribution of the loads. The beam top and corresponding bottom potentiometer recordings were used to calculate the rotations between the sections where the potentiometer ends were connected. This was done by first finding the difference between the recorded

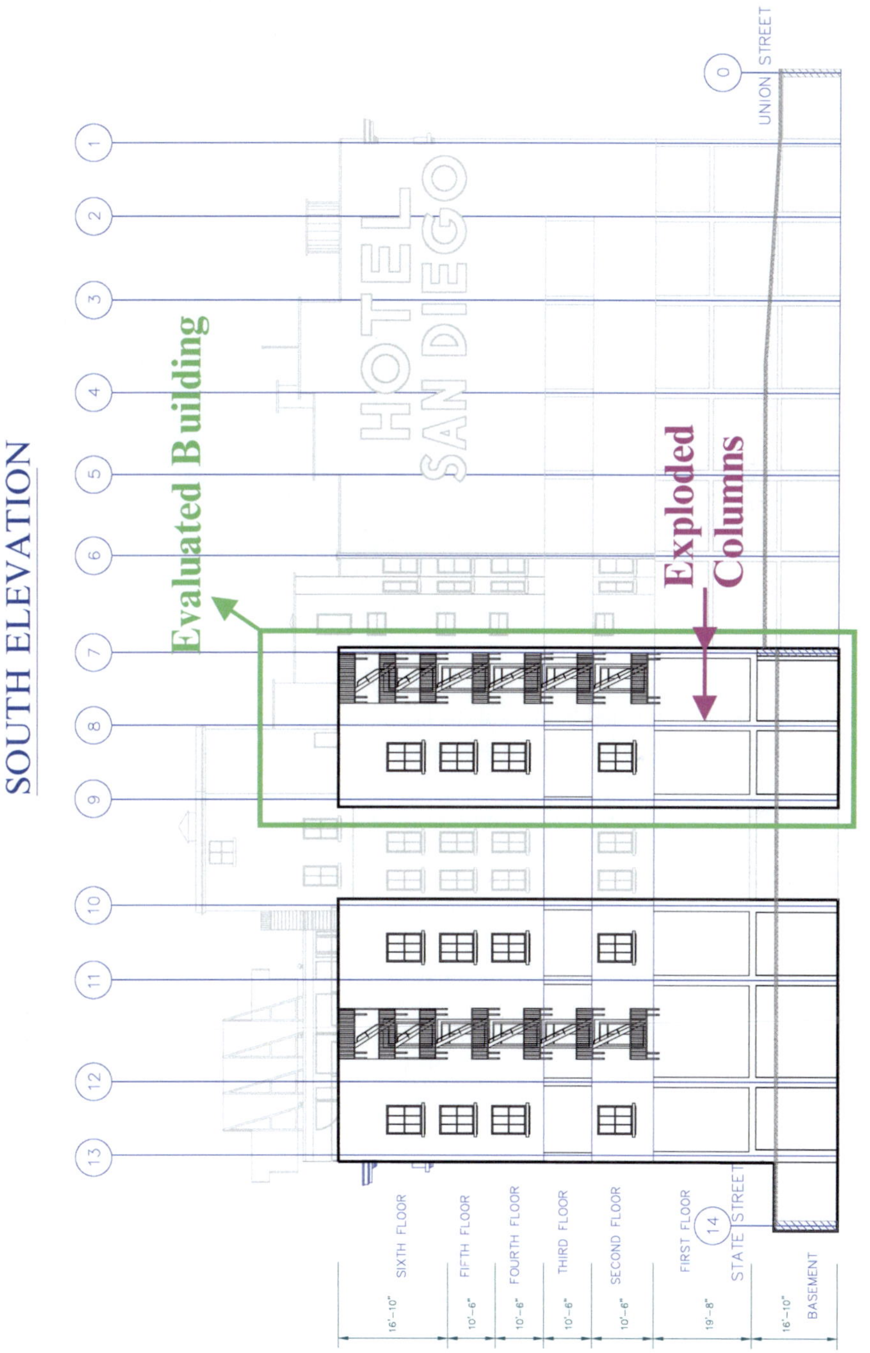

Fig. 11.11 Hotel San Diego

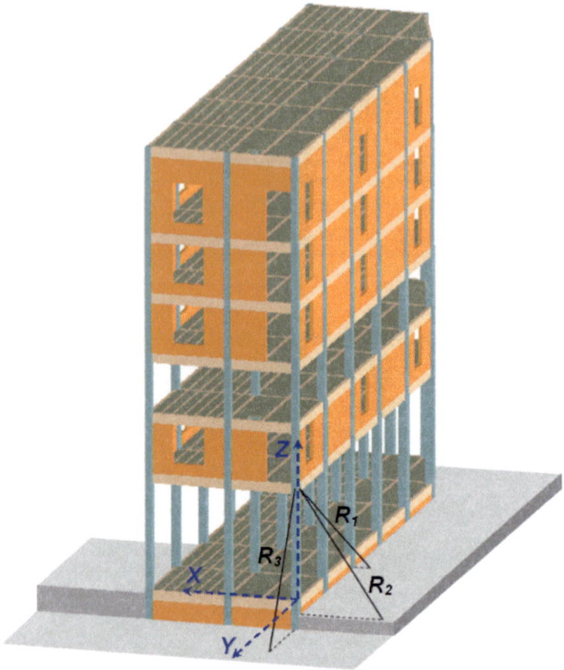

Fig. 11.12 Building model and location of three potentiometers to measure 3D displacements of joint A7 on second floor

deformations at the top and bottom of the beam, and then dividing the value by the distance (along the height of the beam section) between the two potentiometers. The expected deformed shapes between the measured end regions of the second floor beams are shown by dashed lines. As can be seen in the figure, the analytically estimated deformed shape of the beam is in good agreement with the experimentally obtained deformed shape.

Figure 11.14 demonstrates the development of Vierendeel (frame) action in this structure. Vierendeel action can be characterized by relative vertical displacement between beam ends and the corresponding double-curvature deformations of beams. In this structure, such a deformation pattern is developed because of the existence of moment connections and the interaction between beams with columns and infill walls. The double-curvature deformed shape of beams provides the shear force needed to redistribute the loads, following the column removal.

11.4 Conclusions

The experimental and analytical evaluation of progressive collapse resistance of beams in an RC frame structure and the corresponding force–displacement relationship demonstrate three main stages of the response of the beams following the loss of a column. These stages are the initial ascending branch, the descending branch following flexural-axial failure, and the final ascending branch due to catenary

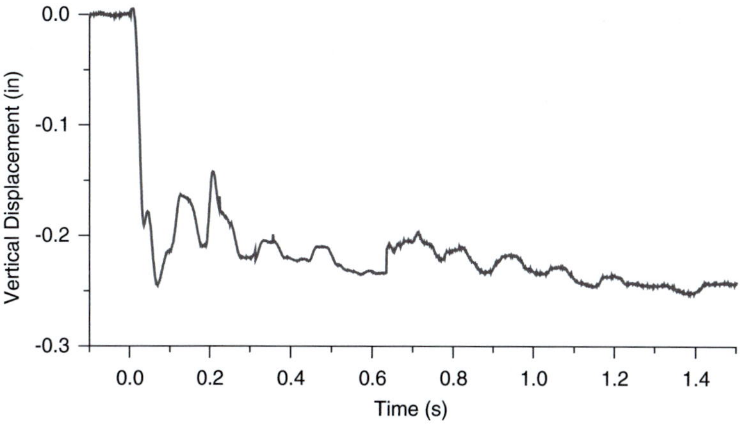

Fig. 11.13 Measured vertical displacement of joint A7 in second floor

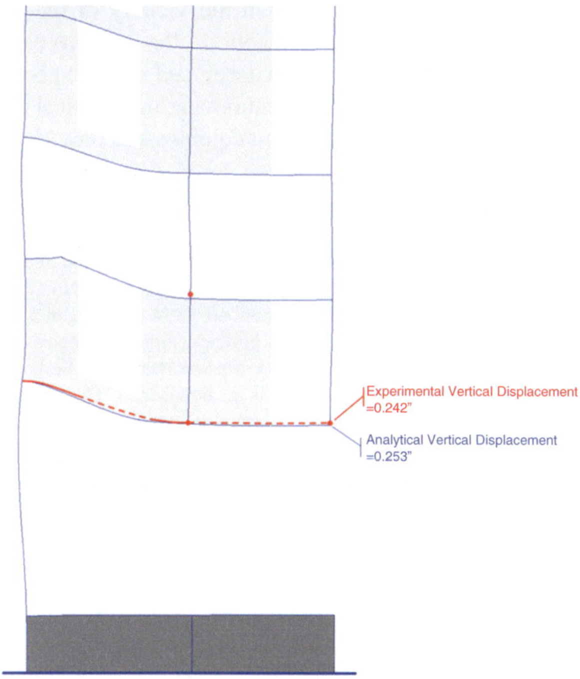

Fig. 11.14 Analytical deformed shape of structure (transverse direction, axis *A*). Experimentally estimated deformed shape is superimposed on second floor. Experimentally estimated location of joint A8 in third floor is also shown

action. Analytical results show that the maximum load resistance capacity of the RC beam is through the flexural and compressive force interaction. It is demonstrated here that accounting for the axial compressive force in the beam capacity is important for reliable estimation of the beam strength.

RC beam growth (tendency to elongate longitudinally) as the beam deforms, cracks, and the steel reinforcing bars yield and its effect on structural response are discussed. Using laboratory test results, experimental data is presented, which clearly demonstrates the growth of a beam as it deforms to bridge over a lost column. It is shown that accounting for the effects of beam growth can be significantly important in evaluating progressive resistance of the structural system following loss of load-bearing elements. As the beam bridging over lost columns deforms and tends to grow, if the structural system is capable of imposing constraint to the beam growth, axial compressive force develops in the beam. The beam axial compressive force enhances its flexural strength and in turn its resistance to progressive collapse.

Vierendeel (frame) action of the transverse and longitudinal frames is identified as the major mechanism for the redistribution of the loads in Hotel San Diego. That is, the beams that connected a joint above a removed column to an adjacent supported joint experienced double-curvature deformation. As a result, the direction of bending moment in these beams changed in the vicinity of the joint above the removed column. Because of moment connections, the columns caused the beams to deform in double curvature, which in turn developed sufficient beam shear forces to redistribute the gravity loads. The maximum measured vertical displacement of the structure after the removal of two adjacent columns was only about ¼ of an inch (6 mm) directly above the removed columns.

Acknowledgments This chapter is based upon research supported by the National Science Foundation Award No. CMMI-0547503, the US Department of Homeland Security under Award No. 2008-ST-061-ED0001, and the General Services Administration Award No. GS09P06KTM0019. The views and conclusions contained in this document are those of the author and should not be interpreted as necessarily representing the official policies, either expressed or implied, of the supporting organizations. The help provided by Drs. Marlon Bazan and Serkan Sagiroglu in the experimental program as well as analytical studies is acknowledged. The author greatly appreciates the support provided by Mark and Douglas Loizeaux (Controlled Demolition Inc) and James Redyke (Dykon Explosive Demolition Corp.); without their help this research was not possible to be completed.

References

ACI 318 (2011) Building code requirement for structural concrete. American Concrete Institute, Farmington Hills, MI

Alsamsam IM, Kamara ME (2004) Simplified design: reinforced concrete buildings of moderate size and height, 3rd edn. Portland Cement Association, Skokie, IL

ASCE/SEI 7 (2010) Minimum design loads for buildings and other structures. Structural Engineering Institute-American Society of Civil Engineers, Reston, VA

Bazan ML (2008) Response of reinforced concrete elements and structures following loss of load bearing elements. PhD Dissertation, Department of Civil and Environmental Engineering, Northeastern University, Boston, MA

Breen JE (1975) Research workshop on progressive collapse of building structures held at the University of Texas at Austin. National Bureau of Standards, Washington, DC

Choi HJ, Krauthammer T (2003) Investigation of progressive collapse phenomena in a multi story building. 11th International symposium on the interaction of the effects of munitions with structures, Mannheim, Germany

Corley WG (2004) Lesson learned on improving resistance of buildings to terrorist attacks. J Perform Constr Facil 18(2):68–78

Corley WG, Mlakar PF, Sozen MA, Thornton CH (1998) The Oklahoma city bombing: summary and recommendations for multihazard mitigation. J Perform Constr Facil 12(3):100–112

Cundall PA (1971) A computer model for simulating progressive, large scale movement in blocky rock system, Proceedings of the international symposium on rock mechanics, Nancy, France, pp 129–136

DOD (2001) Interim antiterrorism/force protection construction standards. Progressive Collapse Design Guidance, Washington, DC

DOD (2010) Design of building to resist progressive collapse, unified facility criteria, UFC 4-023-03. U.S. Department of Defense, Washington, DC

Ellingwood B, Leyendecker EV (1978) Approaches for design against progressive collapse. J Struct Div 104(3):413–423

FEMA 277 (1996) The Oklahoma city bombing: improving building performance through multi-hazard mitigation. Building Performance Assessment Team, Federal Emergency Management Agency, Washington, DC

FEMA 356 (2000) Prestandard and commentary for the seismic rehabilitation of buildings. Federal Emergency Management Agency, Washington, DC

Fenwick RC, Megget LM (1993) Elongation and load deflection characteristics of reinforced concrete members containing plastic hinges. Bull NZNSEE 26(1):28–41

GSA (2003) Progressive collapse analysis and design guidelines for new federal office buildings and major modernization projects. U.S. General Service Administration, Washington, DC

IBC (2003) International building code. International Code Council, Inc., Country Club Hills, IL

Isobe D, Toi Y (2000) Analysis of structurally discontinuous reinforced concrete building frames using the ASI technique. Comp Struct 76:471–481

Itoh M, Yoshida N, Utagawa N, Kondo I (1994) Simulation of blast demolition of reinforced concrete buildings. Proceedings of the 3rd world congress on computational mechanics, Chiba, Japan, vol 2, pp 1152–1153

Kim J, Stanton JF, MacRae GA (2004) Effect of beam growth on reinforced concrete frames. J Struct Eng 130(9):1333–1342

Krauthammer T, Hall RL, Woodson SC, Baylot JT, Hayes JR, Sohn Y (2003) Development of progressive collapse analysis procedure and condition assessment for structures, National workshop on prevention of progressive collapse in Rosemont, IL. Multihazard Mitigation Council of the National Institute of Building Sciences, Washington, DC

MacLeod IA (1990) Analytical modeling of structural systems – an entirely new approach with emphasis on behavior of building structures, Ellis Horwood Series in Civil Engineering. Halsted, England

Meguro K, Hakuno M (1994) Application of the extended distinct element method for collapse simulation of a double-deck bridge. Structural Eng/Earthquake Eng 10(4):175–185

Meguro K, Tagel-Din HS (2002) Applied element method used for large displacement structural analysis. J Nat Disast Sci 24(1):25–34

SAP2000 (2009) Three dimensional static and dynamic finite element analysis and design of structures, analysis reference, version 14.1. Computer and Structures, Inc., Berkeley, CA

Sasani M (2008) Response of a reinforced concrete infilled-frame structure to removal of two adjacent columns. J Eng Struct 30(9):2478–2491

Sasani M, Sagiroglu S (2008) Progressive collapse resistance of hotel San Diego. J Struct Eng 134(3):478–488

Sasani M, Sagiroglu S (2010) Gravity load redistribution and progressive collapse resistance of a 20-story RC structure following loss of an interior column. ACI Struct J 107(6):636–644

Sasani M, Bazan M, Sagiroglu S (2007) Experimental and analytical progressive collapse evaluation of an actual reinforced concrete structure. ACI Struct J 104(6):731–739

Sasani M, Kazemi A, Sagiroglu S, Forest S (2011) Progressive collapse resistance of an actual 11-story structure subjected to severe initial damage. J Struct Eng 137(9):893–902

Shi GH, Goodman RE (1984) Discontinuous deformation analysis, Proceedings of the 25th U.S. symposium on rock mechanics, vol 1, pp 269–277

Sozen MA, Thornton CH, Corley WG, Mlakar PF (1998) The Oklahoma city bombing: structure and mechanisms of the murrah building. J Perform Constr Facil 12(3):120–136

Tosaka N, Kasai Y, Honma T (1988) Computer simulation for falling patterns of buildings, Demolition methods and practice, Tokyo, Japan, pp 395–403

Yarimer E (1989) Demolition by controlled explosion as a dynamical process, structures under shock and impact. Proceedings of the 1st international conference, Cambridge, MA, pp 411–416

Index

A. Shukla et al. (eds.), *Blast Mitigation: Experimental and Numerical Studies*,
DOI 10.1007/978-1-4614-7267-4, © Springer Science+Business Media New York 2014

R

T

W

X

Y

Z

MIX
Papier aus verantwortungsvollen Quellen
Paper from responsible sources
FSC® C105338
FSC
www.fsc.org

If you have any concerns about our products,
you can contact us on
ProductSafety@springernature.com

In case Publisher is established outside the EU,
the EU authorized representative is:
Springer Nature Customer Service Center GmbH
Europaplatz 3, 69115 Heidelberg, Germany

Printed by Libri Plureos GmbH
in Hamburg, Germany